Scaling Concepts in Polymer Physics

by the same author

The Physics of Liquid Crystals

Superconductivity of Metals and Alloys

Scaling Concepts in Polymer Physics

Pierre-Gilles de Gennes

CORNELL UNIVERSITY PRESS
Ithaca and London

First published 1979 by Cornell University Press

LIBRARY OF CONGRESS CATALOGING-IN-PUBLICATION DATA
(For library cataloging purposes only)

Gennes, Pierre G de.
Scaling concepts in polymer physics.

Includes bibliographical references and indexes.
1. Polymers and polymerization. I. Title.
QD381.G45 547'.84 78-21314
ISBN 978-0-8014-1203-5 (cloth : alk. paper)
Library of Congress Catalog Card Number 78-21314

Cornell University Press strives to use environmentally responsible suppliers and materials to the fullest extent possible in the publishing of its books. Such materials include vegetable-based, low-VOC inks and acid-free papers that are recycled, totally chlorine-free, or partly composed of nonwood fibers. For further information, visit our website at www.cornellpress.cornell.edu.

Cloth printing 10

Contents

IV
Incompatibility and Segregation 98

V
Polymer Gels 128

Who hath measured the waters in the hollow of
his hand, and weighed the mountains in scales?

ISAIAH **40,** 12

Preface

The physics of long flexible chains was pioneered by several great scientists: Debye, Kuhn, Kramers, Flory, and so forth. They constructed the basic ideas; those concerning static properties are summarized in Flory's book,[1] and those concerning dynamics in various reviews.[2,3,4] More recently, a second stage in the physics of polymers has evolved, because of the availability of new experimental and theoretical tools. As usual, these new techniques brought about some important changes in our viewpoints.

(i) Neutron diffraction allowed for measurements of polymer conformations at large scales which were not feasible with X-rays. The essential point is that different isotopes give different scattering amplitudes for neutrons.[5] Thus, it became possible to label one chain (replacing, for instance, its protons by deuterons) and to observe it individually in a sea of chemically identical but unlabeled chains. The same operation is not feasible with X-rays, for which the labeling is based on the attachment of heavy atoms to the chain; these atoms make the labeled and unlabeled species very different, and spurious segregation effects always occur. The advent of neutron scattering experiments on labeled species opened up a vast new field; precise data on long-range conformations and correlations became available rapidly.

(ii) Light scattering has traditionally been used for measurements of molecular weights and sizes in dilute solution. This technique, however, was limited and delicate, mainly because of the many spurious sources of scattering (e.g., dust) which were always present. The situation suddenly improved when the *inelastic* scattering of laser light became accessible.

This "photon beat" method[6] allows one to study the dynamics of the scattering centers in a frequency range (1 to 10^6 cycles) which is suitable for the overall motions of polymer chains. Furthermore, all the spurious signals caused by dust particles are easier to eliminate, since large particles are essentially immobile and contribute only to the elastic spectrum.

(iii) A certain refinement also occurred in *theoretical methods*. Functional integrals, Feynman diagrams, and all the techniques of many-body theory were first applied to polymers in the pioneering work of S. F. Edwards.[7] In a different direction, certain *numerical methods*, allowing the study of polymer statistics on simple lattice models, became extremely powerful. The British school used exact summation on short chains, supplemented by clever extrapolation techniques to reproduce the behavior of long chains.[8] Another approach (with a slightly different spectrum of application) was the Monte Carlo method, in which a small (but representative) fraction of all possible conformations in a given problem is generated and sampled.[9] Both techniques have been extremely helpful in elucidating certain geometric laws and in displaying the existence of "characteristic exponents," to which we constantly refer in this book.

In a third stage, a relationship between polymer statistics and phase transition problems was discovered.[10,11] This discovery allowed polymer science to benefit from the vast knowledge which had been accumulated on critical phenomena; a number of remarkably simple *scaling* properties emerged. At this third stage, however, our community is divided; a new theoretical language (heavily loaded with field theoretical concepts) has appeared but has remained essentially unintelligible to most polymer scientists.

The aim of this book is to eliminate this barrier, or at least to reduce it as much as possible. In a series of lectures given between 1975 and 1978, in Paris, Strasbourg, Grenoble, and Leiden, I found that most of the essential concepts of polymer physics can be explained in simple terms and do not require any advanced theoretical education. Thus, I hope to give to my reader a reasonable understanding of certain "universal" properties: scaling laws and characteristic exponents in polymer solutions and melts. All details are systematically omitted.

(i) I ignore numerical coefficients in most formulas, where they would obscure the main line of thought.

(ii) On the experimental side, the discussions are very brief. I do not try to recapitulate all the data on a given problem but simply to select studies in which scaling features are apparent.

(iii) On the other hand, this book is not intended as an introduction for

a young polymer theorist. Theoretical methods are relegated to the last three chapters; and even there my aim is not to provide the reader with the ability to make advanced calculations; more modestly, I would like him (or her) to reach a certain qualitative understanding of the methods—how they work and where they fail. (A much more complete description of polymer theory will be available in a forthcoming book by J. des Cloizeaux and G. Jannink.)

(iv) Certain important areas of polymer physics are not mentioned at all; crystallization kinetics and glass transitions are two glaring examples. Polyelectrolytes are mentioned only briefly. In these areas we do not know whether or not scaling concepts will be really useful.

On the whole, this book is meant for experimentalists in polymer science who wish to incorporate the recent advances into their modes of thinking. Obviously certain difficulties remain even for these readers. In particular there is a general question of language and notation.

(i) I have tried to follow the basic notation of Flory,[1] but I have had to introduce some modifications which correspond to recent trends—for instance, to use a polymerization index (N) rather than a molecular weight (M) as the fundamental object; to eliminate all mention of Avogadro's number; to write thermal energies as T rather than $k_B T$ (i.e., to use energy units for the temperature T, as is done now in most theoretical literature); and so forth. Such modifications, although trivial, will disturb the reader at the beginning, but they represent (I think) a necessary simplification.

(ii) At a more fundamental level, my inclination is always to seek comparisons to other branches of science: conceptually, a single chain in an external field is closely related to a quantum particle, as first found by Edwards; there is a profound analogy between polymer statistics and phase transitions; the gelation problem is related to the general concept of percolation; and so forth. I have tried to explain some of these analogies, without assuming prior knowledge of quantum mechanics or critical phenomena (a summary of critical phenomena is included in Chapter X). One pleasant discovery, when I was teaching polymer statistics, was to find that renormalization groups can be explained in very simple words to polymer chemists; the last chapter describes this approach.

I have also tried to help my readers by carefully selecting references. As explained, I never give a complete historical list on any topic; I quote mainly a few basic reviews which are both clear and accessible. (Most of the polymer literature written before 1965 and relevant to the present text is analyzed in the books mentioned at the beginning of this preface.) For the more recent advances on scaling laws, the majority of my references are French. This is not an expression of nationalistic pride; it just happened that our experimentalists, under the impetus of H. Benoît in Strasbourg and G. Jannink in Saclay, were able to set up at the right time an efficient, cooperative effort for elucidating scaling laws. The present text reflects to a large extent the discussions of this program during the past five or six years.

The laboratories at Strasbourg, Saclay, and Collège de France that joined in this venture have been associated for some time under the acronym STRASACOL, the story of which is summarized in a short note.[12] However, the cooperation has rapidly extended beyond these limits, involving people at Brest, Grenoble, and Chambery, and I hope it expands even further. To all these units I am profoundly grateful, for their eagerness in discussing present research and for their open mind toward new directions. Last but not least, I wish to mention my debt to many friends on the theoretical side: to C. Sadron; to J. des Cloizeaux, G. Sarma, and M. Daoud in Saclay; to F. Brochard and P. Pfeuty in Orsay; to S. F. Edwards in England; and especially to our foreign visitors: S. Alexander, J. Ferry, F. C. Frank, P. Martin, P. Pincus, and W. Stockmayer, who instructed us and corrected many of my mistakes.

Some mistakes certainly remain, and at various points I present very conjectural views. Nevertheless, let us hope that the book will still give a reasonable image of what is universal and what is system dependent in these fascinating systems of mobile entangled chains.

P. G. DE GENNES

Paris

Note to the 1985 printing

The 1985 printing contains a number of alterations. I am particularly indebted to M. Adam, Y. A. Ono, and F. Volino for their close scrutiny of the original text, which allowed for the elimination of many misprints. I have also removed certain serious conceptual errors (on polymer adsorption, on the statistics of contacts). The help of M. F. Jestin and F. David in the organization of the reprint version is also gratefully acknowledged.

REFERENCES

1. P. Flory, *Principles of Polymer Chemistry,* Cornell University Press, Ithaca, New York, 1971.
2. J. D. Ferry, *Viscoelastic Properties of Polymers,* 2nd ed., John Wiley and Sons, New York, 1970.
3. W. H. Stockmayer, *Fluides Moléculaires,* p. 107, R. Balian and G. Weill, Eds., Gordon & Breach, New York, 1976.
4. W. Graessley, *Adv. Polym. Sci.* **16** (1974).
5. W. Marshall, S. Lovesey, *Theory of Neutron Scattering,* Oxford University Press, London, 1971. For the specific problem of deuterated polymers, see F. Boue *et al., Neutron Inelastic Scattering 1977,* Vol. 1, p. 563, International Atomic Energy Agency, Vienna, 1978; and A. Maconnachie, R. W. Richards, *Polymer* **19,** 739 (1978).
6. G. Benedek, *Polarisation, Matière, et Rayonnement,* Société Française de Physics, Presses Universitaires de France, Paris, 1969.
7. S. F. Edwards, *Fluides Moléculaires,* R. Balian and G. Weill, Eds., Gordon & Breach, New York, 1976.
8. C. Domb, *Adv. Chem. Phys.* **15,** 229 (1969).
9. F. T. Wall, J. Erpenbeck, *J. Chem. Phys.* **30,** 634 (1959).
10. P. G. de Gennes, *Phys. Lett.* **38A,** 339 (1972). P. G. de Gennes, *Riv. Nuovo Cimento* **7,** 363 (1977).
11. J. des Cloizeaux, *J. Phys. (Paris)* **36,** 281 (1975).
12. P. G. de Gennes, *J. Polym. Sci., Polym. Lett.* **15,** 623 (1977).

Introduction: Long Flexible Chains

Linear Polymers

This book discusses the statistical properties of long, flexible objects, polymer chains being the fundamental example. The following is a short list of chains which are currently used in physical studies:

$$\ldots-CH_2-CH_2-CH_2-\ldots \quad \text{or} \quad |-CH_2-|_N \qquad \textbf{polyethylene}$$

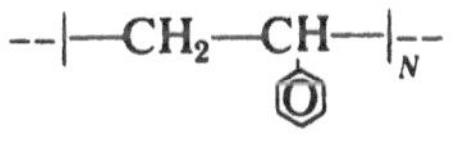

$$--|-CH_2-\underset{\bigcirc}{CH}-|_N-- \qquad \textbf{polystyrene}$$

$$---\left|-CH_2-\underset{\underset{O\swarrow\ \searrow O-CH_3}{C}}{\overset{CH_3}{C}}-\right|_N--- \qquad \textbf{poly(methyl methacrylate)}$$

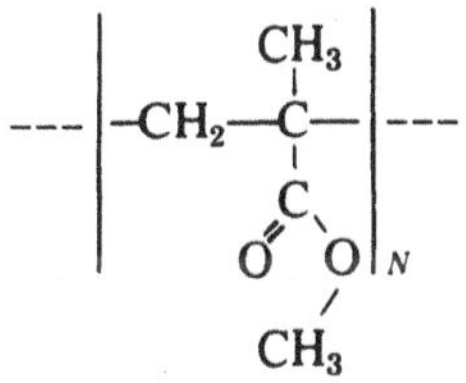

$$---|-CH_2-CH_2-O-|_N--- \qquad \textbf{poly(oxyethylene)}$$

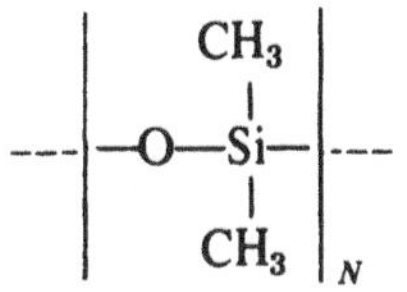

$$---\left|-O-\underset{CH_3}{\overset{CH_3}{Si}}-\right|_N--- \qquad \textbf{poly(dimethyl siloxane)}$$

The number of repeat units, N, in one chain is often called the "degree of polymerization" (DP) and can be amazingly large. (For example, it is possible to reach $N > 10^5$ with polystyrene.) The fabrication of such long chains *without error* in a sequence of 10^5 operations is a remarkable chemical achievement. However, there are many difficulties. Two are particularly important for physical studies: polydispersity and branching.

Polydispersity

Most preparation schemes give chains with a very broad distribution of N values.[1] It is possible, however, to obtain relatively narrow distributions either by physical selection (via precipitations, gel permeation, chromatography, etc.[2]) or through special methods of synthesis, such as anionic polymerization.[3]

Branching

Many parasitic reactions occurring during the synthesis can lead to a chain which is not perfectly linear but which contains branch points. For example, industrial polyethylene has many three-functional branch points of the type

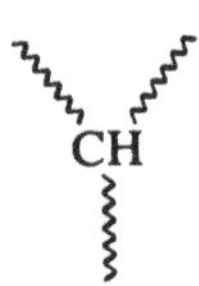

where the zigzag lines represent different chain portions.

When the fraction of branch points in the structure is not too small, these points can be detected by various physical methods, such as infrared spectroscopy. On the other hand, if a long chain has accidentally acquired one or two branch points, it is extremely hard to demonstrate their existence or absence (they show up mainly in certain mechanical studies on concentrated systems, discussed in Chapter VIII).

In some cases we encourage branching. For example, model molecules can be synthesized with the geometry of "stars" or "combs" (Fig. 0.1). More often, branching takes place statistically. It may lead either to tree-like molecules, or, at a higher level, to network structures (discussed in Chapter V). In summary, we can obtain chains that are strictly linear (when N is not too large); we can also insert on a chain a controlled number of branch points.

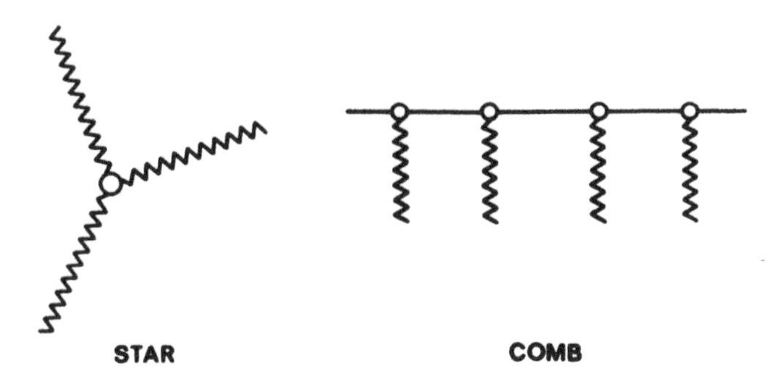

Figure 0.1.

Flexibility

Flexibility can be understood either in a static or in a dynamic sense.

Static flexibility

As a simple example, consider a carbon–carbon chain such as polyethylene. The angle θ between successive C—C bonds is essentially fixed, but when we build up successive units with the carbon atoms ($n - 3$, $n - 2$, $n - 1$) fixed, and add carbon (n), we have one angle φ_n (Fig. 0.2). The energy between successive groups depends on the angle φ_n as shown on Fig. 0.3. There are three minima, corresponding to three principal conformations (called *trans* and *gauche*). In this figure we see two essential energy parameters: 1) the energy difference between minima $\Delta\epsilon$,* and 2) the energy barrier separating the minima ΔE.

For the moment, we focus on $\Delta\epsilon$. When $\Delta\epsilon$ is smaller than the thermal energy T,† we say that the chain is statically flexible. This has striking consequences if we look not at one monomer but at the whole chain. Because the relative weight of *gauche/trans* conformations is of order unity, the chain is not fully stretched. It appears rather as a *random coil* (Fig. 0.4).

Note the difference in magnification between Fig. 0.2 and Fig. 0.4. Fig. 0.2 deals with distances of order 1 Å. Fig. 0.4 deals with hundreds of Angströms.

The case of $\Delta\epsilon < T$ defines a limit of extreme flexibility. If we go to

*In polyethylene $\Delta\epsilon$ (as defined in the figure) is *positive:* the *trans* state is lower in energy than the *gauche* states.

†Recall that we use units where the Boltzmann constant is unity.

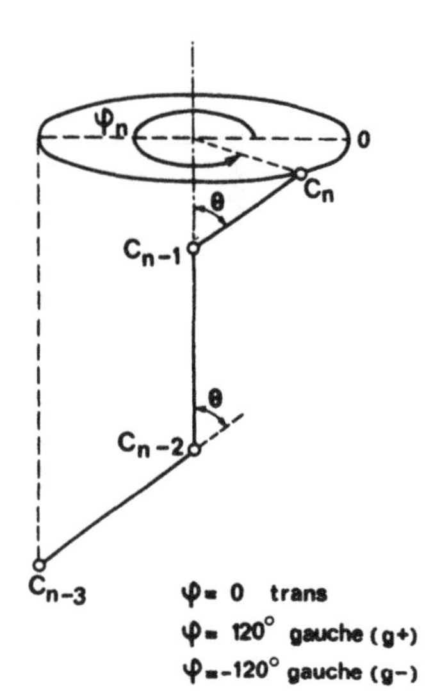

Figure 0.2.

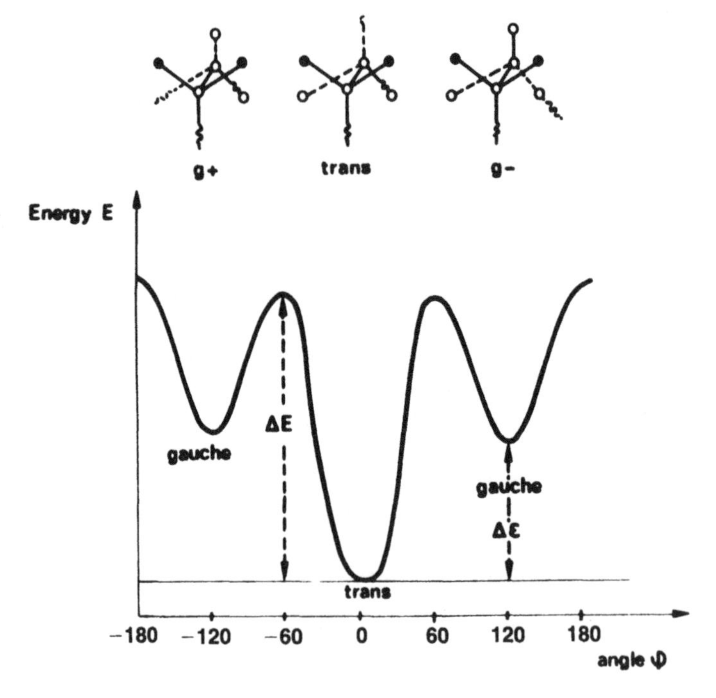

Figure 0.3.

Figure 0.4.

slightly higher values of $\Delta\epsilon/T$, there will be a definite preference for the *trans* state; locally the chain will be rigid. However, if we look at it on a scale which is large enough, it will again appear as a flexible coil. This is illustrated in Fig. 0.5.

More generally, when we ignore details smaller than a certain characteristic length l_p, we see a continuous, flexible chain. The parameter l_p is called the *persistence length* of the chain[4] and can be calculated from the microscopic energies. For the polyethylene chain of Figs. 0.1 and 0.2 l_p is a rapidly increasing function of the energy difference $\Delta\epsilon$

$$l_p = l_0 \exp\left(\frac{\Delta\epsilon}{T}\right) \qquad (\Delta\epsilon > 0)$$

where l_0 is of order a few Angströms.

Whenever l_p is much smaller than the total length L of the chain, we can choose a magnification which is weak, so that the rigid portions (of size $\sim l_p$) are too small to be seen, but which is still strong enough to ensure that the whole chain is not reduced to a point. Then we may say that the molecule is still flexible at large scales. On the other hand, if l_p is larger than the overall chain length, the picture is a rigid rod at all scales.

We see that the essential parameter controlling global flexibility is the ratio

$$x = \frac{l_p}{L} \cong N^{-1} \exp\left(\frac{\Delta\epsilon}{T}\right)$$

Flexible behavior can be observed only at small x.

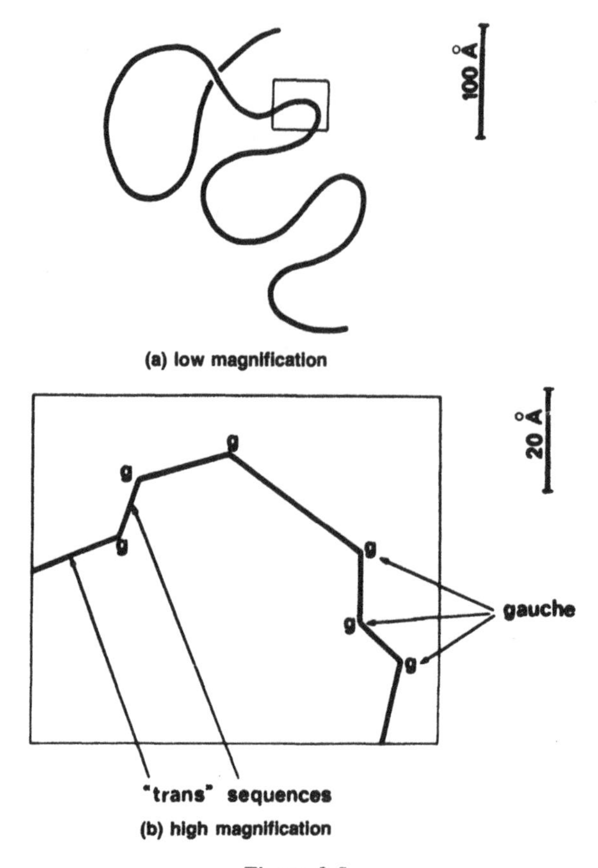

Figure 0.5.

Dynamic flexibility

Successive carbon–carbon links can be in one of two states: *trans* and *gauche*. One important question is related to the time τ_p required for a transition between these two states. This depends mainly on the height ΔE of the barrier separating them in the energy diagram of Fig. 0.3. If ΔE is not much larger than the thermal energy T, the barrier is not important, and the *trans–gauche* isomerization can take place in times $\tau \sim 10^{-11}$ sec. We say then that the chain is dynamically flexible. On the other hand, if the barrier ΔE is high, τ_p becomes exponentially long

$$\tau_p = \tau_0 \exp\left(\frac{\Delta E}{T}\right)$$

It is sometimes useful to call τ_p a persistence time.

Our discussion on spatial scales and static flexibility can be extended to temporal scales and dynamic flexibility. If we are interested in large scale motions of the molecule, involving frequencies ω smaller than $1/\tau_p$, we can still say that the chain is dynamically flexible.

One can find molecules which are flexible from a static point of view but which have high barriers ΔE (with certain flexible backbones carrying bulky side groups). This situation corresponds to a random coil which is essentially *frozen* in one conformation, like a piece of twisted wire. A molecule of this type in dilute solution could be called a "single chain glass," and should have some remarkable mechanical properties.

This book does not discuss any of these rigid molecules. It assumes both static and dynamic flexibility in the strongest form. Then l_p, for example, reduces to a monomer size [currently designated by (a)] and no other characteristic length is involved; this simplification will be helpful.

Global versus Local Properties

Fig. 0.5 illustrates a fundamental distinction between two aspects of polymer science:

(i) Strong magnification or local properties: conformations and motions of one monomer inside the chain, and their dependence on chemical substitutions in the side groups.

(ii) Weak magnification: global properties: dependence of physical properties observables on chain length, on concentration, and on a few basic interaction parameters.

The *local* features are essential whenever we want to choose an optimal polymer for a given practical application. If we want to improve the fabrication of rubbers, we need a good understanding of the local motions of a rubber chain—i.e., how they depend on temperature, the influence of steric constraints between neighboring monomers, and so forth. The experimental methods for local probing of a polymer chain are not very different from those used for small molecules (such as infrared and Raman measurements). Similarly, the theoretical methods are (or will become) related to those which are used for conventional liquids: molecular dynamics, Monte Carlo methods, etc.

The *global* point of view is completely different. Here we try to omit the details of the chain structure as much as possible and to extract simple, universal, features which will remain true for a large class of polymer chains. An example will make this statement more precise: Consider a dilute solution of separate coils in a good solvent. The radius of gyration of one coil R_G depends on the degree of polymerization, N, and we know from Flory that

$$R_G = (\text{constant})\ aN^{\nu} \qquad (0.1)$$

where ν is close to 3/5. What is universal in this law is the exponent ν; it is the same for all coils (in three-dimensional solutions) provided that the solvent is good. What is nonuniversal here is the prefactor. It depends on the detailed monomer structure and on the solvent chosen. If we want to understand the properties of polymer coils in good solvents, the first step is to explain the existence and the value of the exponent ν. The second step is to account for the constant that multiplies a, and this involves delicate studies on local properties. In the present book we are concerned with the first step.

Eq. (0.1) is a good example of a *scaling* law. It tells us that if we double the chain length, the size is increased by a factor 2^{ν}. The theorist using such a scaling law can be compared with the chemist seeking comparisons in homologous series: finding the exact value of R_G for a given chain and solvent is extremely difficult. In a first stage, what we can and must do is to measure R_G for different values of N and compare them. This is the spirit of the present text.

A law such as $R(N)$ above holds only for large N, with flexible chains, and for good solvents. Later we make these statements more precise, but we see already that a scaling law is always defined only in a certain *limit*, which must be specified in each case.

Notation

If we compute a quantity exactly (within a certain model), including all numerical coefficients, we can use an equals sign—i.e., write $A = B$. If we state only a scaling law, ignoring all numerical coefficients but keeping all dimensional factors, we use the symbol, $\cong$ (e.g., $R \cong aN^{3/5}$). If we go to a further reduction and want to stress only the power law involved in $R(N)$, we use the symbol $\sim$ (e.g., $R \sim N^{3/5}$).

REFERENCES

1. P. Flory, *Principles of Polymer Chemistry,* Chaps. III, IV, Cornell University Press, Ithaca, N.Y., 1971.
2. D. D. Bly, *Physical Methods of Macromolecular Chemistry,* B Carroll, Ed., Vol. 2, Marcel Dekker, New York, 1972.
3. P. Flory, *Principles of Polymer Chemistry,* Chap. V, Cornell University Press, Ithaca, N.Y., 1971.
4. T. Birshtein, O. Ptitsyn, *Conformations of Macromolecules,* John Wiley & Sons, New York, 1966.

Part A

STATIC CONFORMATIONS

Various animals attempting to follow a scaling law.

I

A Single Chain

I.1. The Notion of an Ideal Chain

I.1.1. Simple random walks

One of the simplest idealizations of a flexible polymer chain consists in replacing it by a random walk on a periodic lattice, as shown in Fig. I.1. The walk is a sucession of N steps, starting from one end (α) and reaching an arbitrary end point (ω). At each step, the next jump may proceed toward any of the nearest-neighbor sites, and the statistical weight for all these possibilities is the same. The length of one step will be called a.

This description was apparently initiated by Orr in 1947.[1] It is convenient from a pedagogical point of view: all chain properties are easy to visualize. For instance, the entropy $S(\mathbf{r})$ associated with all chain conformations starting from an origin ($\mathbf{r} = 0$) and ending at a lattice point $\mathbf{r}$, is simply related to the number of distinct walks $\mathfrak{N}_N(\mathbf{r})$ going from (0) to ($\mathbf{r}$) in N steps*

$$S(\mathbf{r}) = \ln[\mathfrak{N}_N(\mathbf{r})] \tag{I.1}$$

The main features of the number $\mathfrak{N}_N(\mathbf{r})$ are discussed now. First, the total number of walks is simple to compute; if each lattice site has z neighbors, the number of distinct possibilities at each step is z, and the total number is

*We always use units where Boltzmann's constant k_B is unity.

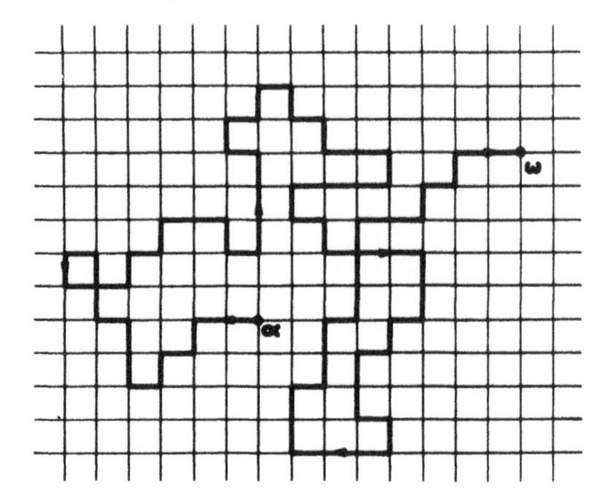

Figure I.1.

$$\sum_{(\mathbf{r})} \mathfrak{N}_N(\mathbf{r}) = z^N \tag{I.2}$$

(where $\sum_{\mathbf{r}}$ denotes a sum over lattice points)

The end-to-end vector $\mathbf{r}$ is the sum of N "jump vectors"

$$\mathbf{r} = \mathbf{a}_1 + \mathbf{a}_2 + \ldots + \mathbf{a}_N = \sum_n \mathbf{a}_n \tag{I.3}$$

where each of the $\mathbf{a}$ terms is a vector of length a with z possible orientations. Different $\mathbf{a}$ vectors have completely *independent* orientations, and this has many consequences:

(*i*) the average square end-to-end distance is *linear in N*

$$\langle \mathbf{r}^2 \rangle = \sum_{nm} \langle \mathbf{a}_n \cdot \mathbf{a}_m \rangle = \sum_n \langle \mathbf{a}_n^2 \rangle = Na^2 \; (= R_0^2) \tag{I.4}$$

since all cross-products vanish. Qualitatively, we shall say that a random walk has a size $R_0 \sim N^{1/2}\, a$.

(*ii*) the distribution function for $\mathbf{r}$, defined by

$$p\,(\mathbf{r}) = \mathfrak{N}_N\,(\mathbf{r}) / (\sum_{\mathbf{r}} \mathfrak{N}_N\,(\mathbf{r})) \tag{I.5}$$

has a gaussian shape as soon as the number of independent jump vectors $\mathbf{a}_n$ is large ($N \gg 1$). For example, if we are in three dimensions

$$p\,(x,y,z) = \text{constant } N^{-1/2} \exp\left(\frac{-x^2}{2\,\langle x^2\rangle}\right) N^{-1/2} \exp\left(\frac{-y^2}{2\,\langle y^2\rangle}\right)$$

$$N^{-1/2} \exp\left(\frac{-z^2}{2\,\langle z^2\rangle}\right) \cong N^{-3/2} \exp\left(\frac{-3\,r^2}{2\,Na^2}\right) \quad \text{(I.6)}$$

The factors $N^{-1/2}$ arise from normalization conditions. We purposely do not write the complete numerical value of the constant in front of eq. (I.6); these constants would obscure our arguments. They can be found in standard textbooks on statistics.[2]

Eq. (I.6) gives a formula for the entropy of the chain at fixed elongation

$$S(\mathbf{r}) = S(0) - \frac{3\,r^2}{2\,R_0^2} \quad \text{(three dimensions)} \quad \text{(I.7)}$$

The entropy decreases when the elongation increases. It is often convenient to rewrite eq. (I.7) in terms of free energy

$$F(\mathbf{r}) = E - TS$$

In the Orr model the energy E is a constant (independent of the chain conformation), and we have simply

$$F(\mathbf{r}) = F(0) + \frac{3\,Tr^2}{2\,R_0^2} \quad \text{(I.8)}$$

This is a fundamental formula, giving the "spring constant" of an ideal chain. We return to it in eq. (I.11) and use it frequently.

I.1.2. More general models for ideal chains

The model in Fig. I.1 is crude but convenient. More accurately, it is possible to build up the chain by successive steps, taking into account all valence angles, the correct weights for *trans*/*gauche* conformations (or their generalization) and even statistical deviations from the ideal *trans* or *gauche* states. This type of calculation is described fully in the second book by P. Flory.[3]

The crucial approximation involved in this "progressive buildup" amounts to taking into account only the interactions between each unit (n) and its neighbor ($n+1$) [or possibly between (n) and ($n+1$, $n+2$, $n+p$)

with p fixed and finite]. Let us accept this for the moment. We can define the "backbone" of the chain by a sequence of vectors $\mathbf{b}_1 + \mathbf{b}_2 + \ldots + \mathbf{b}_N = \mathbf{r}$, each of the vectors $\mathbf{b}_i$ linking two consecutive monomers. In the Orr model these vectors are now correlated. For example, the average

$$\langle \mathbf{b}_n \cdot \mathbf{b}_m \rangle = \gamma_{nm} \tag{I.9}$$

is nonzero even for $m \neq n$. It is, however, a decreasing function of the chemical interval $|m - n|$, and it decays exponentially at large $|n - m|$. Thus the correlations are of finite range. We now show that, in this case, the global properties are not affected seriously.

Let us put g consecutive vectors $\mathbf{b}$ into one *subunit*. In Fig. I.2 we show the case for $g = 3$. If g is much larger than the range of the correlations c_{nm}, the new vectors $\mathbf{c}$ will be uncorrelated, and we face the problem of N/g independent variables $\mathbf{c}_1, \mathbf{c}_2, \ldots$, leading again to gaussian statistics provided that N/g is large; this is what we call ideal chain behavior. The mean square end-to-end distance is linear in N

$$\langle r^2 \rangle = \frac{N}{g} \langle c^2 \rangle = Na^2 \tag{I.10}$$

where $a = (\langle c^2 \rangle / g)^{1/2}$ is now an effective length per monomer. Thus, whatever the microscopic structure of the chain, *if we take into account only the interactions between neighboring units on the chemical sequence, we always get an ideal chain if N is large enough.*

The single (but important) weak point in this approach is the neglect of interactions between monomers n and m with $|n - m|$ very large. Fig. I.3 shows an interaction which is omitted. When these "large loop interactions" are included, the chain is *not* gaussian. We discuss this extensively later in this chapter.

Figure I.2.

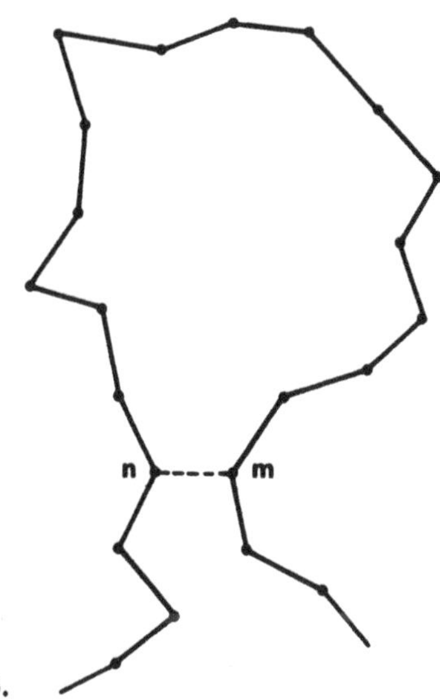

Figure I.3.

I.1.3. Ideal chains under external actions

It is of interest to study the response of a chain to external perturbations. With an ideal chain, this response is particularly easy to derive. We are concerned here with two main situations: pulling and squeezing.

PULLING A CHAIN AT BOTH ENDS (Fig. I.4)

We apply forces **f** and $-\mathbf{f}$ at both ends and ask for the average elongation $\langle \mathbf{r} \rangle_f$ of the chain. For an f that is not too large the answer is derived from the "spring constant equation" (I.8). The force **f** is $\partial F/\partial \mathbf{r}$ taken at $\mathbf{r} = \langle \mathbf{r} \rangle_f$, and thus

$$\langle \mathbf{r} \rangle_f = \mathbf{f} \frac{R_0^2}{3\,T} \qquad \text{(I.11)}$$

Eq. (I.9) holds whenever $\langle \mathbf{r} \rangle$ is much smaller than Na (chain not fully stretched). This corresponds to $f \ll T/a$.

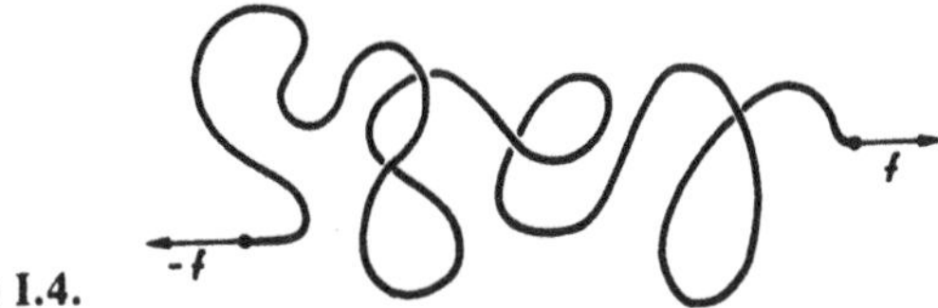

Figure I.4.

We rederive eq. (I.11) here through a scaling argument, which is good training for later problems. This derivation is based on the following points:

(*i*) Since the tension **f** is the same all along the chain, the elongation $\langle \mathbf{r} \rangle$ must be a linear function of N.

(*ii*) We expect $\langle \mathbf{r} \rangle$ to depend only on **f**, on temperature, and on the unperturbed size $R_0 = N^{1/2} a$. This leads to

$$|\langle \mathbf{r} \rangle| \cong R_0 \left(\frac{f R_0}{T} \right)^x$$

where x is fixed by requirement (i)—namely, $R_0^{(1+x)} \sim N$. Thus $x = 1$, and the elongation is a linear function of the force.

Eq. (I.11) is the basis of rubber elasticity, and we shall use it often.

Exercise: consider an ideal chain carrying charges $\pm e$ at both ends (e is one electron charge). What will be its relative elongation in a field $E = 30{,}000 \text{V/cm}$?

Answer: we have $r/R_0 \cong R_0 eE/3\,T$. Take $N = 10^4$ and $a = 2$ Å (giving $R_0 = 200$ Å). The voltage drop on a length R_0 is $3.10^4 \times 2.10^{-6} = 0.06$ V. At room temperature $T = 1/40$ eV and thus $r/R_0 \sim 0.8$.

AN IDEAL CHAIN TRAPPED IN A TUBE

The chain is captured in a cylindrical tube of diameter $D \ll R_0$ (Fig. 1.5). On the other hand, we want $D \gg a$, so that the chain still retains some

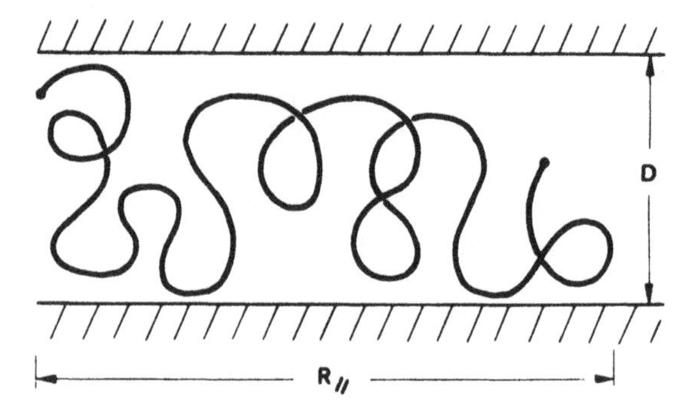

Figure I.5.

lateral wiggling. We assume that the tube walls repel the chain strongly (no trend towards adsorption).

We ask first, what is the length of tube ($R_{\parallel}$) occupied by the chain? The answer is $R_{\parallel} = R_0$; that is, confinement does not affect the components of the random walk parallel to the tube axis.

Second, we discuss the energy required to squeeze the chain, starting from a dilute solution in the same solvent and assuming that chain entropy is the only significant factor (no long-range van der Waals force in the tube). We try to estimate the reduction in entropy ΔS due to confinement:

(i) The leading term in ΔS will be a linear function of N.

(ii) ΔS is dimensionless and depends only on the length ratio R_0/D.

This leads to $\Delta S = -(R_0/D)^y \sim N^{y/2}$, and from (i) we must have $y = 2$. The corresponding free energy is

$$F \cong T \frac{R_0^2}{D^2} \qquad \text{(I.12)}$$

The argument holds equally for a confinement in a slit or in a hollow sphere; only the coefficients differ. They have been computed first by Cassasa and co-workers[4,5] (see Chapter IX for more details).

WEAK ADSORPTION OF AN IDEAL CHAIN

The situation is represented in Fig. I.6. The chain sticks slightly to the wall and has large loops extending up to an average distance D. Exact calculations on this system have been performed in the past.[6,7,8] Here we present a simple scaling argument that relates D to the strength of the adsorption.[9] The starting point is a free energy per chain of the form

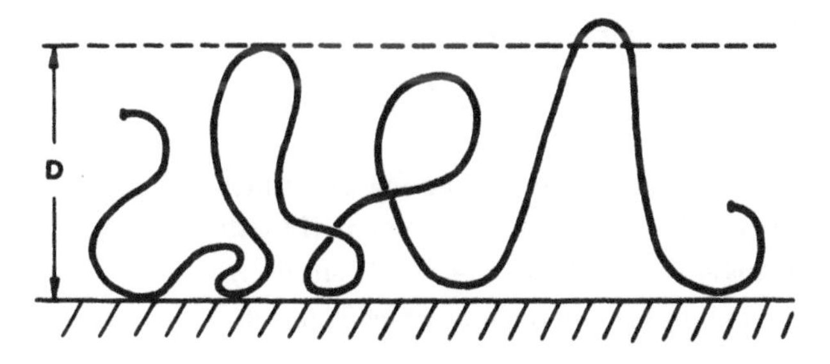

Figure I.6.

$$F \cong T\frac{R_0^2}{D^2} - T\delta f_b N \tag{I.13}$$

The first term is the confinement energy (eq. I.12), and the second term describes the contact interactions with the surface; $T\delta$ is the effective attraction seen by a monomer adsorbed at the surface (a balance between an attractive energy and a loss of entropy), and f_b is the fraction of bound monomers. Since the monomer density is spread over a thickness D, we expect

$$f_b \cong a/D \tag{I.14}$$

Inserting this in eq. (I.13) and minimizing the sum with respect to D, we reach a thickness

$$D \cong a\delta^{-1} \qquad (\delta \ll 1,\ D \ll R_0) \tag{I.15}$$

and a free energy of binding

$$F \cong -TN\delta^2 \tag{I.16}$$

The conditions required for the adsorption of *separate* chains are never realized in practice, but they provide a useful framework for future discussions of many chain adsorption.

I.1.4. Pair correlations inside an ideal chain

A pair correlation function $g(\mathbf{r})$ may be defined as follows. We pick one monomer at random in the chain, and we place it at the origin. Then we ask, what is the number density of other monomers at a distance $\mathbf{r}$ from the first, and we average the result over all choices of the first monomer.

The Fourier transform of $g(\mathbf{r})$

$$g(\mathbf{q}) = \int g(\mathbf{r}) d\mathbf{r} e^{i\mathbf{q}\cdot\mathbf{r}}$$

is directly measured in many scattering experiments (light, X-rays, neutrons), $\mathbf{q}$ being the scattering wave vector. (In terms of wavelength λ and scattering angle θ we have $q = 4\pi\lambda^{-1}\sin\theta/2$.)

The function $g(\mathbf{r})$ has an integral which is just the total number of monomers per chain N

$$\int g(\mathbf{r}) d\mathbf{r} = N = g(\mathbf{q} = 0)$$

The functions $g(\mathbf{r})$ and $g(\mathbf{q})$ obey simple scaling rules:

$$g(\mathbf{r}) = \frac{N}{R_0^{\,3}}\,\tilde{g}\left(\frac{r}{R_0}\right)$$

where $\tilde{g}$ is an universal function.

The structure of $g(\mathbf{q})$ for ideal chains was discussed first by Debye,[10] and thus we call $g(\mathbf{q})$ the *Debye function* $g_D(\mathbf{q})$.

Focusing on the limit $\mathbf{r} \ll R_0$, we can reach the form of $g(\mathbf{r})$ by a simple argument. In a sphere of radius r we have a certain number of monomers n, related to r by the random walk scaling law: $na^2 \sim r^2$. The function $g(\mathbf{r})$ scales like the density of monomers in the sphere:

$$g_D(\mathbf{r}) \cong n/r^3 \cong \frac{1}{a^2 r} \qquad (r \ll R_0) \tag{I.17}$$

The exact coefficient is displayed in Fig. I.7; for its complete derivation see Chapter IX. The Fourier transform of $1/r$ is $4\pi/q^2$, and the scattering function is

$$g_D(q) = \frac{12}{q^2 a^2} \qquad (qR_0 \gg 1) \tag{I.18}$$

It is not easy to measure this $g_D(q)$ on dilute chains directly; in light scattering q is too small, and in X-rays or neutron experiments the signals from dilute systems are weak. However, the result [eq. (I.18)] will be useful for more complicated systems.

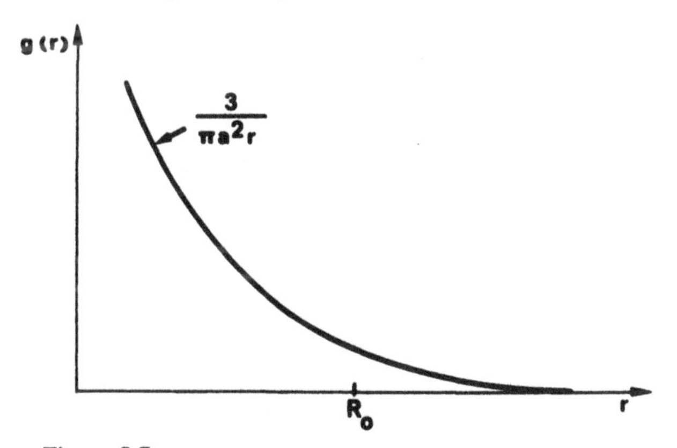

Figure I.7.
Pair correlation between all monomers in an ideal chain. The correlations decrease like $1/r$ at distances r, smaller than the chain size R_0. They fall off sharply for $r > R_0$.

I.1.5. Summary

Ideal chains are characterized by: 1) gaussian statistics; 2) size proportional to $N^{1/2}$; 3) large domain of linear relation between force and elongation; and 4) scattering law of the q^{-2} type. We now see how these properties are altered when we switch from ideal to real chains.

I.2. A "Real" Chain in a Good Solvent

I.2.1. The main experiments

The size of real chains in dilute solutions can be determined by various standard experimental methods:

(i) Measurements on scattered light intensity versus angle give us the radius of gyration R_G.[3]

(ii) More simply, a study of the viscosity η of dilute solutions measures a certain hydrodynamic radius R_η[11]

$$\eta = \eta_S \left[1 + 2.5 \frac{c}{N} \frac{4\pi}{3} R_\eta^3 \right] \qquad (c \to 0) \tag{I.19}$$

Here η_S is the solvent viscosity, and c is the concentration; we do not define it by weight but rather as *a number of monomers per unit volume.* Similarly c/N is the number of chains per cm^3. The numerical factors in eq. (I.19) correspond to a rigid sphere of radius R_η. On the experimental side this provides an excellent determination of R_η. However the interpretation of R_η is delicate. We return to this question in Chapter VI.

(iii) Photon beat measurements give us the diffusion coefficient D_0 for a single coil. This coefficient may be related to another effective radius R_D, defined through the Stokes relation for a sphere

$$D = \frac{T}{6\pi\eta_S R_D} \tag{I.20}$$

Summarizing a vast literature, we may say that the light scattering experiments (i) give a radius $R_G \sim N^{0.60}$ while the hydrodynamic studies (ii) and (iii) give a slightly weaker power $R \sim N^{0.55}$ or $N^{0.57}$. This discrepancy reflects some subtle corrections involved in dynamical experiments and is discussed in Chapter VI.

I.2.2. Numerical data on self-avoiding walks

We see that the direct data on coils are not quite conclusive. It is then helpful to return to theoretical calculations. There do exist rather accurate numerical studies on *real chains on a lattice*. The chain is still represented by a random walk as in Fig. (I.1), but the main difference is that now this walk can never intersect itself. We call it a *self-avoiding walk* (SAW).

The mathematical properties of simple random walks are trivial, but the mathematical properties of SAWs are complex. Two numerical methods have been used to study the SAWs:

(i) Exact counting of walks for finite N (typically up to $N \sim 10$) plus extrapolation methods allowing us to extend the results toward $N \rightarrow \infty$.[12]

(ii) Monte Carlo methods, where the computer generates a certain (manageable) fraction of all SAWs of N steps and performs averages on these.[13]

All these studies have been performed on three-dimensional lattices and in other dimensionalities, d. The case for $d = 1$ corresponds to chains along a line and is simple. The case for $d = 2$ may physically correspond to chains adsorbed at an interface. Higher dimensionalities ($d = 4, 5 \ldots$) are also of interest for the theorist, although they do not correspond to realizable systems. One important advance (during the past 10 years) has been to recognize the interest of discussing any statistical problem in arbitrary dimensions and to classify systems according to their behavior as a function of d. Thus, we shall often keep d as a parameter in our discussion of polymer chains.

The results of numerical studies on SAWs are usefully summarized in a recent review by McKenzie.[12] Our presentation, however, is slightly different since the physical meaning of the essential exponents has become more apparent in the recent years.

The total number of SAWs of N steps has the asymptotic form (at large N)

$$\mathfrak{N}_N\,(tot) = \text{constant}\ \tilde{z}^N\, N^{\gamma-1} \tag{I.21}$$

The first factor $\tilde{z}^N$ is reminiscent of the z^N which we had for ideal chains, but $\tilde{z}$ is somewhat smaller than z. For the three-dimensional simple cubic lattice, $z = 6$ and $\tilde{z} = 4.68$. The second factor, $N^{\gamma-1}$, is more unexpected and will be called the enhancement factor. The exponent γ *depends only on the dimensionality, d:*

$$\text{for all three-dimensional lattices } \gamma = \gamma_3 \cong 7/6 \tag{I.22a}$$

$$\text{for all two-dimensional lattices } \gamma = \gamma_2 \cong 4/3 \tag{I.22b}$$

We say that γ is a universal exponent; this is in contrast to $\tilde{z}$, which does depend not only on d but also on the particular lattice chosen (e.g., face-centered cubic/simple cubic). Note that for $d = 1$, $\mathfrak{N}_N\,(tot) = 2$, independently of N. Thus $\tilde{z}_1 = 1$ and $\gamma_1 = 1$.

The end-to-end distance r has a mean square average which we shall call R_F^2, and which scales as

$$R_F \cong a\, N^{\nu} \tag{I.23}$$

Here ν is another universal exponent ($\nu_3 \cong 3/5$, $\nu_2 \cong 3/4$, $\nu_1 = 1$)

The distribution law for $\mathbf{r}$ depends on $\mathbf{r}$ *only through the ratio* r/R_F

$$p_N(r) = \frac{1}{R_F{}^d} f_p\left(\frac{r}{R_F}\right) \quad (a \ll r \ll Na) \tag{I.24}$$

The prefactor $\frac{1}{R_F{}^d}$ is required to ensure the normalization

$$\int p_N(\mathbf{r})d\mathbf{r} = 1$$

The general structure of the reduced distribution $f_p(x)$ is shown in Fig. I.8 for $d = 3$. There is a very strong drop at large x

$$\lim_{x \to \infty} f_p(x) = \exp(-\,x^{\delta})\, f_1(x) \tag{I.25}$$

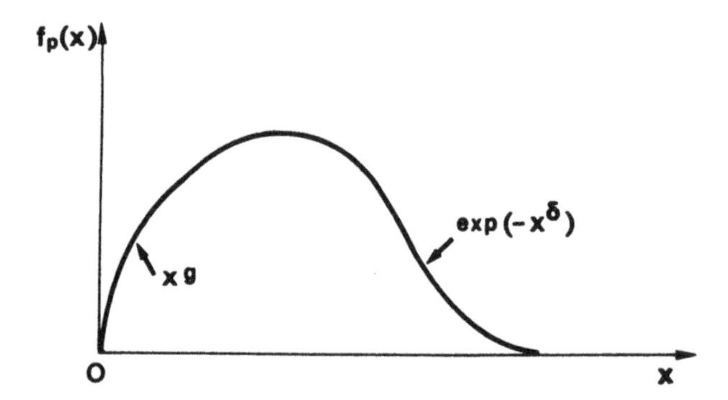

Figure I.8.

Distribution of the end-to-end distance r in a self-avoiding walk of N steps; x is equal to r/R_F, where R_F is the root mean square value.

where f_1 varies as a power of x. The exponent δ controls most of the chain properties for strong stretching and is given by:[14,15,16]

$$\delta = (1 - \nu)^{-1} \tag{I.26}$$

We present a simplified derivation of eq. (I.26) later in this chapter (see eq. (I.47) and the discussion following it).

At small x, f_p decreases sharply; it is exceptional for a self-avoiding walk to return close to its starting point

$$\lim_{x \to 0} f_p(x) = \text{constant } x^g \tag{I.27}$$

In three dimensions $g = g_3 \cong 1/3$. We relate g to other exponents below.

Let us consider the SAWs that return to a terminal site adjacent to the origin (Fig. I.9). In closing the $\alpha - \omega$ link we may say that each of these SAWs is associated with a closed polygon of $N + 1$ edges (and self-avoiding). The number of such polygons is of the form

$$\mathfrak{N}_N (r = a) \cong \bar{z}^N \left(\frac{a}{R_F}\right)^d \tag{I.28}$$

The factor R_F^{-d} is natural since the terminal points ω of *all* SAWs of N steps are spread over a d-dimensional volume R_F^d. What is remarkable in eq. (I.28) is the absence of the enhancement factor ($N^{\gamma-1}$) which was present in $\mathfrak{N}_N$ (tot) [eq. (I.21)]. This absence also reflects the difficulty for a SAW to return near its starting point.

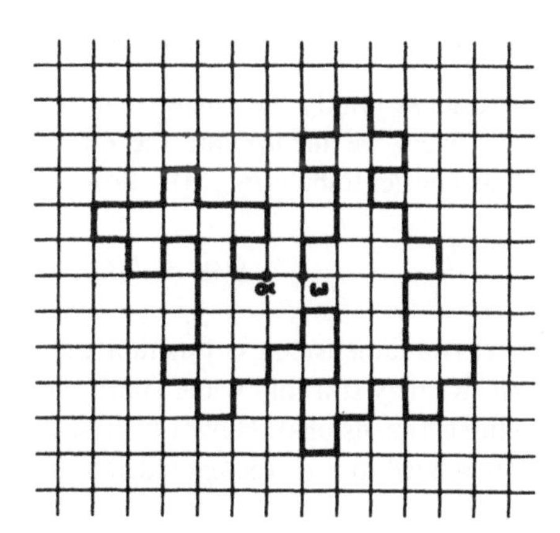

Figure I.9.

Eq. (I.28) is proved later by two independent methods (Chapters X and XI). If we accept it for the moment, we can predict simply what is the exponent g in eq. (I.27). The distribution function $p_N(\mathbf{r})$ taken for a terminal point adjacent to the origin ($r = a$) is from eqs. (I.24 and I.27)

$$p_N(a) \cong \frac{1}{R_F{}^d}\left(\frac{a}{R_F}\right)^g = \frac{1}{R_F{}^d} N^{-\nu g} \tag{I.29}$$

On the other hand, it is (by definition) related to $\mathfrak{N}_N(a)$ [eq. (I.28)]

$$p_N(a) = a^{-d} \frac{\mathfrak{N}_N(a)}{\mathfrak{N}_N(tot)} = \frac{1}{R_F{}^d} N^{1-\gamma}$$

Comparing this with eq. (I.29), we obtain:

$$g = \frac{\gamma - 1}{\nu} \tag{I.30}$$

a result first derived by des Cloizeaux.[17]

A REMARK ON HIGHER DIMENSIONALITIES

We have presented numerical data concerning $d = 3$, 2, and 1 (the latter being trivial). What would happen for larger d? The answer is simple: for $d > 4$, all exponents return to the ideal chain value ($\nu = 1/2$, $\gamma = 1$). This did not show up very clearly in the early numerical work but is a general theorem and is explained in Section I.3.2.

I.2.3. Correlations inside a swollen coil

Let us discuss briefly the changes in the pair correlation function $g(\mathbf{r})$ that occur when we incorporate the effects of excluded volume. First, $g(\mathbf{r})$ and its Fourier transform $g(\mathbf{q})$ follow simple scaling laws. For instance

$$g(\mathbf{q}) = N\tilde{g}(qR_F)$$

where $\tilde{g}(x)$ is a dimensionless function and $\tilde{g}(0) = 1$.

Second, we may still follow the approach of Chapter I and write $g(\mathbf{r}) = n/r^3$ (in three dimensions). However, now the number n of units inside the radius r is related to r by the excluded volume exponent $n^{3/5}\, a \sim r$. This gives

$$g(r) \cong \frac{1}{r^{4/3}\, a^{5/3}} \; (r < R_F) \; (d = 3) \tag{I.31a}$$

i.e., a more rapid decrease than for ideal coils. The Fourier transform is

$$g(q) \cong \frac{1}{(qa)^{5/3}} \; (qR_F > 1) \; (d = 3) \tag{I.31b}$$

These power laws were derived first by S. F. Edwards.[18] They have been verified directly on dilute chains with X-rays.[19] They have also been checked by neutron scattering experiments on semi-dilute systems (see Section III.2.5).

I.2.4. Summary

Real chains in good solvents have the same universal features as self-avoiding walks on a lattice. These features are described by two "critical exponents," γ and ν. All other exponents of interest can be expressed in terms of these two. The exponent γ is related to chain entropy, and the exponent ν is related to chain size. A real chain has a size ($R_F \sim N^\nu$), which is much larger than an ideal chain ($R_0 \sim N^{1/2}$). For three dimensions the exponent ν is very close to 3/5.

I.3. The Flory Calculation of the Exponent ν

I.3.1. Principles

Long ago, Flory devised a simple and brilliant scheme for computing the exponent ν, which gives excellent values for all dimensionalities.[20] We briefly describe his method and the approximations involved. The starting point is a chain, with a certain unknown radius R and an internal monomer concentration

$$c_{int} \cong \frac{N}{R^d} \tag{I.32}$$

(Note that we present the argument for an arbitrary dimensionality d).

There is a certain repulsive energy in the chain due to monomer monomer interactions. If c is the local concentration of monomers, the repulsive energy per cm^3 is proportional to the number of pairs present—i.e., to c^2. We write it (per unit volume) as:

$$F_{rep} = \frac{1}{2} Tv\,(T)c^2 \tag{I.33}$$

where v has the dimension of a (d dimensional) volume and is positive. We call v the excluded volume parameter. [In the Flory notation $v = (1-2\chi)\,a^d$ where a^d is the monomer volume and χ is an interaction parameter. For good solvents $\chi < 1/2$ and $v > 0$.]

One essential approximation is to replace the average of c^2 (inside the coil) by the square of the average

$$\langle c^2 \rangle \rightarrow \langle c \rangle^2 \sim c_{int}{}^2 \tag{I.34}$$

Eq. (I.34) is typical of a *mean field* approach: all correlations between monomers are ignored. The overall repulsive energy after integration over a volume R^d, scales as:

$$F_{rep}\big|_{tot} \cong Tv(T)c_{int}{}^2\, R^d = Tv\frac{N^2}{R^d} \tag{I.35}$$

This tends to favor large values of R (i.e., to swell the chain). However if the distortion is too large, the chain entropy becomes too small, and this is unfavorable. Flory includes this through an elastic energy term derived from the ideal chain result [eq. (I.8)]

$$F_{el} \cong T\frac{R^2}{Na^2} \tag{I.36}$$

Eq. (I.36) is also a very strong approximation; as shown later, the spring constant of a real chain is much smaller than that suggested by eq. (I.36). However, let us accept eqs. (I.35) and (I.36) and add them:

$$\frac{F}{T} \cong v\frac{N^2}{R^d} + \frac{R^2}{Na^2} \tag{I.37}$$

Eq. (I.37) has a minimum for a well defined radius $R = R_F$. Omitting all numerical coefficients, we find

$$R_F{}^{d+2} \cong va^2N^3 \tag{I.38}$$

or $R_F \sim N^\nu$ with*

*Eq. (I.39) was written by Flory for $d = 3$. For general d, it was first quoted by M. Fisher, *J. Phys. Soc. Japan* **26** (Suppl.) 44 (1969).

$$\nu = \frac{3}{d+2} \tag{I.39}$$

Eq. (I.39) is amazingly good; it gives the correct value for $d = 1$ ($\nu_1 = 1$). The values for $d = 2$ and $d = 3$ are within a percent of the most accurate numerical results.[12,21] For most practical applications the Flory formula can be considered exact.

I.3.2. Chains are ideal above four dimensions

Eq. (I.39) tells us that $\nu = 1/2$ for $d = 4$. This is precisely the ideal chain exponent. We can understand this better if we return to the repulsive energy [eq. (I.35)]. We expect $R > R_0$, and thus the repulsive energy is at most of order

$$F_{rep.\ max} \cong vT \frac{N^2}{R_0^d} \cong T \frac{v}{a^d} N^{2-d/2} \tag{I.40}$$

while the elastic energy [eq. (I.36)] is at least of order T. We see then that the ratio

$$\frac{F_{rep}}{F_{el}} \leqslant N^{2-d/2} \tag{I.41}$$

For dimensionalities of $d > 4$ we conclude that repulsions between monomers represent only a weak perturbation; the local concentration in an ideal chain is so low that excluded volume effects become negligible.

The idea of calculating the effects of repulsions by *perturbation* methods (treating the excluded volume v as infinitesimally small) is relatively old.[22] When this is done, to first order in v, one finds*

$$\frac{R - R_0}{R_0} \cong \frac{F_{rep.\ max}}{T} = (\text{constant})\ \zeta + 0\ (\zeta^2)$$

$$\zeta \cong \frac{v}{a^d} N^{2-d/2} \tag{I.42}$$

Thus the real, dimensionless, expansion parameter is ζ. When ζ is small, the chain is ideal. When ζ is large, the chain shows strong excluded volume effects. (For intermediate ζ values a precise interpolation formula

*For the most simplified models R/R_0 is a function of ζ only. This point will be discussed more in Chapter XI.

has been worked out by Domb and Barrett.[23]) Note that for the usual case $d = 3$, the parameter

$$\zeta \cong \frac{v}{a^3} N^{1/2} \tag{I.42'}$$

is always large for large N; eq. (I.42) has a very limited range of validity. The self-consistent method of Flory is clearly much more powerful, but the characteristic parameter ζ will be of frequent use in this book.

I.3.3. Why is the Flory method successful?

It is important to realize that the self-consistent calculation of eqs. (I.35, I.36) benefits from a remarkable cancellation of two errors:

(i) The repulsive energy is enormously overestimated when correlations are omitted.

(ii) The elastic energy is also largely overestimated; if we think for example of the end-to-end elongation of the chain, since the distribution function $p_N(r)$ [eq. (I.24)] is a function of (r/R_F) only, this implies that the entropy at fixed r is also a function of r/R_F only. Finally the elastic energy should be written $Tr^2/R_F{}^2$ rather than Tr^2/R_0^2. Again this brings in a large reduction.

As often happens in self-consistent field calculations (e.g., in the Hartree theory of atoms) the two errors *(i)* and *(ii)* cancel each other to a large extent. Many post-Flory attempts, which tried to improve on one term, *(i)* or *(ii)*, leaving the other unaltered, led to results that were poorer than eq. (I.39).

In fact, another problem exists in chain statistics, where the self-consistent method does not benefit from the same cancellations. This is the case of a charged chain (polyelectrolyte) for which a self-consistent approach was attempted very early.[24,25] Here the neglect of correlations is less serious because most of the repulsion comes from very distant monomers. Thus point (i) is improved, but point (ii) remains weak; the net result is a formula for ν in charged systems which gives incorrect values for $3 < d < 6$.[26] We return to this problem in Chapter XI.

I.4. Constrained Chains

We now turn to a discussion of real chains in good solvents, when external constraints are applied. The basic situations are listed in Section I.1. in connection with ideal chains. We shall see that all exponents are modified strongly by excluded volume effects, and that most of them can be related directly to the exponent ν. To simplify the notation, we set $\nu = 3/5$ (the Flory value) for three-dimensional systems.

I.4.1. A chain under traction (Fig. I.4)

The external energy due to the force f, when the end-to-end distance is $\mathbf{r}$, is simply $-\mathbf{f}\cdot\mathbf{r}$. Thus, we may write a partition function for the chain in the form

$$Z = \int d\mathbf{r}\; p_N(\mathbf{r}) \exp(\mathbf{f}\cdot\mathbf{r}/T) \tag{I.43}$$

and using the results from Section (I.2) on $p_N(\mathbf{r})$, we can compute all averages involving $\mathbf{r}$.[12] Here, however, we use a simpler approach due to Pincus.[27] The only characteristic lengths entering into eq. (I.43) are: 1) the Flory radius, $R_F \cong aN^{3/5}$, and 2) the length $\xi_p = T/f$.

Let us now consider the elongation $\mathbf{r}(\mathbf{f})$. We may write

$$|\langle\mathbf{r}\rangle| = R_F\, \varphi_r \left(\frac{R_F}{\xi_p}\right) = R_F\, \varphi_r\,(x) \tag{I.44}$$

where φ_r is a dimensionless function. For small f we expect $|\langle\mathbf{r}\rangle|$ to be linear in x, and thus $\varphi_r\,(x \rightarrow 0) \cong x$

$$|\langle\mathbf{r}\rangle| \cong \frac{R_F{}^2}{T} f \qquad (fR_F < T) \tag{I.45}$$

Note that $\langle\mathbf{r}\rangle$ is not linear in N at small f. This means that the tension f is transmitted not only through the backbone (as in the ideal case) but also through contacts between certain pairs of monomers (n, m) (with $|n-m|$ large).

Consider now the limit of large tensions ($x \gg 1$). What happens here can be idealized as shown in Fig. I.10. The chain breaks up into a series

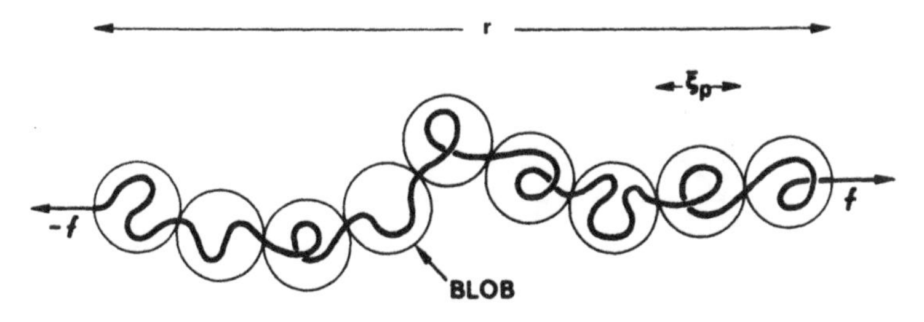

Figure I.10.

of "blobs" each of size ξ_p. Inside a blob (i.e., for spatial scales $r < \xi_p$) the force f (measured by the dimensionless number fr/T) is a weak perturbation. Thus, each blob retains the local correlations of a Flory chain, but at larger scales $r > \xi_p$ we have a string of independent blobs.

The number of monomers per blob, g_p, is related to ξ_p by the Flory law of real chains [eq. (I.39)], giving

$$\xi_p \cong a\, g_p^{3/5}$$

or

$$g_p = \left(\frac{T}{af}\right)^{5/3} \qquad \text{(I.46)}$$

and the total number of blobs is N/g_p. The chain elongation is then

$$|\langle \mathbf{r}\rangle| \cong \frac{N}{g_p}\xi_p \cong Na\left(\frac{fa}{T}\right)^{2/3} \qquad \left(\frac{fa}{T} \ll 1\right) \qquad \text{(I.47)}$$

Eq. (I.47) deserves some discussion. We see that a real chain has an elastic response which is significantly more nonlinear than an ideal chain. This appears on the plot of $\varphi(x)$ shown qualitatively in Fig. I.11.

The high f limit could have been obtained directly on the scaling form [eq. (I.44)] by imposing the restriction that $|\langle \mathbf{r}\rangle|$ becomes linear in N at high f. The reason for this linearity is that at high f, separate blobs do not interact; thus, we return to an ideal string of blobs.

Eq. (I.47) allows us to derive the exponent δ defined in connection with the strongly stretched limit [eq. (I.25)]. At large r the probability distribu-

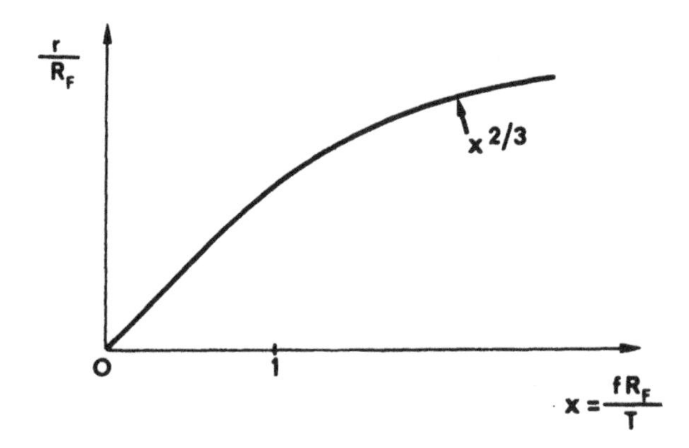

Figure I.11.

tion is essentially proportional to exp $-(r/R_F)^\delta$, and the entropy for fixed elongation $S(\mathbf{r})$ has the form

$$S(\mathbf{r}) = \text{constant} + \ln p_N(\mathbf{r})$$
$$= \text{constant}\ (-)\left(\frac{r}{R_F}\right)^\delta \qquad \text{(I.48)}$$

The corresponding elastic free energy is $-TS$, and the overall energy is

$$F_{tot} = T\left(\frac{r}{R_F}\right)^\delta - fr$$

The physically realized elongation corresponds to the minimum of F:*

$$f \cong \frac{T}{R_F}\left(\frac{r}{R_F}\right)^{\delta-1} \qquad \text{(I.49)}$$

Comparing eq. (I.49) with (I.47) we see that $\delta = 5/2$ (when $\nu = 3/5$). Keeping a more general value of ν would lead to eq. (I.26).

Apart from the longitudinal elongation $\langle \mathbf{r} \rangle$ (parallel to f) it is of interest to ascertain the lateral spread of the chain $r_\perp$ in strong elongation. The projection of the string of blobs on a plane normal to $\mathbf{f}$ is an ideal string, and thus

$$\langle r_\perp^2 \rangle \cong \frac{N}{g_p}\,\xi_p^2 \cong Na^2\left(\frac{T}{fa}\right)^{1/3} \qquad (fR_F > T) \qquad \text{(I.50)}$$

Thus the chain not only elongates but also shrinks in its lateral dimensions.

No experimental verifications of the laws [eqs. (I.47, I.50)] seem to be available at present. For the future, studies on strong distortions in flows of dilute solutions, and also in gels, may become relevant.

I.4.2. Squeezing a real chain in a tube

In one dimension, excluded volume effects are very strong. Thus it is of interest to consider a chain trapped in a thin tube of diameter $D \ll R_F$ (but D still larger than a). Situations of this sort may become available in the future. What is the length of tube $R_\parallel$ occupied by the chain? What is the energy required to squeeze the chain in?

Let us start with the length $R_\parallel$; it must have the scaling form

*The mathematically inclined reader will recognize that this describes a saddle point integration in Eq. (I.43).

$$R_{\|} = R_F \, \Phi_{\|} \, (R_F/D) \tag{I.51}$$

where $\Phi_{\|}(x) \rightarrow 1$ for $x \rightarrow 0$ (thick tube) and $\Phi_{\|}(x) \rightarrow x^m$ when $x \rightarrow \infty$ (thin tube). To determine the exponent m, we notice that for a thin tube we have a one-dimensional problem, and $R_{\|}$ must therefore be a *linear* function of N. Since $R_F \sim N^{\nu_3}$ this requirement means that

$$\begin{aligned} N^{\nu_3(1+m)} &\cong N \\ m = \nu_3^{-1} - 1 &= 2/3 \end{aligned} \tag{I.52}$$

Thus the formula for the length of the chain is[28]

$$R_{\|} \cong Na \left(\frac{a}{D}\right)^{2/3} \qquad (a \ll D \ll R_F) \tag{I.53}$$

Note that $R_{\|}$ is larger than R_F. The chain is extended by squeezing, and this behavior is very different from an ideal chain. Further, the concentration inside the chain is interesting. It scales according to:

$$c_{int} \cong \frac{N}{D^2 \, R_{\|}} \sim \frac{1}{a^3} \left(\frac{a}{D}\right)^{4/3} \tag{I.54}$$

and is independent of N.

Another derivation of eq. (I.53) is based on a "blob" picture. The chain behaves as a sequence of blobs of diameter D. Inside each blob the effects of the boundaries are weak. The number g_D of monomers per blob is still given by the three-dimensional law: $g_D^{3/5} = D/a$. Successive blobs act as hard spheres and pack into a regular one-dimensional array. Thus $R_{\|} = N/g_D \, D$ in agreement with eq. (I.53).

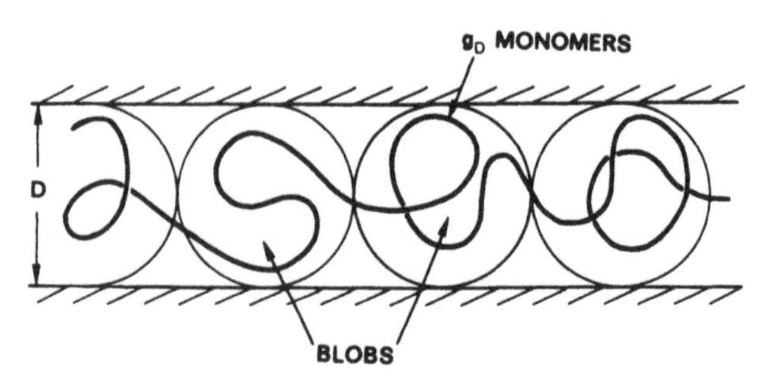

Figure I.12.

Let us now turn to the *confinement energy*. In the strong confinement limit ($D < R_F$) we see from Fig. I.12 that the energy must be linear in N; doubling N simply doubles the number of blobs. On the other hand, the energy must be of the form

$$F_{conf} \cong T\, \varphi_F \left(\frac{R_F}{D}\right) \cong T\, \varphi_F\,(x)$$

$$\lim_{x\to\infty} F_{conf} \cong Tx^n \qquad \text{(I.55)}$$

Thus R_F^n must be linear in N, and $n = 5/3$.

$$F_{conf} = TN \left(\frac{a}{D}\right)^{5/3} \qquad \text{(I.56)}$$

Note first the difference in behavior from the ideal chain [eq. (I.12)]. The confinement energy (at given D and N) is larger for the real chain. Note also the analogy between eq. (I.56) and the energy for strong elongation of a chain in free space [eq. (I.48)] $1/T\,(F_{conf}) \cong (R_{\parallel}/R_F)^{5/2}$. Thus $R_{\parallel}$ plays the role of the total elongation $\langle \mathbf{r} \rangle$ in the Pincus problem.

GENERALIZATIONS

This analysis can be extended to chains that are squeezed in slits and to other geometries provided that the confining object is characterized by a single length D. One such case has been recently studied by numerical methods.[29,30] This corresponds to a two-dimensional lattice, where we allow the chains to explore only a finite strip of width D. Then a similar argument suggests $R_{\parallel} \sim Na\,(a/D)^{1/3}$; this dependence on N and especially on D seems well confirmed by the data.

REFERENCES

1. W. J. Orr, *Trans. Faraday Soc.* **43**, 12 (1947).
2. *Selected Papers on Noise and Stochastic Processes*, N. Wax, Ed., Dover, New York, 1954.

3. P. Flory, *Statistics of Chain Molecules,* Interscience Publishers, New York, 1969.
4. E. F. Cassasa, *J. Polymer Sci.* **B5,** 773 (1967).
5. E. F. Cassasa, Y. Tagami, *Macromolecules* **2,** 14 (1969). E. F. Cassasa, *Macromolecules* **9,** 182 (1976).
6. R. Rubin, *J. Chem. Phys.* **43,** 2392 (1965). R. Rubin, *J. Res. Nat. Bur. Std.* **70B,** 237 (1966).
7. C. Hoeve, S. Di Marzio, P. Peyser, *J. Chem. Phys.* **42,** 2558 (1965). C. Hoeve, *J. Polym. Sci.* **C30,** 361 (1970). C. Hoeve, *J. Polym. Sci.* **C34,** I (1971).
8. A. Silberberg, *J. Chem. Phys.* **46,** 1105 (1967). A. Silberberg, *J. Chem. Phys.* **48,** 2835 (1968).
9. P. G. de Gennes, *J. Phys. (Paris)* **37,** 1445 (1976).
10. P. Debye, *J. Phys. Colloid Chem.* **51,** 18 (1947).
11. W. Stockmayer, "Dynamics of Chain Molecules," in *Fluides Moléculaires,* R. Balian and G. Weill, Eds., Gordon & Breach, New York, 1976.
12. For reviews on these methods as applied to polymers see: C. Domb, *Adv. Chem. Phys.* **15,** 229 (1969). D. S. McKenzie, *Phys. Rept.* **27C** (2) (1976).
13. F. T. Wall, S. Windwer, P. J. Gans, "Monte Carlo Methods Applied to Configurations of Flexible Polymer Molecules," in *Methods of Computational Physics,* Vol. 1, Academic Press, New York, 1963.
14. M. E. Fisher, *J. Chem. Phys.* **44,** 616 (1966).
15. M. E. Fisher, R. J. Burford, *Phys. Rev.* **156,** 583 (1967).
16. D. McKenzie, M. Moore, *J. Phys.* **A4,** L82 (1971).
17. J. des Cloiseaux, *Phys. Rev.* **A10,** 1665 (1974).
18. S. F. Edwards, *Proc. Phys. Soc. (London)* **85,** 613 (1965).
19. K. Okano, E. Wada, H. Hiramatsu, *Rep. Prog. Polym. Sci. Japan* **17,** 145 (1974).
20. P. Flory, *Principles of Polymer Chemistry,* Chap. XII, Cornell University Press, Ithaca, N.Y., 1971.
21. J. C. Le Guillou, J. Zinn Justin, *Phys. Rev. Lett.* **39,** 95 (1977).
22. See, for example, H. Yamakawa, *Modern Theory of Polymer Solutions,* Harper & Row, New York, 1972.
23. C. Domb, A. J. Barrett, *Polymer* **17,** 179 (1976).
24. J. Hermans, J. Overbeek, *Rec. Trav. Chim.* **67,** 761 (1948).
25. W. Kuhn, D. Kunzie, A. Katchalsky, *Helv. Chim. Acta* **31,** 1994 (1948).
26. P. Pfeuty, R. M. Velasco, P. G. de Gennes, *J. Phys. (Paris) Lett.* **38,** 5 (1977).
27. P. Pincus, *Macromolecules* **9,** 386 (1976).
28. M. Daoud, P. G. de Gennes, *J. Phys. (Paris)* **38,** 85 (1977).
29. F. T. Wall, W. A. Seitz, J. C. Chin, P. G. de Gennes, *Proc. Nat. Acad. Sci. U.S.* **75,** 2069 (1978).

30. M. Barber, A. Guttman, K. Middlemiss, G. Torrie, S. Whittington, *J. Phys. (London)* **A11,** 1833 (1978).

II

Polymer Melts

II.1.

Molten Chains are Ideal

Dilute polymer chains in a good solvent are swollen; the size R_F increases as $N^{3/5}$ rather than $N^{1/2}$. When we squeeze the chains together and reach a concentrated solution or a melt, we might expect the situation to become even more complicated because the interactions between monomers are much stronger. Actually the correct conclusion is different. In a dense system of chains each chain is gaussian and ideal. This was first understood by Flory,[1] but the notion is so unexpected that it took a long time to reach the community of polymer scientists. Here we give two presentations of this ''Flory theorem'' and discuss fine correlation effects between chains in the melt.

II.1.1. A self-consistent field argument

Consider a dense system of identical chains (Fig. II.1). Let us focus on one chain, which we shall call the ''white chain'' (the other chains are ''black''), and study the repulsive potential U experienced by one white monomer. This U is essentially proportional to the local concentration c of monomers, which is shown in Fig. II.1b. The concentration c has two components.

The concentration of white monomers (and the corresponding potential U_w) is peaked around the center of gravity of the white molecule. On the slopes of this peak there is a force $-\,\partial U_w/\partial x$ pointing outward. In the

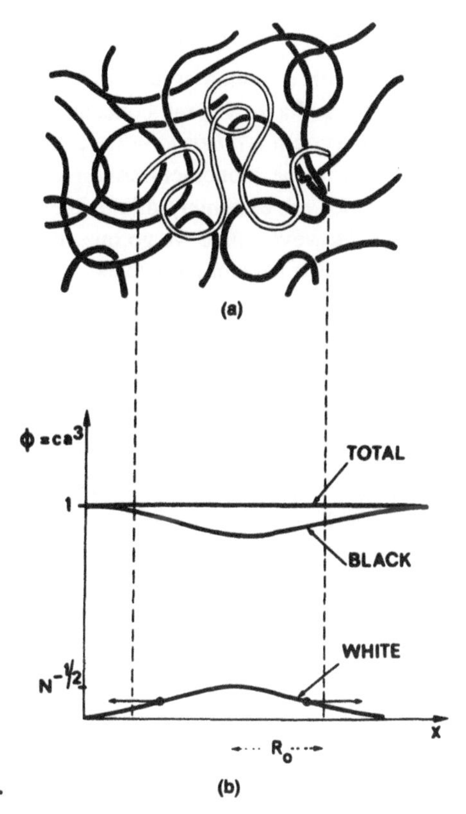

Figure II.1. (b)

single-chain problem, it is this force which is responsible for swelling and nonideal behavior. However, the concentration profile for the black chains has a trough in the same region. This is necessary since in a molten polymer, the fluctuations of the *total* concentration (or density) are very weak.

Therefore, the black potential creates an *inward* force. This force exactly equals the force caused by the white monomers since U_{tot} (like c_{tot}) is constant in space: $\partial U_{tot}/\partial x \equiv 0$. The chain experiences no force and remains ideal.

The above argument is useful and valid in three dimensions. However, it applies only if the whole idea of a *self-consistent field* makes sense. Clearly, there are cases where it fails; for example, in one dimension, the chains must be completely stretched (at all concentrations). What happens then in two dimensions? The answer can be derived from the detailed (local) discussion below.

II.1.2. Screening in dense polymer systems

Let us start with a molten system of chains (degree of polymerization N) to which we add a very weak concentration of solvent molecules. We assume that the system is *athermal*—i.e., the solvent-monomer interaction is exactly equal to the monomer-monomer interaction. Of course, if the solvent molecules are extremely dilute, they behave as a gas of independent particles. However, here we go one step further to consider the interactions between two solvent molecules.

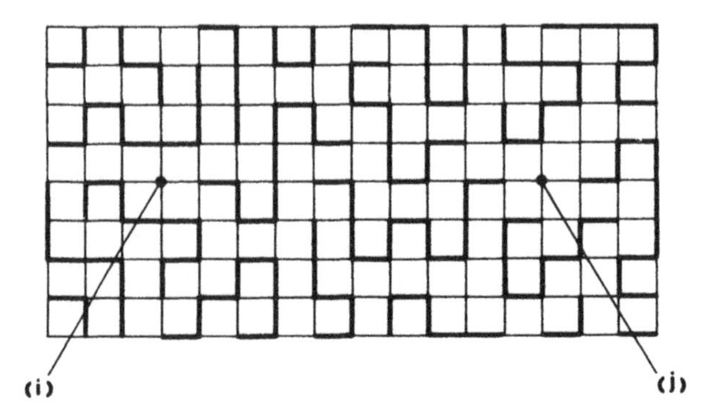

Figure II.2.

The situation is shown on a lattice model in Fig. II.2. The two solvent units block sites (i) and (j) of the lattice, and the chains fill all other sites. There is a certain (huge) number of ways of filling the allowed sites with chains, and the logarithm of this number defines the total entropy S_{ij}. For large separations R_{ij} the entropy S_{ij} goes to a well-defined limit S_∞. When R_{ij} becomes smaller and comparable with the single chain radius $R_0 = N^{1/2} a$, the entropy S_{ij} changes* as shown qualitatively in Fig. II.3. No rigorous calculation of S_{ij} is available at present, but an approximate form can be derived from methods to be described in Chapter IX. The results are as follows.

When averaged over all sites j different from i, the shift ΔS_{ij} is *positive*. Our reader may acquire a qualitative feeling for this sign by inspection of a (comparatively) simple case—namely, dimers ($N = 2$) on a square lattice. If we look at two *distant* units (i) and (j) (Fig. II.4a), we see that each has four neighbors where a dimer molecule is restricted in orientation; starting on any of these neighbors the dimer can point only in three directions (α β γ). On the other hand, if (i) and (j) are adjacent, there are only six sites

*Note that Fig. II.3 applies only for dimensionalities $d = 2, 3, \ldots$ but not for $d \sim 1$ (S_{ij} is meaningless for $d = 1$).

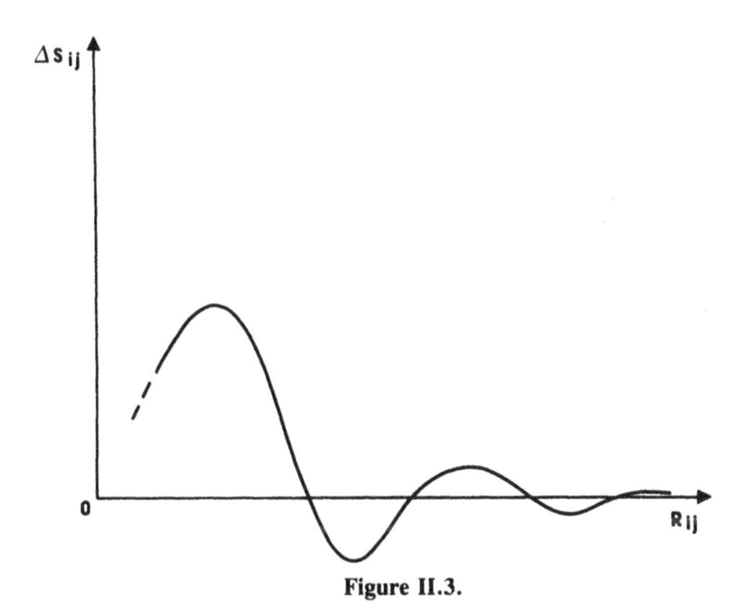

Figure II.3.

(rather than eight) which have this reduction in allowed orientations. Thus, the entropy should be larger in Fig. II.4b. The above argument does not give a precise calculation of the entropy, but it is a good clue to its sign.

Returning to the general form of S_{ij}, we may also translate it into an effective interaction U_{ij} between sites (i) and (j) putting

$$\exp\left(-\frac{U_{ij}}{T}\right) = \Delta S_{ij} \qquad (i \neq j) \tag{II.1}$$

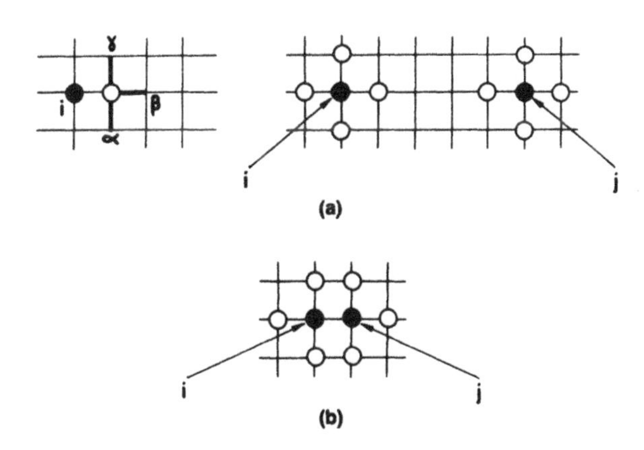

Figure II.4.

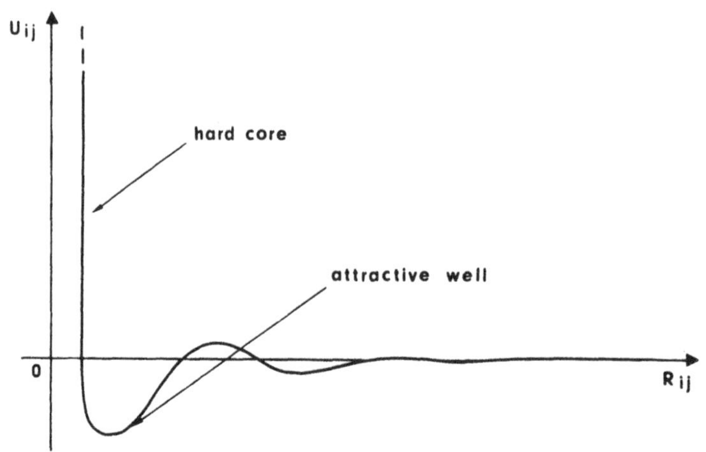

Figure II.5.

For $i = j$ we must take U_{ij} as repulsive and infinitely large because we do not allow two solvent molecules on the same site. The resulting aspect of U_{ij} is shown in Fig. II.5. At finite R_{ij}, the interaction oscillates in sign, but it is predominantly attractive. At $R_{ij} > R_0$ the interaction drops to zero.

The repulsive core is what we might call the bare interaction; it would be present even in the absence of polymer chains on surrounding sites. The attractive part is due to the chains and tends to reduce or "screen out" the bare interaction. This notion of screening was first introduced—in a somewhat different context—for semi-dilute solutions by S. F. Edwards.[2]

To measure the importance of the screening effect, the main parameter of interest is the *second virial coefficient* A_{2S} of the solvent–solvent interaction. If we define a solvent fraction Φ_S (the average number of solvent units per site) and a solvent osmotic pressure Π_S, the latter has an expression of the form

$$\frac{a^3 \Pi_S}{T} = \Phi_S + A_{2S}\,\Phi_S^2 + O\,(\Phi_S^3) \qquad \text{(II.2)}$$

and this defines A_{2S}. On a microscopic level, A_{2S} is related to the pair correlation function g_{ij} between two solvent units

$$g_{ij} = \begin{cases} e^{\Delta S_{ij}} & (i \neq j) \\ 0 & i = j \end{cases} \qquad \text{(II.3)}$$

and we have

$$\begin{aligned} 2A_{2S} &= \sum_j (g_{ij} - 1) \\ &= -1 + \sum_{j \neq i} [exp\,(\Delta S_{ij}) - 1] \end{aligned} \qquad \text{(II.4)}$$

Having defined A_{2S} in detail, we can now state a fundamental *screening theorem* (in three dimensions)

$$\lim_{N\to\infty} A_{2S} = 0 \tag{II.5}$$

Physically this means that for long chains (large N) there is complete cancellation between the attractive and repulsive parts of the interaction U_{ij}—i.e., a dilute solution of monomers in the melt is *ideal*.

We give some justification for eq. (II.5) in Chapter IV, where polymer–solvent systems are discussed in more detail. In particular, with the simplest model, due to Flory and Huggins, one finds for all N [see Chapter III, eq. (III.16)]:

$$2A_{2S} = \frac{1}{N} \tag{II.6}$$

Thus, for large N, the direct repulsion between solvent molecules is very efficiently screened out. We accept for the moment the result [eq. (II.6)] and use it to discuss chain conformations.

II.1.3. One long chain among shorter chains

Having stated the screening theorem, we can now return to a deeper discussion of the conformation of one particular chain in the melt. It is interesting to generalize the problem slightly and to assign a degree of polymerization N_1 to this particular chain, all others having the degree of polymerization N. This generalization is interesting because we already have a feeling for two limiting behaviors:

(i) If $N_1 = N$, we expect the chain to be ideal (in three dimensions).

(ii) If N_1 is large but N is unity, we return to the problem of a single chain in a good solvent: the chain is swollen. The question is then: when do we cross from regime (i) to regime (ii)?

The Flory approach to this problem[1] is simply to consider that all monomers of the long chain interact between themselves through the screened interaction; for the excluded volume parameter v of Chapter I we must substitute the virial coefficient A_{2S}. More precisely we have

$$v = 2A_{2S}\, a^d = a^d \frac{1}{N} \tag{II.7}$$

where a^d is the volume of the unit cell in the lattice (d is always the dimensionality).

We can now write the dimensionless parameter ζ [introduced in eq. (I.42)] which tells us whether the chain (N_1) is ideal or not.

$$\zeta \cong \frac{v}{a^d} N_1^{2-d/2} = N_1^{2-d/2}/N \tag{II.8}$$

Consider first the usual *three-dimensional* case, $d = 3$. Then we have

$$\zeta = N_1^{1/2}/N \tag{II.9}$$

and we see the following limits:

(i) If N_1 is comparable with N, ζ is of order $N^{-1/2}$ and is small—i.e., the chain is ideal.

(ii) If N decreases and becomes a somewhat smaller than N_1, the chain stays ideal up $N = N_1^{1/2}$. Then it begins to swell.*

For *two dimensions*

$$\zeta = N_1/N \tag{II.10}$$

is large whenever $N_1 > N$ and is of order unity when $N_1 \sim N$. We conclude that for $d = 2$ the chains in a "melt" are *not quite ideal.* This point does not seem to have been noticed in the literature.

At first one might be inclined to use eq. (II.10) also for one short chain among larger chains ($N_1 < N$). However, in this limit, the simple recipe [eq. (II.7)] breaks down. Because the screening radius ($R_0 = N^{1/2}\, a$) is now larger than the size of the chain of interest ($\sim N_1^{1/2}\, a$), it is not possible to replace U_{ij} of eq. (II.1) by a point interaction with one coefficient A_{2S}. (A more detailed study suggests that ζ stays of order unity at all $N_1 < N$).

II.1.4. Mixed chains versus segregated chains

Some years ago there existed among polymer scientists some dispute between two opposite views on polymer melts as represented in Fig. II.6 a, b). In the standard model (a) the chains interpenetrate very strongly.

*Neutron experiments qualitatively confirm these features; see R. Kirste, B. Lehnen, *Makromol. Chem.* **177**, 1137 (1976). Quantitatively there remain some unexplained discrepancies.

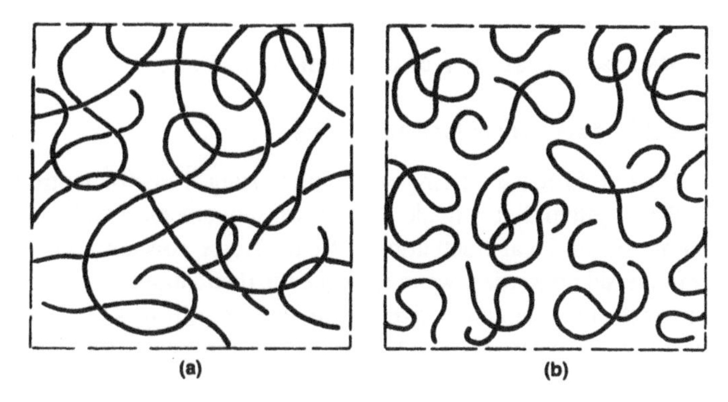

Figure II.6.

In the segregated model (b) different chains overlap only slightly. What is the correct picture? The answer depends on dimensionality.

For three dimensional situations, neutron experiments probing a few labeled (deuterated) chains in melt of identical (hydrogenated) chains, have shown quite convincingly that the chains are ideal and gaussian as expected from the Flory theorem.[3,4,5] The radius is $R_0 = N^{1/2} a$, and the local concentration due to one labeled chain is of order $N/R_0^3 \sim N^{-1/2} a^{-3}$. This local concentration is small; this implies that there are many chains overlapping to build up the total concentration (a^{-3}) in the melt, and that Fig. II.6a applies.

Very few data are available at the moment on "two-dimensional melts" or analog systems on computers. However, we expect that here model (b) is closer to the real situation. Since the perturbation parameter ζ is of order unity, the chains still have a radius comparable with R_0. Thus the two-dimensional concentration associated with one chain is $c = N/R_0^2 \sim a^{-2}$. A single chain is enough to build up a c value comparable with the total concentration, and the chains must be somewhat segregated. It is to be hoped that future experiments using chains on adsorbed layers (or chains trapped in lamellar systems such as lipid + water), will be able to probe these questions.

II.1.5 Summary

In three-dimensional polymer melts, the chains are essentially ideal and move freely. In two dimensions they should be slightly swollen and strongly segregated.

II.2.
Microscopic Studies of Correlations in Melts

II.2.1. Necessity of labeled species

If we perform a scattering experiment (using X-rays or neutrons) on a dense chain system that is made of identical monomers, we obtain some information on the local fluctuations of density, or concentration. If we look for universal properties, we must probe the fluctuations at wavelengths much larger than the monomer size. (More precisely, if q is the scattering wave vector,* we want $qa \ll 1$.) However, at this scale, the density in a melt is essentially fixed—i.e., there is no scattering intensity.

Thus, to obtain interesting data at small q, we must label the chains. When working with X-rays, it is tempting to insert *heavy atoms* at definite positions along the chain (e.g., at both ends). However, this does not work very well because of the strong chemical perturbation introduced by these atoms; very often they tend to segregate, building up micellar (or more complex) structures which have little to do with the original, unperturbed melt.

A cleaner situation has been achieved with *deuterium labeling* in neutron scattering. Fortunately, deuterium and hydrogen have very different coherent scattering amplitudes for thermal neutrons. Thus, it is possible to achieve an isotopic labeling which is efficient for neutrons but leaves the system nearly unperturbed.†

Experiments of this type have been carried out on partially deuterated polystyrenes (using quenched phases from the melt) by J. P. Cotton and co-workers.[6] These experiments give us precise information on the local correlations between chains in a polymer melt. Also, because of the simplicity introduced by the Flory theorem, this is one of the few cases where the scattering diagrams can be computed accurately.[7] The method is described in Chapter IX. Here we present the results only, in qualitative terms.

II.2.2. The correlation hole

Consider, for example, the case where each chain is labeled at *one end* only: the first monomer of each chain is deuterated. We take one such

*In terms of the beam wavelength λ and scattering angle θ we have $q = 4\pi\lambda^{-1} \sin \theta/2$.

†There is, however, an ultra-weak difference in interactions between normal and deuterated monomers, which has been considered recently by Buckingham. This may lead to some segregation effects in chains of very high molecular weight. We return to this point in Chapter IV.

chain in the melt and put its labeled unit at the origin. Now we ask for the distribution S_{11} (**r**) of labeled ends at distance **r** from the origin. There is, of course, a delta function at the origin (self-correlation), but the most interesting part, for $r \neq 0$, is shown in Fig. II.7.

(*i*) At large distances $S_{11}(\mathbf{r})$ is equal to the average density of labels in the melt. If we define density as a number per site on the lattice, the label density is just $1/N$ (since there are N lattice sites occupied by one chain, of which one only carries a label).

(*ii*) At smaller distances ($a < R_0$) $S_{11}(r)$ is *reduced*. When we have fixed one label at the origin 0, we know that one chain is inside a sphere of radius $\sim R_0$ near 0. Thus, the density allowed for other chains is reduced, and the density of other labels also drops.

(*iii*) At the origin $S_{11}(\mathbf{r})$ has a delta function peak corresponding to the fixed labeled unit.

Qualitatively, we may say that $S_{11}(\mathbf{r})$ shows a *correlation hole* of width R_0 and of average depth $\sim 1/N\ a/R_0 \sim N^{-3/2}$. From an experimental point of view, what is directly measured is the Fourier transform of $S_{11}(\mathbf{r})$

$$S_{11}(\mathbf{q}) = \frac{1}{a^3} \int d\mathbf{r} \exp(i\mathbf{q}\cdot\mathbf{r})\, S_{11}(\mathbf{r}) \qquad \text{(II.11-12)}$$

(where the unit cell volume a^3 has been introduced to make $S_{11}(q)$ dimensionless).

The structure of $S_{11}(q)$ for a chain labeled at one end is shown in Fig. II.8 (from Ref. 7). Note that for $q \rightarrow 0$ the scattering intensity vanishes. For long wavelength fluctuations, the label density ρ_1 is simply ρ/N, and the total density ρ does not fluctuate in our model; there is no scattering.

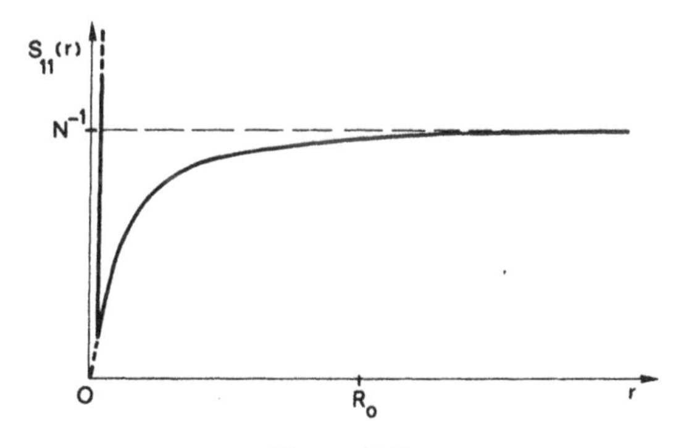

Figure II.7.

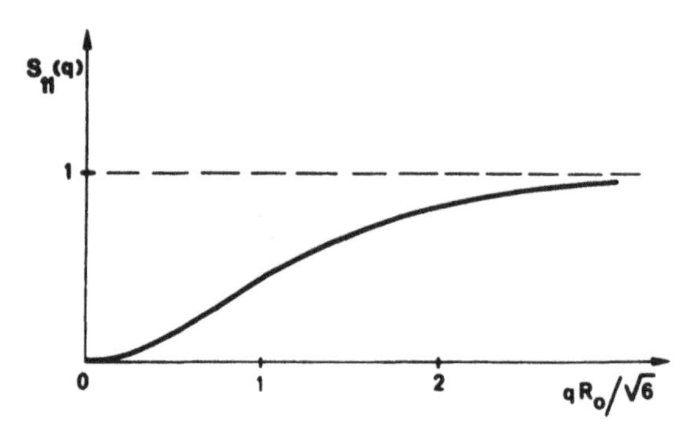

Figure II.8.

This corresponds to the following sum rule in direct space:

$$\int S_{11}(\mathbf{r})d\mathbf{r} = 0 \qquad \text{(II.13)}$$

In more general terms, this sum rule tells us that the depletion in the correlation hole corresponds exactly to one monomer unit

$$\frac{1}{a^d}\int_{r>0}\left(-\,S_{11}(\mathbf{r}) + \frac{1}{N}\right)d\mathbf{r} = 1 \qquad \text{(II.14)}$$

At large q, all interference effects between different labels drop out; the intensity $S_{11}(\mathbf{q})$ reduces to the value for a single unit, and if the unit is point-like, the intensity is constant $S_{11}(\mathbf{q}) \rightarrow 1$.

II.2.3. More general sequences

We now consider the more usual case where the chains are labeled not at one point but over a finite fraction of their length. A typical sequence is represented in Fig. II.9 (the deuterated portions are marked D). To understand what happens, it is convenient to think first of a short D portion ($N_D \ll N$). Then, at most q values, the D portion can be treated as a point, and the scattered intensity $I(q)$ is very similar to Fig. II.8 except for

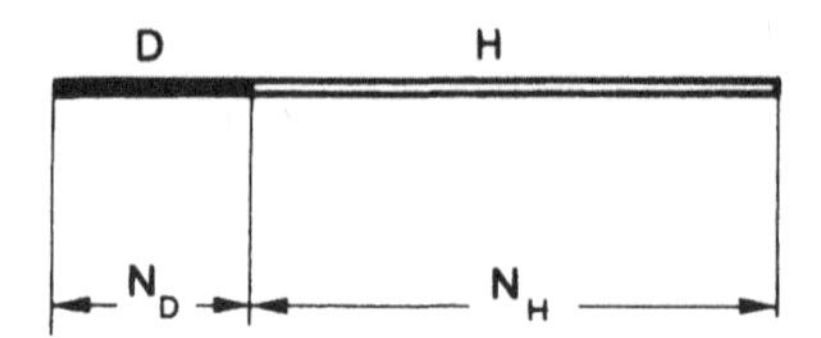

Figure II.9.

the normalization. If we measure $I(q)$ per labeled monomer, we have in this region:

$$I(q) = N_D\, S_{11}(q) \qquad (qR_D < 1) \tag{II.15}$$

However, eq. (II.15) will break down when qR_D becomes larger than 1 (R_D being the natural size of the labeled portion: $R_D = N_D^{1/2}\, a$).

In this relatively high q region, interference between different D coils becomes negligible, and $I(q)$ behaves essentially like the intensity due to a single coil of N_D units. This is discussed in Chapter I [eq. (I.18)] and gives

$$I(q) = g_D(q) = \frac{12}{q^2a^2} \qquad (qR_D \gg 1)$$

The net result is shown in Fig. II.10. The intensity now has a broad maximum at some intermediate wave vector q_m. When peaks of this type were first observed on partly labeled chains, there was a natural temptation to ascribe them to some sort of local segregation of the deuterated species. In fact, as shown in Fig. IX.4 there is very good agreement between the data and precise calculations on the correlation hole (described in Chapter IX). No special segregation effect needs to be involved (at least for the molecular weights $\sim 10^5$ which have been used in practice).

If we increase the length of the D portion, $1/R_D$ decreases and the peak first moves slightly to the left. The overall intensity also decreases since when $N_D = N$, we are left with no intensity at all. When $N_H = N - N_D$

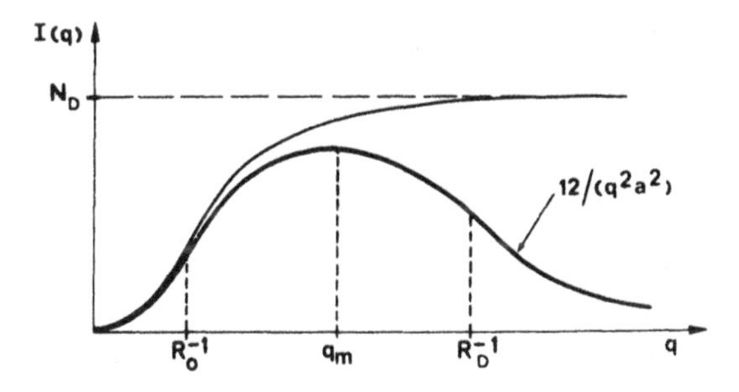

Figure II.10.

Qualitative plot of scattering intensity versus scattering wave vector in a neutron experiment using a melt of polymer chains, each chain being deuterated in one portion of its length as shown in Fig. II.9.

becomes small, it is convenient to reverse our view and to think of the scattering as being due to H portions in a sea of deuterated material. Then Fig. II.10 still holds provided that we replace N_D (and R_D) by N_H (and $R_H = N_H^{1/2}a$)

II.2.4. The correlation hole in two dimensions

Let us return to the correlation function S_{11} between the "heads" of all chains in a molten system. In the three-dimensional world this has the general appearance shown in Fig. II.7. What would happen if we could measure it in two dimensions (e.g., with chains adsorbed on a flat surface)?

Our prediction is that the correlation hole is much deeper. The sum rule [e.g. (II.14)] still holds, but the domain of integration is now an *area* of order $R_0^2 \sim Na^2$. Thus we must have

$$\frac{1}{a^2}\left(\frac{1}{N} - S_{11}\right) Na^2 \sim 1$$

$$\frac{1}{N} - S_{11} \sim \frac{\text{constant}}{N}$$

Thus the depletion is a finite fraction of the maximum concentration. This is another version of the discussion in Section II.1.4, showing that segregation is important for two-dimensional systems.

II.2.5. Mixtures of labeled and unlabeled chains

In the above discussion, we assumed that *all* chains were partly labeled, each with the same sequence of labels. Now we turn to a different experiment, where a fraction f of the chains is *completely* deuterated and the remaining fraction, $1 - f$, is entirely hydrogenated (we always assume that the deuterated and normal species are entirely identical in their interactions).

When f is small, we measure the scattering due to individual chains. This is the limit used by various workers to prove that chains in a melt are gaussian.[3,4,5] The scattering intensity per site (in the lattice model) is

$$I(q) = fg_D(q) \qquad (f \to 0) \tag{II.16}$$

where $g_D(q)$ is the Debye function defined in Chapter I.

When f is larger, we know $I(q)$ by a simple argument. Since the intensity involves only pair correlations, it must be a polynomial in f of order 2

$$I(q) = fg_D(q) + f^2h(q) \tag{II.17}$$

where $h(q)$ is yet unknown. However, we must impose the condition that for $f = 1$ we get no scattering (no fluctuations in concentration). This means that $h(q) = - g_D(q)$ and thus

$$I(q) = f(1 - f)g_D(q) \tag{II.18}$$

Eq. (II.18) is useful for the experimentalist because it allows him to study single-chain properties by using f values of order 1/2—i.e., with signals that are much higher than in the $f \rightarrow 0$ limit.

Eq. (II.18) has another interesting feature. We know that the intensity at $q = 0$ is related simply to the osmotic compressibility $(df/d\Pi_L)$ of the gas of labeled chains

$$I(q = 0) = Tf \frac{df}{d\Pi_L} \tag{II.19}$$

At $q = 0$ the Debye function $g_D(0)$ is equal to N. Inserting this value into eq. (II.18) and integrating, we find a small f expression

$$\frac{\Pi_L}{T} = \frac{f}{N} + \frac{f^2}{2N} + \ldots \tag{II.20}$$

This is exactly what we would expect from the monomer–monomer second virial A_{2S} of eq. (II.6). The total of pair-wise interactions between two chains is $N^2 A_{2S}$, and we may write, in terms of the *chain* density f/N

$$\frac{\Pi_L}{T} = \frac{f}{N} + \left(\frac{f}{N}\right)^2 A_{2S} N^2 + \ldots \tag{II.21}$$

Since $2A_{2S} = N^{-1}$, eqs. (II.20) and (II.21) coincide.

II.2.6. Summary

In a melt, each monomer is surrounded by a correlation hole, inside of which the concentration of monomers from other chains is slightly reduced. This correlation hole has a size comparable with the overall size of one chain. It gives rise to remarkable patterns in neutron scattering by partly deuterated chains. (The detailed scattering laws are given in Chapter X.) Experiments on partly labeled chains give us a precise check on the entire picture and prove that three-dimensional chains are strongly intertwined in a melt. The two-dimensional situation would be quite different and should be studied in the future.

REFERENCES

1. P. Flory, *J. Chem. Phys.* **17,** 303 (1949). For recent discussions on this subject, see the proceedings of the Symposium on the Amorphous State, *J. Macromol. Sci.* **B12** (1, 2) (1976).
2. S. F. Edwards, *Proc. Phys. Soc. (London)* **88,** 265 (1966).
3. D. Ballard *et al., Eur. Polymer J.* **9,** 965 (1973).
4. J. P. Cotton *et al., Macromolecules* **7,** 863 (1974).
5. R. Kirste *et al., Polymer* **16,** 120 (1975).
6. F. Boue *et al., Neutron Inelastic Scattering 1977,* Vol. 1, p. 563, International Atomic Energy Agency, Vienna, 1978.
7. P. G. de Gennes, *J. Phys. (Paris)* **31,** 235 (1970).

III

Polymer Solutions in Good Solvents

From Chapter I we know the properties of dilute (nonoverlapping) coils in a good solvent: they are swollen, with a size $R_F = N^{3/5}a$. At the other end, we know from Chapter II the limit of very concentrated solutions or melts: the chains are essentially ideal, with a size $R_D = N^{1/2}a$, and they interpenetrate each other very strongly.

What happens in between? For a long time, the only interpolation method available was a mean field theory due to Flory[1] and Huggins.[2] However (as realized early by these authors) the mean field idea is not adequate at low and intermediate concentrations. The question remained obscure for a long time. It has now been clarified experimentally (mainly through neutron scattering experiments; see Fig. III.1) and theoretically (by certain manipulations as described in Chapter X). Fortunately, the final picture is simple and can be explained without referring to abstract theory.

III.1. The Mean Field Picture (Flory-Huggins)

III.1.1. Entropy and energy in a lattice model

Again it is convenient to represent the polymer chains as random walks on a lattice, each lattice site being either occupied by one (and only one) chain monomer or by a "solvent molecule," as shown in Fig. III.2. We denote the fraction of sites occupied by monomers as Φ. This term is related to the concentration c (number of monomers per cm^3) by $\Phi = ca^3$,

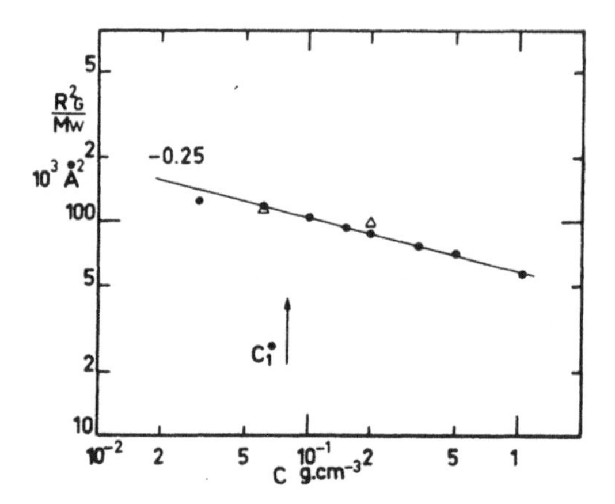

Figure III.1.

Square of gyration radius R_G for one labeled polystyrene chain in a polystyrene solution of concentration c. Above $c = c^*$, $R_G^2(c)$ decreases with a well-defined exponent. At high c, R_G returns to the ideal chain value. After Daoud *et al.*, *Macromolecules* **8**, 804 (1975).

where a^3 is the volume of the unit cell in the cubic lattice. Clearly the lattice model is somewhat artificial, but it does not lose any essential feature of the problem, and it provides a convenient framework to describe solutions at *all* concentrations.

The free energy F for this model has two components: an entropy term describing how many arrangements of chains can exist on the lattice for a given Φ, and an energy term describing the interactions between adjacent molecules.

In the mean field approximation the *entropy* S has a simple structure:

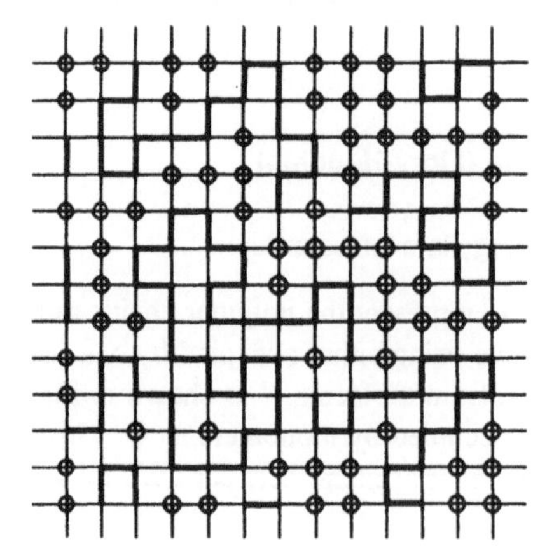

Figure III.2.

$$- S|_{site} = \frac{\Phi}{N} \ln \frac{\Phi}{N} + (1 - \Phi) \ln (1 - \Phi) \tag{III.1}$$

The first term is related to the translational entropy of the chain (Φ/N is the chain concentration in dimensionless units). The second term may be similarly conceived as the translational entropy of the solvent molecules (of volume fraction $1 - \Phi$). An interesting check on eq. (III.1) is obtained for $N = 1$. Then eq. (III.1) becomes the standard entropy for a collection of independent systems, each with two states "full" (probability Φ) or "empty" (probability $1 - \Phi$).

Note that eq. (III.1) is an approximation; in the real problem, neighboring sites are *not* independent. (We return to this later.)

A technical modification. Instead of considering the full entropy $S(\Phi)$, it is often convenient to focus our attention on a slightly different object, the *entropy of mixing* S_{mix}. This is defined by the difference between $S(\Phi)$ and the weighted average of the entropies of pure polymer [$S(1)$] and pure solvent [$S(0)$]

$$S^{mix}(\Phi) = S(\Phi) - \Phi S(1) - (1-\Phi)\, S(0)$$

The subtraction described above is interesting, because it eliminates certain trivial terms. All contributions to $S(\Phi)$ which are independent of Φ, or linear in Φ, drop out from S_{mix}. In what follows we shall perform a similar transformation for all thermodynamic functions of the mixture.

For the particular form in eq. (III.1) the only change obtained by going from S to S_{mix} is to eliminate a term $\Phi/N \ln 1/N$ which is linear in Φ

$$- S^{mix}|_{site} = \frac{\Phi}{N} \ln \Phi + (1-\Phi)\ln(1-\Phi) \tag{III.1a}$$

The *energy* term E contains, in general, three terms that describe

$$\begin{aligned} &\text{monomer–monomer interactions:} && \frac{T}{2}\chi_{MM}\,\Phi^2 \\ &\text{monomer-solvent interactions:} && T\,\chi_{MS}\,\Phi\,(1 - \Phi) \\ &\text{solvent–solvent interactions:} && \frac{T}{2}\chi_{SS}\,(1 - \Phi)^2 \end{aligned} \tag{III.2}$$

However, we do *not* need three constants because all terms in the free energy per site which are independent of Φ, or linear in Φ, drop out when we consider E_{mix}.

The truly interesting feature is the *curvature* of the plot $E(\Phi)$: only one parameter is relevant. In fact we can write:

$$\frac{1}{T} E_{mix/site} = \chi\Phi\,(1 - \Phi) + \text{constant} + \text{terms linear in } \Phi \qquad \text{(III.3)}$$

with

$$\chi = \chi_{MS} - \frac{1}{2}(\chi_{MM} + \chi_{SS}) \qquad \text{(III.4)}$$

We call χ the Flory interaction parameter. It is dimensionless, and it depends on temperature, pressure, etc. Good solvents have a low χ, while poor solvents have a high χ (we see later that the borderline corresponds to $\chi = 1/2$). The case $\chi = 0$ corresponds to a solvent which is very similar to the monomer. In our lattice model this is the case where the free energy comes entirely from the entropy associated with various chain patterns on the lattice. In such a case temperature has no effect on structure, and we say that the solvent is "athermal." Athermal solvents are a particularly simple example of good solvents. Most of the scaling laws of the present chapter will be written only for athermal solvents.

In most cases the parameter χ is *positive*. This is because the interactions (MM, MS, SS) are mainly van der Waals attractions, which are essentially proportional to the *product* of the electronic polarizabilities (α) for both molecules.[3] Thus we have

$$\begin{aligned} \chi_{MM} &= -\,k\alpha_M^2 \\ \chi_{SS} &= -\,k\alpha_S^2 \\ \chi_{MS} &= -\,k\alpha_S\alpha_M \end{aligned} \qquad \text{(III.5)}$$

where k is positive (the van der Waals interactions being attractive). The net result is

$$\chi = +\frac{k}{2}(\alpha_S - \alpha_M)^2 > 0 \qquad \text{(III.6)}$$

Of course the simple estimate of eq. (III.5) can be corrected by special bonding effects, steric corrections, and so forth, but the trend toward positive χ is quite general. The temperature dependence of χ is a delicate matter. If the monomer-monomer local correlations were independent of

T, we would expect the interactions to be independent of T, and χ to vary like $1/T$. However, reality is much more complex: χ is an increasing function of T in a number of cases.

Having introduced the interaction parameter χ, we can now discuss the overall free energy. It is obtained by adding eqs. (III.1) and (III.3)

$$\frac{1}{T} F_{mix}\Big|_{site} = \frac{\Phi}{N} \ln \Phi + (1 - \Phi) \ln (1 - \Phi) + \chi\Phi(1 - \Phi) \text{ (mean field)} \qquad \text{(III.7)}$$

In the following paragraphs we discuss the main consequences of this basic formula. In this chapter we restrict our attention to good solvents (small χ). The situation for poor solvents is investigated in Chapter IV.

III.1.2. Low concentrations

Expanding the regular terms in the free energy [eq. (III.5)] at small Φ we find

$$\frac{F_{mix}}{T}\Big|_{site} = \frac{\Phi}{N} \ln \Phi + \frac{1}{2} \Phi^2 (1 - 2\chi) + \frac{1}{6} \Phi^3 + \ldots \qquad \text{(III.8)}$$

Note that the linear terms in Φ have dropped out because we are considering only the free energy of mixing (as explained on p. 71). We may interpret the $\Phi^2/2$ term as defining an effective pair interaction between dilute monomers. Note that the coefficient $1 - 2\chi$ contains two contributions: one part (-2χ) which is related to interactions between adjacent sites, and another part (1) which expresses the existence of steric repulsions between monomers at short distances (in the lattice model, two monomers are *not* allowed on the same site).

Eq. (III.8) is one example of the low concentration expressions which can be proposed for the free energy. A slightly more general version, which is also of use, can be written independently of any lattice model:

$$\frac{F}{T}\Big|_{cm^3} = \frac{c}{N} \ln c + \frac{1}{2} vc^2 + \frac{1}{6} w^2c^3 + \ldots \qquad \text{(III.9)}$$

where c is the number of monomers per cm^3. Here v is the excluded volume parameter, following Edwards.[4] Using $\Phi = ca^3$, we find the following correspondence

$$v = a^3 (1 - 2\chi) \tag{III.10}$$

The value of v is positive for good solvents and negative for poor solvents. For an athermal solvent $v = a^3$.

Note that in the Flory-Huggins approach [eq. (III.8)] the coefficient w is fixed ($w = a^3$) while it remains as a free parameter in eq. (III.9). The second formulation is slightly more general, but it is restricted to small Φ values. The lattice model has the advantage of providing a description at all Φ values.

III.1.3. Osmotic pressures

Let us consider first the osmotic pressure Π of the macromolecules in the solution. This is defined by an operation where we change the solution volume (by adding more solvent: $V_{tot} \rightarrow V_{tot} + \Delta V$) while keeping fixed the number of monomers present (ν_m)

$$-\Pi = \left.\frac{\Delta F_{tot}}{\Delta V}\right|_{\nu_m} \tag{III.11}$$

The term ν_m is related to V_{tot} by $\nu_m = \Phi V_{tot}/a^3$, and the total free energy is $F_{tot} = F_{site}\, V_{tot}/a^3$. Inserting this into eq. (III.9), we get

$$a^3 \Pi = -\frac{\partial(F_{site}/\Phi)}{\partial(1/\Phi)} = \Phi^2 \frac{\partial}{\partial \Phi}(F_{site}/\Phi) \tag{III.12}$$

and using the Flory-Huggins form [eq. (III.7)], we arrive at

$$a^3 \frac{\Pi}{T} = \frac{\Phi}{N} + \ln\left(\frac{1}{1-\Phi}\right) - \Phi - \chi\Phi^2 \tag{III.13}$$

The discussion of eq. (III.11) breaks up into three parts:

(i) For $\Phi \rightarrow 0$ we have a perfect gas law $\Pi/T = c/N$ where c/N is the number of *chains* per cm^3.

(ii) For $1/N \ll \Phi \ll 1$ we may still expand Π in powers of Φ, and the dominant term is the quadratic part

$$\frac{a^3 \Pi}{T} = \frac{\Phi}{N} + \frac{1}{2}(1 - 2\chi)\,\Phi^2 \cong \frac{1}{2}(1 - 2\chi)\,\Phi^2 \tag{III.14}$$

(iii) For dense systems ($\Phi \rightarrow 1$) the pressure tends to diverge logarithmically.

Thus, provided that the solvent is good ($1 - 2\chi > 0$), mean field theory predicts a crossover from ideal gas behavior to a strongly interacting behavior at $\Phi \sim 1/N$. Beyond this point the osmotic pressure should become proportional to Φ^2. We shall see that, in fact, all these predictions are qualitatively wrong.

Let us now discuss the *dense limit*. This is of some interest in connection with our discussion of screening in molten polymers. Here we consider the osmotic pressure of the *solvent* molecules Π_S. At first sight Π_S is an absurd concept; to measure it directly we would need an osmometer which is transparent for the chains but impermeable for the solvent. However Π_S is a useful theoretical intermediate for studying solvent–solvent interactions.

The starting formula for Π_S is quite similar to eq. (III.12). Introducing the solvent fraction $\Phi_S = 1 - \Phi$ we have

$$a^3\Pi_S = \Phi_S{}^2 \frac{\partial}{\partial \Phi_S}(F_{site}/\Phi_S)$$

giving

$$\frac{a^3\Pi_S}{T} = \Phi_S - \frac{1}{N}[\Phi_S + \ln(1 - \Phi_S)] - \chi\Phi_S{}^2 \qquad \text{(III.15)}$$

At low Φ_S we may expand and get

$$\frac{a^3\Pi_S}{T} = \Phi_S + \Phi_S{}^2\left(\frac{1}{2N} - \chi\right) + \dots \qquad \text{(III.16)}$$

Of particular interest is the athermal case* which was chosen in our discussion of screening in Section II.1.2. Setting $\chi = 0$ we arrive at a second virial coefficient A_{2S} between solvent monomers which is very small: $A_{2S} = 1/2N$.

Eq. (III.16), for dense systems, has a sounder basis than eq. (III.14), for dilute systems. The reason is that for dense systems the chains are nearly ideal, and the mean field description is then nearly exact.

We could extend the above discussion to many other thermodynamic properties, such as the heats of mixing, etc. However, as regards concepts, the osmotic pressures Π and Π_S are the basic tools.

Another thermodynamic law is worth mentioning. The chemical potential μ_S of the solvent is proportional to the osmotic pressure of the solute Π.[1] This is shown in eq. (III.11) since the change in volume ΔV involved

*The discussion of Section II.1.2. is directed toward a system of chains filling the lattice, which is intrinsically athermal.

in the definition of Π is simply a change in the number of solvent molecules $\Delta\nu_S = \Delta V/a^3$. Thus

$$\mu_S = \frac{\Delta (F_{tot})}{\Delta\nu_S} = -\Pi a^3 \tag{III.17}$$

III.1.4. Critique of mean field theory

As mentioned the mean field calculation neglects certain correlations between adjacent (and even more distant) monomers. For example, when we write a Φ^2 term for the monomer–monomer interactions, we really mean a certain average $\langle\Phi^2\rangle$ which may be quite different from $\langle\Phi\rangle^2$. In a good solvent the monomers tend to avoid each other.

The mean field calculation replaces the monomer–monomer interactions by a certain self-consistent potential which is uniform in space;* such a potential cannot induce any swelling of the chains (as explained in Section II.1). Thus, the mean field for polymer solutions is intrinsically associated with ideal chains. This is clearly not acceptable at low concentrations.

These statements are classical and give no real clue on how to improve the situation. It turns out, however, that scaling arguments will work.

III.2. Scaling Laws for Athermal Solvents

From now on in this chapter, we assume that the solvent is very good (athermal): numerically, the Flory parameter χ is much smaller than 0.5. This removes one parameter from the discussion, and simplifies all scaling laws.

III.2.1. The overlap threshold c^*

A fundamental distinction exists between dilute polymer solutions where the coils are separate (Fig. III.3a) and more concentrated solutions where the coils overlap (Fig. III.3c). At the overlap threshold ($c = c^*$) the coils begin to be densely packed. Clearly this threshold is not sharp; it is more properly defined as a region of crossover between regimes (a) and (c), but the scaling properties of c^* are essential. We expect c^* to be comparable with the local concentration inside a single coil. In a very good (athermal) solvent this implies:

*This is very different from the Flory calculation for a single chain where the self-consistent potential is localized in a small region of space and creates finite forces on the chain.

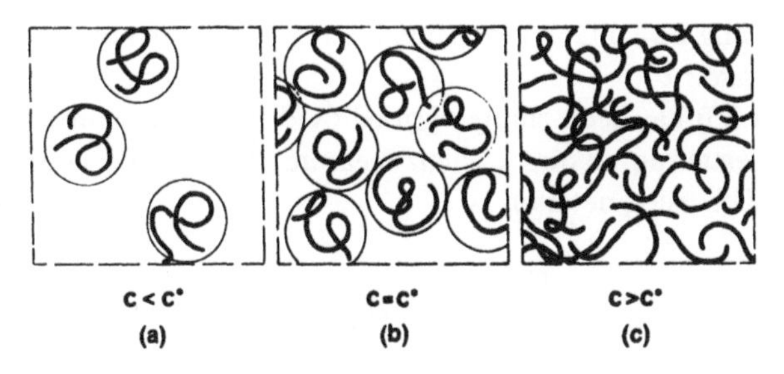

Figure III.3.

Crossover between dilute and semi-dilute solutions: (a) dilute, (b) onset of overlap, and (c) semi-dilute.

$$c^\star \cong N/R_F{}^3 = a^{-3}\, N^{1-3\nu} = a^{-3}\, N^{-4/5} \qquad \text{(III.18)}$$

In terms of the polymer fraction Φ we may define the corresponding threshold $\Phi^\star \sim N^{-4/5}$. Note that for large N, $\Phi^\star$ is very small; for example, with $N = 10^4$ (corresponding to molecular weights of order 10^6 for current polymers) we expect $\Phi^\star \sim 10^{-3}$.

III.2.2. The dilute regime

When $c < c^\star$, we have a dilute system of coils. In the zeroth approximation, we can treat these as a perfect gas, with a number of coils per cm^3 equal to c/N and an osmotic pressure of

$$\Pi = \frac{c}{N}T \qquad (c \rightarrow 0) \qquad \text{(III.19)}$$

In an improved approximation, we have to take into account the interactions between coils; in a good solvent two coils tend to repel each other. Flory has shown that in this regime the coils behave essentially like *hard spheres* of radius $\sim R_F$.[1] This implies an equation of state of the form

$$\frac{\Pi}{T} = \frac{c}{N} + \text{constant}\left(\frac{c}{N}\right)^2 R_F{}^3 + O\left(\frac{c}{N}\right)^3 \qquad \text{(III.20)}$$

Here $R_F{}^3$ appears as the second virial coefficient between coils.

Why do the coils behave as hard spheres? If we force two coils to overlap strongly, many contacts must occur between them. Each contact brings an energy of order T, and the overall overlap energy is many times T; the corresponding Boltzmann exponential is very small, and the coils refuse to interpenetrate.*

Eq. (III.20) has been amply verified by direct measurements of Π and by light scattering studies (which measure the osmotic compressibility $dc/d\Pi$). The usual notation is

$$\frac{\Pi}{T} = \frac{c}{N} + A_2\, c^2 + \dots \qquad \text{(III.21)}$$

and the N dependence of A_2 has the form

$$A_2 \cong R_F^{\,3}\, N^{-2} \sim N^{-1/5} \qquad \text{(III.22)}$$

We now have some sophisticated field theoretic calculations of the coefficient in eq. (III.22) for three-dimensional systems.[5]

III.2.3. Semi-dilute solutions

We now consider solutions where the coils do overlap but where the polymer fraction Φ is still low

$$\Phi^{\star} \ll \Phi \ll 1 \qquad \text{(III.23)}$$

The two inequalities are compatible at high N because $\Phi^{\star}$ is then very small. This semi-dilute regime is of great interest for two reasons:

(i) It corresponds to a large fraction of the Φ interval in high polymer solutions.

(ii) Since Φ is small, the monomer–monomer interactions can be described very simply (much like the interactions in an imperfect, dilute, gas); we need only *one interaction constant,* such as the excluded volume parameter v of eqs. (III.9, III.10). This is in contrast to the situation $\Phi \sim 1$, where, in principle, we need all of the theory of liquids to obtain "exact" results, using realistic interaction potentials and highly complex numerical calculations. In the current jargon, we may say that semi-dilute solutions have a high degree of universality.

How do we predict the thermodynamic properties of semi-dilute solutions? The basic notion is a scaling law for the osmotic pressure, proved by des Cloizeaux for one specific example.[6] This scaling law is a natural generalization of eq. (III.20) and reads

*Note that this statement holds in one, two, and three dimensions but not at $d = 4, 5$, etc. Two overlapping ideal chains in d dimensions have a number of contacts $\sim N^2\,(a/R_0)^d \cong N^{2-d/2}$, and this is small for $d > 4$.

$$\frac{\Pi}{T} = \frac{c}{N} f_\Pi \left(\frac{cR_F^3}{N}\right) = \frac{c}{N} f_\Pi \left(\frac{c}{c^\star}\right) \qquad \text{(III.24)}$$

where the function $f_\Pi(x)$ is dimensionless and has the following limiting properties:

(i) At small x (dilute solutions) f_Π is expressed as $f_\Pi = 1 + \text{constant}\ x + \ldots$

(ii) At large x (semi-dilute solutions) all thermodynamic properties must reach a limit which depends on c but which becomes independent of the degree of polymerization N. Physically this means that local energies, entropies, etc. are controlled entirely by the concentration c; the local properties are not different for a solution of chains of N monomers each or a single chain that fills the whole vessel ($N \to \infty$).

This requirement is very strong; because eq. (III.24) has the prefactor c/N, to eliminate all dependence on N, the function $f_\Pi(x)$ *must* behave like a simple power of x

$$\lim_{x\to\infty} f_\Pi(x) = \text{constant}\ x^m = \text{constant} \left(\frac{\Phi}{\Phi^\star}\right)^m = \text{constant}\ \Phi^m\ N^{4m/5} \qquad \text{(III.25)}$$

(where we have used eq. (III.18) for $\Phi^\star$). In terms of Φ and N, this gives

$$\frac{a^3\,\Pi}{T} = \text{constant}\ \Phi^{m+1}\ N^{4\,m/5\,-\,1}\ (\Phi \gg \Phi^\star) \qquad \text{(III.26)}$$

and since we want Π independent of N, we must have $m = 5/4$.

Then the osmotic pressure follows what we call the des Cloiseaux law:

$$\frac{\Pi\, a^3}{T} = \text{constant}\ \Phi^{9/4} \quad \text{(semi-dilute)} \qquad \text{(III.27)}$$

Eq. (III.27) has been confirmed to some extent by osmometric and light scattering data.[7] It is important to note the difference from the mean field prediction (eq. III.14) which is $\Pi \sim \Phi^2$. This difference represents a *correlation effect*. In the semi-dilute regime Π measures the number of contacts between monomers. If we neglect correlations, this number per site is $\sim \Phi^2$, but correlations reduce it by an extra factor, $\Phi^{1/4}$. Since Φ can be as low as 10^{-3} (while still belonging to the semi-dilute regime), the correlation factor may be of order 1/10, and is thus important.

A similar factor should occur in all properties which measure local correlations in semi-dilute solutions with good solvents: heats of mixing, hypochromicity in optical spectra,* etc. Further experiments along these lines are desirable.†

III.2.4. The correlation length

The above presentation of scaling for thermodynamic properties is direct but not illuminating. A much better picture of what happens in semi-dilute solutions can be obtained if we investigate spatial properties. Consider the solution shown in Fig. III.4. When photographed at a certain time, this looks very much like a network with a *certain average mesh size* ξ.

We can use neutron scattering,[7] to measure ξ but we can also use a simple idea first suggested by H. Benoît.‡ This amounts to adding a small number of inert spheres with diameter $D \sim$ 50–100 Å. When $D < \xi$, we expect the spheres to move easily, with a friction coefficient which is essentially related to the viscosity of the pure solvent. When $D > \xi$, the spheres are trapped—the effective viscosity controlling their friction is closer to the viscosity of the entangled solution.[8]

Let us now construct the scaling form of ξ in the semi-dilute regime for a good (athermal) solvent. This is based on two requirements:

(i) For $\Phi > \Phi^\star$ the network structure on the scale ξ will depend only on concentration and not on the degree of polymerization N (the chains being much longer than the mesh size).

(ii) For $\Phi \sim \Phi^\star$ where we have coils in contact (but not yet interpenetrating) the mesh size must be comparable with the size of one coil R_F.

These two requirements lead to the form

$$\xi(\Phi) = R_F \left(\frac{\Phi^\star}{\Phi}\right)^{m_\xi} \qquad (\Phi > \Phi^\star) \qquad \text{(III.28)}$$

where the exponent m_ξ must be such that the powers of N from $R_F(\sim N^{3/5})$ and from $\Phi^\star$ $(\sim N^{-4/5})$ cancel. This means that $m_\xi = 3/4$

$$\xi(\Phi) \cong a\Phi^{-3/4} \qquad (\Phi^\star \ll \Phi \ll 1) \qquad \text{(III.29)}$$

Thus, the mesh size decreases rapidly with concentration. Note an inter-

*See C. Destor, D. Langevin, F. Rondelez, *J. Polym. Sci. (Polym. Lett.)* **16,** 229 (1978).

†An interesting experiment with some chains carrying a fluorescent group and others carrying a quencher was done by Y. Kirsh *et al., Europ. Polymer J.* **11,** 495 (1975).

‡Private communication, 1975.

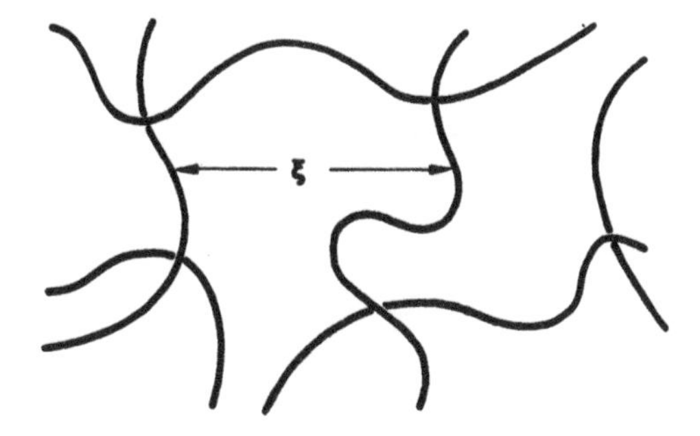

Figure III.4.

esting scaling relationship between the osmotic pressure (eq. III.27) and the correlation length [eq. (III.29)]

$$\Pi \cong T/\xi^3 \qquad (\Phi^\star \ll \Phi \ll 1) \tag{III.30}$$

III.2.5. The notion of blobs

We now focus on one particular chain in the semi-dilute solution; this could be, for example, one chain labeled by deuteration with all the other chains being normal. We may visualize it as a succession of units or "blobs" of size ξ (Fig. III.5). Inside one blob, (from the definition of the mesh size) the chain does not interact with other chains. Thus, inside one blob we must still have correlations of the excluded volume type. This implies that the number of monomers per blob (g) is related to ξ by the law of swollen coils

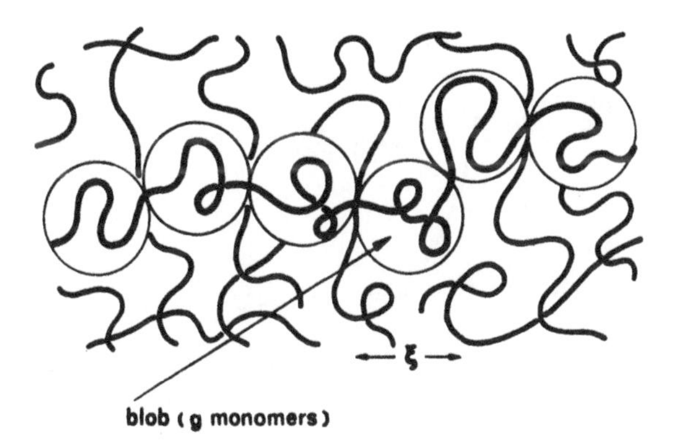

Figure III.5.

$$\xi \cong a g^{3/5}$$

$$g = \left(\frac{\xi}{a}\right)^{5/3} = \Phi^{-5/4} \qquad \text{(III.31)}$$

Note that

$$g = c\,\xi^3 \qquad \text{(III.32)}$$

as can be seen immediately from eq. (III.29). Eq. (III.32) says that the solution is essentially a *closely packed system of blobs*. If we take the blobs as the basic units, we are led back to the molten chain problem of Chapter II. We know that the Flory theorem holds: the chains are ideal on a large scale. Their mean square end-to-end size can be estimated from the ideal chain formula for N/g blobs of size ξ

$$R^2(\Phi) = \frac{N}{g}\xi^2$$

$$\cong N a^2\,\Phi^{-1/4} \qquad (\Phi^\star \ll \Phi \ll 1) \qquad \text{(III.33)}$$

an equation derived first by Daoud[7,9] and verified with reasonable accuracy by neutron experiments on polystyrene solutions (see Fig. III.1). Eq. (III.33) could have been obtained directly by a scaling argument, writing $R = R_F\,(\Phi^\star/\Phi)^{m_R}$ and imposing the condition $R \sim N^{1/2}$, but the derivation based on blobs is more illuminating.

III.2.6. Correlation functions

Our statements on blobs can be made more precise in terms of correla-

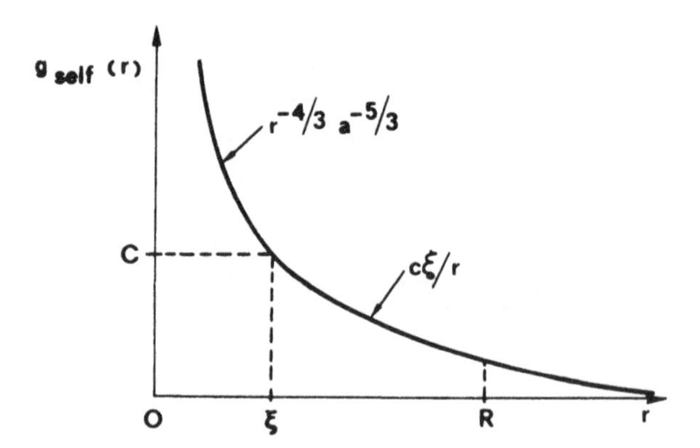

Figure III.6.

tion functions. Consider first the correlation function $g_{self}(r)$ *inside one labeled chain,* which has been studied in great detail on polystyrene solutions by Farnoux[10] (Fig. III.6). At a short distance $(r < \xi)$ $g_{self}(r)$ is identical to the correlation function of a single swollen chain and follows the Edwards law (eq. I.31). However, when $g(r)$ becomes smaller than the ambiant concentration c, the effects of surrounding chains begin to be felt, and the chain becomes ideal. Thus $g_{self}(r)$ decreases as $1/r$ at $r > \xi$. More precisely

$$g_{self}(r) = c\,\xi/r = \frac{g}{\xi^2 r} \tag{III.34}$$

The reader may check that the two laws cross over smoothly at $r = \xi$. Ultimately when r becomes larger than the overall size R, the correlation $g_{self}(r)$ drops sharply to zero.

Imagine a chain which is labeled at one end by a fluorescent molecule and at the other end by an optical trap. The trap is an efficient quencher of fluorescence only if both partners are in close contact. We put one such chain into the solution; all the other chains are normal. This situation should then allow us to measure the probability E of contact between the two ends as a function of Φ. Experiments of this type have been proposed recently.[11] The scaling law for E must have the form

$$E = p_N(a)\left(\frac{\Phi}{\Phi^\star}\right)^{m_p} \tag{III.35}$$

where $p_N(a) \sim N^{1-\gamma-3\nu}$, is the probability of contact between ends for a single chain (eqs. I.29, I.30). In three dimensions $p_N(a) \sim N^{-2}$ to a good approximation. The exponent m_p must be such that $E \sim N^{-3/2}$ (as it should be for a chain which is ideal at large scales). Taking $\Phi^\star = N^{-4/5}$, this gives

$$m_p = 5/8 \tag{III.36}$$

More generally, the end-to-end distance $\mathbf{r}$ for one chain in the solution has the probability distribution $p_N(\mathbf{r})$ shown in Fig. III.7.

At small distances $(r < \xi)$ we have a power law increase, based on the same exponent g which appeared for a single chain [eq. (I.27)]

$$p_N(r) = \left(\frac{r}{\xi}\right)^g p_N(\xi) \quad (r < \xi) \tag{III.37}$$

At larger distances we have a gaussian behavior

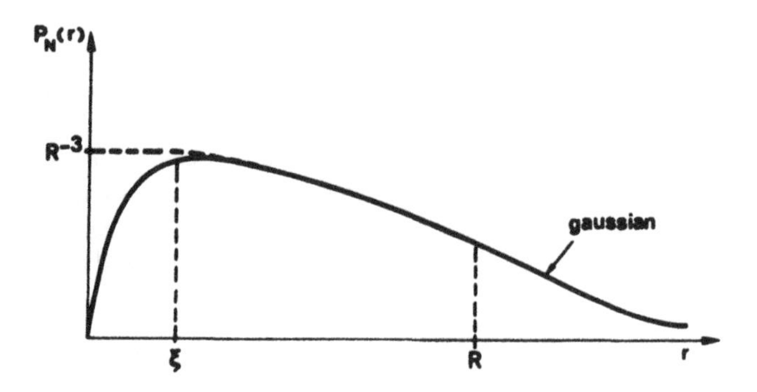

Figure III.7.
Distribution of the end-to-end distance for one chain in a semi-dilute solution. Note the reduced values in the short range region ($r<\xi$).

$$p_N(r) \cong p_N(\xi) \exp(-3r^2/2R^2) \qquad (r > \xi) \tag{III.38}$$

and the normalization condition ($\int p_N \, dr = 1$) then imposes the restriction

$$p_N(\xi) \cong \frac{1}{R^3} \tag{III.39}$$

It is easy to check that when we set $r = a$ in eq. (III.37) we recover eqs. (III.35, III.36) for the probability of contact.

Until now we have discussed the correlations inside one labeled chain. It is also of interest to analyze the correlations between *all pairs* of monomers which are measured by X-ray (or neutron) scattering without any labeling. This gives a pair correlation

$$g(\mathbf{r}) = \frac{1}{c} \left[\langle c(0)c(\mathbf{r})\rangle - c^2\right] \tag{III.40}$$

through its Fourier transform $g(\mathbf{q})$. At short distances $r < \xi$, $g(r)$ is still dominated by correlations inside the same chain

$$g(\mathbf{r}) = g_{self}(\mathbf{r}) \qquad (r < \xi) \tag{III.41}$$

However, at larger distances $g(\mathbf{r})$ becomes much smaller than g_{self}. The reason is that at large distances we can describe the system as a molten system of blobs, with very little fluctuations in density and thus very little scattering. Although no complete proof has been written, there are reasons to believe that in this region $g(\mathbf{r})$ follows a simple Ornstein-Zernike form

$$g(\mathbf{r}) \cong c\frac{\xi}{r}\exp(-r/\xi) \tag{III.42}$$

In terms of Fourier transforms this gives

$$g(q) \cong \frac{c\xi}{q^2 + \xi^{-2}} \tag{III.43}$$

a law which is followed well by the Farnoux data.[7,10] Note that for $q = 0$, $g(q) \rightarrow c\,\xi^3 = g$, the number of monomers in one blob.

Eq. (III.43) satisfies the compressibility sum rule

$$g(q = 0) = T\frac{\partial c}{\partial \Pi} \tag{III.44}$$

as can be checked from eq. (III.27) for the osmotic pressure Π.

III.2.7. Screening in semi-dilute solutions

There is an analogy between eq. (III.42) for the correlations and the Debye-Hückel law for screened coulomb interactions in an electrolyte. The notion of screening was first introduced by Edwards for polymer solutions.[12] His argument may be presented as follows. We assume that one monomer is fixed at point O in the solution, and we ask for the excess concentration $g(\mathbf{r})$ which will result from this condition, at a distance r from O.

(*i*) If our chains were ideal and independent, the only contributions to $g(\mathbf{r})$ would come from monomers belonging to the same chain: they would then be given by the Debye function $g_D(\mathbf{r})$ introduced in Chapter I. At distances larger than a but smaller than the chain size R_0, $g_D(\mathbf{r})$ decreases like $1/r$:

$$g_D(\mathbf{r}) = \frac{3}{\pi a^2 r}$$

Note that g_D satisfies the analog of a Laplace equation

$$\nabla^2 g_D = -\frac{12}{a^2}\delta(\mathbf{r})$$

(*ii*) But in fact our chains repel each other, and $g(\mathbf{r})$ will differ from $g_D(\mathbf{r})$. The difference will be described here in terms of a potential $U(\mathbf{r})$ acting on all monomers. Because the monomer density is increased near the origin, $U(\mathbf{r})$ will be high in this region. At longer distances it will decay down to a constant value U_∞, describing the bulk solution. (We shall omit U_∞ in the following discussion, since it is a constant potential, giving no interesting spatial effects.)

If we know $U(\mathbf{r})$, how do we calculate the resulting concentration profile $g(\mathbf{r})$? An approximate answer is written below:

$$g(\mathbf{r}) - g_D(\mathbf{r}) = -\int g_D(\mathbf{r} - \mathbf{r}') \frac{cU(\mathbf{r}')}{T} \, d\mathbf{r}'$$

It may be explained as follows: locally at point $\mathbf{r}'$, the potential $U(\mathbf{r}')$ shifts the monomer concentration from c to:

$$c \exp(-U/T) \cong c - cU(\mathbf{r}')/T$$

This local concentration shift then implies shifts in neighboring regions since the monomers are not independent, but are linked in the form of chains. At this stage Edwards assumes that the chains are ideal: then the concentration shift around point $\mathbf{r}'$ is still given by a Debye function $g_D(\mathbf{r} - \mathbf{r}')$. The total function $g(\mathbf{r})$ is a sum of such contributions from all points $\mathbf{r}'$.

The equation for g may in fact be written more compactly if we apply the Laplacian operator on both sides—the result being:

$$\frac{a^2}{12} \nabla^2 g(\mathbf{r}) = \frac{cU(\mathbf{r})}{T} \quad (\mathbf{r} \neq 0)$$

(*iii*) There remains for us to write a condition of self-consistency for the repulsion potential U. It is natural to assume that $U(\mathbf{r})$ is proportional to the local chain concentration $g(\mathbf{r})$

$$U(\mathbf{r}) = Tv \, g(\mathbf{r})$$

where v is a coefficient, with the dimensions of a volume, which we call the excluded volume parameter. (It will be discussed more fully in Chapter IV.) For the athermal solvents of interest here, it is enough to say that $v \sim a^3$.

Inserting this condition into the former equation for $g(\mathbf{r})$ we arrive at an equation of the Debye-Hückel form:

$$\nabla^2 g = \xi_E^{-2} g \quad (\mathbf{r} \neq 0)$$

where ξ_E is defined by

$$\xi_E^{-2} = \frac{12cv}{a^2}$$

We shall call ξ_E the Edwards correlation length.

The solution of the equation for g must be chosen such that, at short distances, g returns to its unperturbed value g_D. (At small r the correlations are always dominated by effects involving a single chain, and, in the Edwards approximation, each chain in ideal). Then the solution is of the form shown in Fig. III.8

$$g(\mathbf{r}) = \frac{3}{\pi a^2 r} \exp(-r/\xi_E)$$

For $v \cong a^3$ we may then write

$$g(\mathbf{r}) \cong c \frac{\xi_E}{r} \exp(-r/\xi_E)$$

in full agreement with eq. (III.42). We note that the correlations die out in a length ξ_E. The same property holds for the potential $U(\mathbf{r})$: we say that

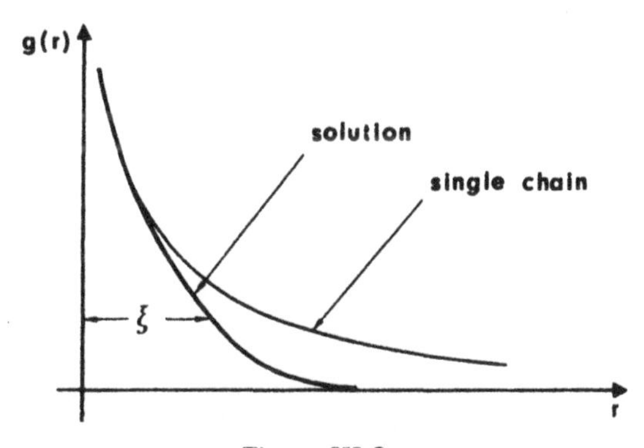

Figure III.8.

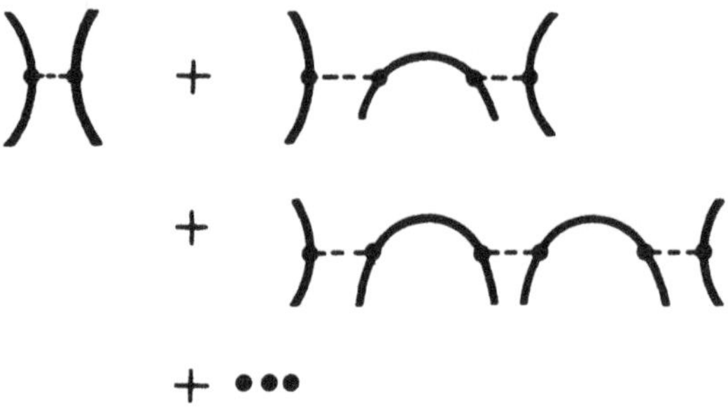

Figure III.9.
Calculation of the effective interaction between two monomers in a semi-dilute solution, following the Edwards approximation. Chains are represented by continuous lines, the interactions by dotted lines. The length of the dotted lines is short, but the effective interaction can be mediated by intermediate chains. In the present approximation, only linear sequences of interactions are retained.

$U(\mathbf{r})$ is *screened*—the interaction between two objects in solution is reduced by the presence of other chains in between (see Fig. III.9). This notion of screening has already appeared in our discussion on melts in Chapter II, and is fundamental.

Technically, the original Edwards calculation is not quite perfect. It was based on ideal chains and ignored correlations—i.e., it does not give the correct power for ξ. However, if the calculation is rewritten with *blobs* as the basic units, then the chains can be treated as nearly ideal, and the correct powers result. This revised calculation shows that the screening length scales like the mesh size ξ. More generally, all characteristic lengths which are independent of N in the semi-dilute solution must have the same scaling property, and thus can differ from the mesh size only by a numerical factor.

III.3. Confined Polymer Solutions

III.3.1. A semi-dilute solution in contact with a repulsive wall

The physical situation is represented in Fig. III.10 (a,b). We assume that the wall is repulsive; the simplest (athermal) model corre-

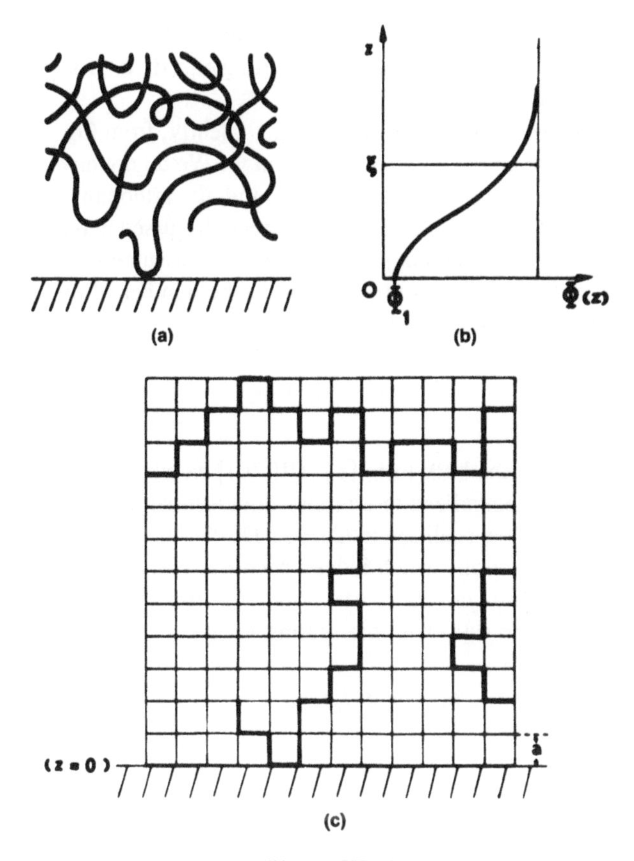

Figure III.10.

sponding to this condition is a cubic lattice occupying the half-space $z >= 0$ (Fig. III.10c), and it will have layers at $z = 0$, $z = a$, etc., but negative z values are forbidden.

The wall depletes the concentration $\Phi(z)$ at finite distance z—up to one screening length ξ. At distance $z > \xi$ all perturbations are screened out, and $\Phi(z)$ recovers the bulk values (Φ) as shown in Fig. III.10b.

(i) *Interfacial energy of the wall/solution system.* The osmotic pressure in the bulk is $\Pi \cong T\xi^{-3}$ [eq. (III.30)], and the concentration is reduced on a thickness ξ. This imposes a contribution to the interfacial free energy of the form:

$$A - A_0 \cong T\xi^{-3}\,\xi \cong T\xi^{-2} \qquad \text{(III.45)}$$

where A_0 is the interfacial free energy between wall and pure solvent. Eq. (III.45) is due to Joanny and Leibler.[13] Inserting eq. (III.29) for ξ we see that $A - A_0$ scales like $\Phi^{3/2}$.

(*ii*) *Concentration in the first layer* Φ_1. Clearly, from Fig. III.10b, Φ_1 is much smaller than the bulk concentration Φ. We do not have yet a complete theory of the relation between Φ_1 and Φ, although some related problems in the theory of magnetism have been discussed in the literature.[13] Here we present a simple conjecture on Φ_1. We assume that the osmotic pressure Π of the solution is proportional to the number of direct contacts between monomers and surface, i.e. to Φ_1

$$\Pi \cong T a^{-3} \Phi_1$$

Equating this to the bulk form [eq. (III.30)] we obtain $\Phi_1 \cong \Phi^{9/4}$. The argument can be generalized to arbitrary dimensionalities *d;* for $d = 4$, it lends to $\Phi_1 \cong \Phi^2$; we shall see in Chapter IX that this agrees with a mean field type of calculation, which indeed becomes correct at $d = 4$.

(*iii*) *Concentration profile*. At a distance z from the wall, we expect to find a concentration $\Phi(z)$ with the scaling form

$$\Phi(z) = \Phi f_\Phi(z/\xi)$$

$$f_\Phi(x) \cong \begin{cases} 1 & (x \gg 1) \\ x^m & (x \ll 1) \end{cases}$$

Writing that $\Phi(z) \approx \Phi_1$ for the first layer ($z = a$), we are then led to the following value for the exponent m:

$$m = \nu^{-1} = 5/3 \qquad \text{(III.46)}$$

Our main conclusion is that experiments measuring the direct contacts between chains and a repulsive wall would be most interesting: for instance if the wall surface contains fluorescent groups, while the chains act as optical traps.

A practical difficulty may be related to the existence of *long range van der Waals forces*. When a monomer immersed in a solvent is at a distance z from the wall, it experiences an attractive potential[3] that decreases as z^{-3}. Then, even if the monomers in the *first* layer experience repulsion, the region of thickness ξ near the wall might have its concentration increased rather than decreased, and the situation would be seriously modified. To avoid this, it is best to use a solvent whose polarizability is as identical to the monomer's as possible; in many cases, the solvent can be the monomer itself.

III.3.2. A semi-dilute solution in a cylindrical pore[14]

This situation is shown in Fig. III.11. The pore has a diameter D smaller than the natural size R_F of the chains. [The precise shape of the cross-section (circular, square, etc.) is not important for our scaling arguments.] Physically, this might be achieved in ternary solutions: lipid + water + polymer, where the lipid tends to make a hexagonal phase with long, parallel tubes.

We take the wall to be repulsive as in the preceding section. From Chapter I we know the behavior of a single chain in this situation: it occupies a region of length $R_{\parallel} \sim aN\,(a/D)^{2/3}$ [eq. (I.53). What happens if we increase the concentration of chains?

We expect to find a dilute regime, with coils that behave very much like a one-dimensional gas of hard rods, each of length $R_{\parallel}$. If the average concentration is c, the number of chains per unit length is $(\pi/4)\,(c/N)\,D^2$. The fraction of the tube which has a chain in it is

$$\psi = \frac{\pi}{4}\,\frac{c}{N}\,D^2\,R_{\parallel} \cong \frac{c}{c_1^*} \qquad \text{(III.47)}$$

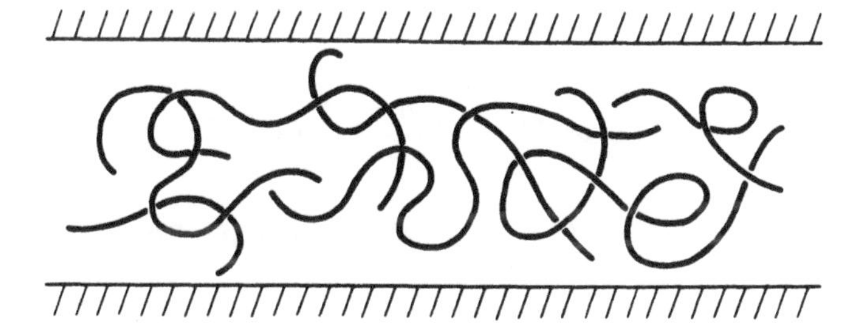

Figure III.11.

A semi-dilute solution trapped in a cylindrical pore with repulsive walls (no adsorption). Note the depletion layer near the wall.

where the overlap threshold is defined by

$$c_1^\star \cong \frac{N}{D^2 R_\parallel} \cong \frac{1}{a^3}\left(\frac{a}{D}\right)^{4/3} \qquad (R_F \gg D \gg a) \qquad \text{(III.48)}$$

The osmotic pressure in this dilute regime would be

$$\Pi = \frac{cT}{N}\frac{1}{1-\psi} \qquad \psi < 1 \qquad \text{(III.49)}$$

where the factor $(1 - \psi)^{-1}$ accounts for steric hindrance between different "rods." When $\psi \to 1$, however, the rods will begin to interpenetrate, since they are not completely "hard," and a different viewpoint is required.

The transition from $c < c_1^\star$ to $c > c_1^\star$ is very different from what happens in three dimensions. The point is that $c_1^\star$ is independent of N. Thus, if we tried to define a correlation length ξ_1 by a formula similar to eq. (III.29), writing

$$\xi_1 = R_\parallel \left(\frac{c^\star}{c}\right)^{m_\xi}$$

with an unknown exponent m_ξ, we would find that *no* value of m_ξ can lead us to a ξ_1 which is independent of N.

The physical answer is different. As soon as $c > c_1^\star$, the three-dimensional correlation length ξ becomes smaller than the tube diameter. To see this, start with eq. (III.28):

$$\xi = a(\Phi)^{-3/4} = a\left(\frac{\Phi_1^\star}{\Phi}\right)^{3/4} (\Phi_1^\star)^{-3/4}$$

and insert the value of $\Phi_1^\star = c_1^\star a^3$ from eq. (III.48). The result is

$$\xi = D\left(\frac{\Phi_1^\star}{\Phi}\right)^{3/4} \qquad \text{(III.50)}$$

Thus, for $\Phi > \Phi^\star$ we have blobs which are smaller than D, and all local correlation properties return to their three-dimensional value. The osmotic pressure is still given by eq. (III.27)

$$\Pi = Ta^{-3}\,\Phi^{2.25} \qquad (1 \gg \Phi \gg \Phi_1{}^{\star} \qquad \text{(III.51)}$$

and there is no simple crossover between eq. (III.49) and eq. (III.51). The osmotic pressure is qualitatively shown in Fig. III.12.

We shall now discuss briefly the *conformation of one chain* in the overlapping regime ($\Phi > \Phi_1$). To simplify the discussion we shall in fact restrict our attention to large concentrations $\Phi = 1$, corresponding to $\xi = a$. The reason for this choice is that the most interesting effects occur at scales $\geqslant D$, while $\xi < D$ in the overlapping regime. Even with this simplification, the discussion is rather delicate, and the conclusions reached by the present author in Ref. 14 were incorrect. An improved version has been developed recently,[15] and will be summarized here.

We start from a melt of chains and confine it in a tube of diameter D. When D is large, we are dealing with a three-dimensional system, and we know from Chapter II that the chains are ideal, with a size $R_0 = N^{1/2}a$. Let us now decrease D at fixed N, and reach the situation where $D < R_0$. Each chain is then confined to a linear dimension D for directions normal to the tube axis. Along the tube axis, the chain spans a certain length $R_{\parallel}$. We shall now define two essential parameters controlling the chain conformation.

(*i*) *The internal filling fraction* corresponding to N monomers spread in a volume $\sim D^2 R_{\parallel}$ is

$$\Phi_{int} \cong Na^3/(D^2 R_{\parallel}) \qquad \text{(III.52)}$$

In particular, when the chain is still equivalent to an ideal random walk in the direction of the tube axis, this filling fraction is

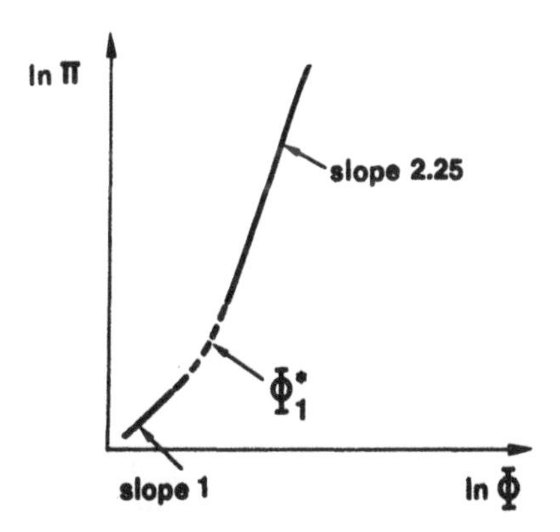

Figure III.12.

$$\Phi_{int} \rightarrow \Phi^{o}_{int} = Na^3/(D^2R_0) = N^{1/2}(a/D)^2 \tag{III.53}$$

(*ii*) *The perturbation parameter* ζ, introduced in eq. (I.42), tells us when the interactions inside one chain become important ($\zeta > 1$). However, ζ must incorporate two effects which were absent in Chapter I. First, we must recall that the interactions are screened out in a melt [eq. (II.7)]

$$v \rightarrow \tilde{v} = vN^{-1} \tag{III.54}$$

Second, we now have an internal filling fraction Φ_{int} which is increased by the confinement; it is given by eq. (III.52). In a perturbation calculation, the repulsion potential on one monomer is proportional to $T\tilde{v}\,\Phi_{int}a^{-3}$, and the repulsive energy for the chain is $T\zeta \cong NT\tilde{v}\,\Phi_{int}a^{-3}$. We see then that

$$\zeta \cong \Phi_{int} \tag{III.55}$$

Having defined the essential parameters, we can now return to eqs. (III.52, 53) and reach the following conclusions.

(*i*) When $D > N^{1/4}\,a$, it is consistent to assume that the chain size $R_{\parallel}$ is equal to the ideal value of R_0: when we do this, we find that Φ_{int} and ζ are small as required.

(*ii*) When $D < N^{1/4}\,a$, the chain cannot remain ideal, since this would lead to values $\Phi_{int} > 1$ which are not acceptable. We expect Φ_{int} to increase monotonically when D is decreased (at constant N). Thus in regime (*ii*) Φ_{int} must reach its maximum allowed value $\Phi_{int} \rightarrow 1$. Using eq. (III.52) this leads to

$$R_{\parallel} \cong Na^3D^{-2} \tag{III.56}$$

In regime (*ii*) there is not much overlap between consecutive chains: they lie in sequence very much like jammed automobiles in a one-lane tunnel. (Two adjacent chains overlap only in a small "terminal region" of linear dimensions $\gtrsim D$.) Ultimately when D decreases down to a, the chain becomes fully stretched ($R_{\parallel} \rightarrow Na$).

We find here a very striking difference between the confined single chain (in a good solvent) described by eqs. (I.51–56), and the confined melt. For the single chain, the longitudinal dimension $R_{\parallel}$ is increased by confinement, as soon as $D < R_F$. For the melt, the length $R_{\parallel}$ is not modified when D becomes smaller than R_0, but only when D reaches the much smaller value $(R_0 a)^{1/2}$.

Let us now return to the *semi-dilute case* and discuss briefly the following problem: we have a pore of diameter D in equilibrium with a bulk solution. The average concentration in the bulk is fixed at a certain value Φ_B. what is the average concentration in the pore Φ? Clearly, when Φ_B is very small, Φ depends critically on the pore diameter

$$\Phi \cong \Phi_B \exp\left(-\ F_{conf}/T\right) \tag{III.57}$$

where F_{conf} is the confinement energy for a single coil, given by eq. (I.56) for $D < R_F$.

On the other hand, when $\Phi_B = 1$ we also expect $\Phi = 1$: a melt penetrates all pores of size $D \geqslant a$. The matching curve between these two limits is qualitatively shown in Fig. III.13.

When do we cross over from the weak penetration regime of eq. (III.57) to the strong penetration regime for concentrated systems? A detailed answer to this question could be obtained from a study of the chain chemical potential, but the essential features may be reached more simply. Let us start from a tube with a diameter D somewhat larger than the correlation length ξ_B in the bulk. Then we expect to have a depletion layer of thickness ξ_B near the tube walls, as described in Section III.3.1. The perturbing effects of the wall do not extend further than ξ_B.

Since we assumed $D \gtrsim 2\xi_B$, this means that the central portion of the tube is unaffected, and reaches a concentration Φ equal to the bulk value. On the other hand, if $D \lesssim 2\xi_B$ the depletion layers occupy all the tube volume, and $\Phi \ll \Phi_B$. Thus we are led to the following rule: upon increasing the bulk concentration Φ_B, penetration in the pore occurs when the bulk correlation length $\xi_B = a\,\Phi_B^{-3/4}$ becomes comparable to the pore diameter. If we call Φ_p the threshold for penetration, we predict $\Phi_p^{-3/4} \cong D/a$ or:

$$\Phi_p \cong (a/D)^{4/3} \cong \Phi_1\bigstar \tag{III.58}$$

Experiments using well-calibrated tubes to confirm (or infirm) the law [eq. (III.58)] are currently underway.*

*D. Cannell, private communication, 1979.

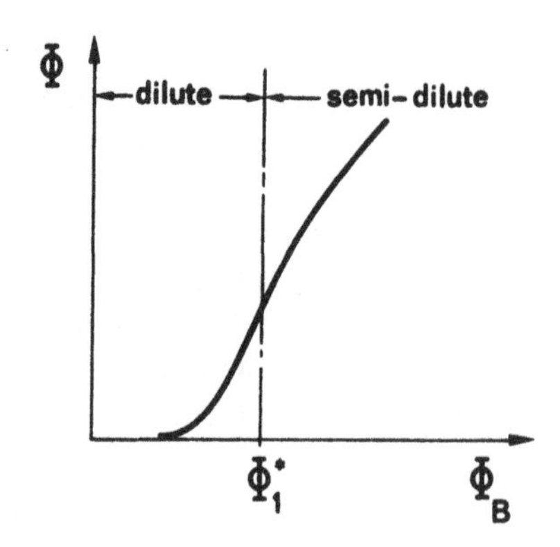

Figure III.13.

SUMMARY

Near a repulsive wall a polymer solution shows a depletion layer of thickness ξ. In a thin tube of diameter D, the solution can exist in two very distinct regimes: dilute and semi-dilute. In the latter case all local properties are similar to those of a bulk solution, but the chain can still be stretched along the tube axis.

Remark. The problem of chains confined in a slit (two-dimensional confinement) is also of interest; it may be achieved in solutions of polymers + lipid + water (lamellar phases). The predicted behavior of the chains is very different from what we found in a tube.[14]

REFERENCES

1. P. Flory, *Principles of Polymer Chemistry,* Chap. XII, Cornell University Press, Ithaca, New York, 1971.
2. M. Huggins, *J. Phys. Chem.* **46,** 151 (1942). M. Huggins, *Ann. N.Y. Acad. Sci.* **41,** 1 (1942). M. Huggins, *J. Am. Chem. Soc.* **64,** 1712 (1942).
3. *Molecular Forces,* Pontifical Academy of Science, Ed., North-Holland, Amsterdam, 1967.
4. S. F. Edwards, *Fluides Moléculaires,* R. Balian and G. Weill, Eds., Gordon & Breach, New York, 1976.
5. D. J. Burch, M. Moore, *J. Phys.* **A9,** 435 (1976). L. Schafer, T. Witten, *J. Chem. Phys.* **66,** 2121 (1977). L. Schafer, T. Witten, *Proc. STATPHYS, Annals Israel Phys. Soc.* **13,** (2), 974 (1978).
6. J. des Cloizeaux, *J. Phys. (Paris)* **36,** 281 (1975).
7. M. Daoud, *et al. Macromolecules* **8,** 804 (1975). See also K. Okano, E. Wada, Y. Taru, H. Hiramatsu, *Rep. Prog. Polym. Sci. Japan* **17,** 141 (1974).

8. D. Langevin, F. Rondelez, *Polymer* **19,** 875 (1978).
9. M. Daoud, Ph.D. Thesis, Paris, 1976. Available from Laboratoire Léon Brillouin, CEN Saclay, 91 Gif s/ Yvette, France.
10. B. Farnoux, *Annales de Phys.* **I,** 73 (1976).
11. I. Ohmine, R. Silbey, J. Deutch, *Macromolecules* **10,** 862 (1977).
12. S. F. Edwards, *Proc. Phys. Soc.* **88,** 265 (1966).
13. J. F. Joanny, L. Leibler, P. G. de Gennes, *J. Polym. Sci.* (in press).
14. M. Daoud, P. G. de Gennes, *J. Phys. (Paris)* **38,** 85 (1977). F. Brochard, S. Daoudi, *Macromolecules* **11,** 751 (1978).
15. F. Brochard, P. G. de Gennes, *J. Phys. (Paris) Lett.* (in press).

IV

Incompatibility and Segregation

IV.1. General Principles and Questions

IV.1.1. The trend toward segregation

In discussing the interaction between a polymer and a solvent in Section III.1 we were led to define the interactions in terms of a Flory parameter χ; the energy of interaction per site was of the form

$$F_{int} = T \chi \Phi_A \Phi_B$$

where Φ_A and Φ_B are the volume fractions of the two portions ($\Phi_A + \Phi_B = 1$). A positive χ implies that species A and B tend to *lower the energy* when they separate into two phases: one (A-rich) has a small Φ_B; the other phase (B-rich) has a small Φ_A. Thus, for both phases the product $\Phi_A \Phi_B$ is small and the energy is low.

As pointed out in Section III.1, a positive χ is the most frequent case, at least whenever van der Waals interactions are dominant. This also occurs more often when the A and B are small molecules rather than when A (or B, or both) are polymers. However, the consequences are more drastic in the latter case: large molecules have extremely strong segregation effects.

Consider two immiscible solvents (1) and (2) that are in contact and add one large polymer chain of N units. If we bring the chain from solvent 1 to solvent 2, each monomer experiences a change in free energy ΔF_{12} (containing both an energy and an entropy contribution). Positive ΔF_{12}

values mean that the chains prefer solvent 1, and vice-versa. For dilute chains, the ratio of chain concentrations in the two solvents is

$$\rho \equiv \frac{c_2}{c_1 + c_2} = \frac{\exp(-N\Delta F_{12}/T)}{1 + \exp(-N\Delta F_{12}/T)} \qquad \text{(IV.1)}$$

The essential feature is the presence of the factor N in the exponent; if we plot ρ as a function of $\Delta F_{12}/T$, we have an extremely sharp step.

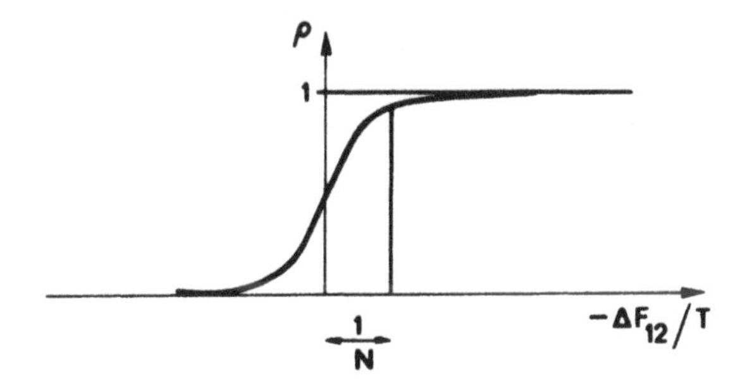

Figure IV.1.

Even when the two solvents are very similar (ΔF_{12} small compared with T), the chains will usually show a *strong preference toward one of the solvents*. These are classical considerations in liquid–liquid extraction. A recent example of interest is the case of *one* solvent which can exist in two different liquid states: the isotropic and the nematic state. In the nematic state the solvent molecules tend to be parallel.[1] It is of some interest to dissolve macromolecular chains in the nematic state. However, for the few cases studied,[2,3] there is a small ΔF_{12} favoring solution in the isotropic state. Typically $\Delta F_{12}/T \sim 1/40$, and this implies that chains of more than ca. 40 units cannot be dissolved in the nematic phase.

Returning to our phase separation problem, if our system separates into two phases, we may consider them as the solvents of eq. (IV.1), and we see that a given type of chain will tend to go entirely into one of the two phases. Segregation effects will be very strong—too strong. First we must list cases where some compatibility can be maintained and where demixing transitions can be observed.

IV.1.2. Cases of partial compatibility

(a) The simplest case is obtained if we try to mix *two types of polymers* A + B, with chain lengths N_A and N_B. In most cases if χ is positive and the interaction per chain is very large, there is no miscibility between A and B at any reasonable temperature.

However, there are a few cases where compatibility is apparently observed. A detailed list and discussion have been given by Krause.[4] Here we quote four systems in which miscibility can occur:

(i) Mixtures of polystyrene (PS) and poly(vinyl methyl ether) (PVME) have a Flory interaction parameter χ which has been measured separately (through vapor pressure measurements on ternary mixtures with benzene).[5] The resulting χ depends strongly on concentration—an anomalous feature—but it appears to be negative at 30°C for all mixtures below 70% PS.

(ii) If a polymer species A is "doped" with a small fraction of positively charged side groups, while B is doped with negative groups, compatibility may occur.*

(iii) A case of interest could be obtained by mixing a pure polymeric species U

$$(A) = U\ U\ U\ U$$

with a statistical U V copolymer

$$(B) = (U_{1-x}\ V_x) = U\ U\ V\ U\ U\ V\ U\ U\ U\ U\ V \ldots$$

We are interested mainly in making the difference between (A) and (B) small—i.e., in *small x values*. Then some compatibility can be maintained. It is important that the chemical sequence of B be statistical; in particular, the number of V units (xN_B) must be much larger than 1 (although x is small). Then it is possible to show (using the methods of Chapter X) that the V units will not tend to segregate locally by building up micelles or other organized structures. Micelles seem to be compatible only with well-defined periodicities in the chemical sequence of the (B) chains.

(iv) Another interesting case may occur with mixtures of hydrogenated and deuterated species. In the preceding chapters we considered such isotopic mixtures as ideal, and this is indeed an excellent approximation for an N that is not too large. However, there is a small interaction parameter χ between (H) and (D) monomers which has been estimated recently by Buckingham.[6] It appears that χ may be positive, and of order 10^{-4} to 10^{-3}, depending on the chemical species under study. This would then mean that demixing could occur if the molecular weight were very high ($N \sim 10^4$). If this is confirmed, it may lead to an interesting set of neutron experiments.

*S. Djadour, R. N. Goldberg, H. Morawetz, *Macromolecules* **10,** 1015 (1977).

(b) Another important case is obtained with simple polymer solutions in *poor solvents*—i.e., when the corresponding Flory parameter χ becomes larger than 0.5, or when the excluded volume $v(T) = a^3(1 - 2\chi)$ becomes negative. This situation has been studied extensively during the past 30 years; it occurs in many simple systems, the classical example being polystyrene dissolved in cyclohexane. For this system, the temperature Θ at which $v(T)$ vanishes is $\sim$ 37°C and is thus in a convenient range.

(c) A third group of interesting segregation effects is obtained with ternary mixtures: polymer A + polymer B + solvent S. It is particularly convenient to have S be a good solvent for both A and B, but to have also a very strong repulsion between A and B. In Flory-Huggins language we have three volume fractions Φ_A, Φ_B, Φ_S, and the interaction energy is a quadratic function of these fractions.

$$\frac{1}{T} F_{int/site} = \frac{1}{2} \sum_{ij} \chi_{ij}^{dir} \Phi_i \Phi_j \qquad \text{(IV.2)}$$

where the symbol χ^{dir} stands for direct interaction, and $\sum_{ij}$ is over the three components. However, the Φ terms are related by the condition

$$\sum_i \Phi_i = 1 \qquad \text{(IV.3)}$$

and this allows us to reduce F_{int} to a quadratic function of two variables (plus linear and constant terms which we drop as explained in Section III.1). A quadratic function of two variables has three independent coefficients—i.e., we need three parameters to characterize the interactions. In practice, we prefer to do this in a symmetrical way; we transform the diagonal term in eq. (IV.2) as follows

$$\frac{1}{2} \chi_{AA}^{dir} \Phi_A^2 = \frac{1}{2} \chi_{AA}^{dir} \Phi_A (1 - \Phi_B - \Phi_S) \text{ etc.}$$

Then we drop the linear terms and arrive at an interaction

$$\frac{1}{T} F_{int/site} = \chi_{AB} \Phi_A \Phi_B + \chi_{AS} \Phi_A \Phi_S + \chi_{BS} \Phi_B \Phi_S \qquad \text{(IV.4)}$$

where the three χ parameters are defined in terms of the direct coefficient χ^{dir} by equations similar to eq. (III.4)

$$\chi_{AB} = \chi_{AB}^{dir} - \frac{1}{2}(\chi^{dir}_{AA} + \chi^{dir}_{BB}) \text{ etc.} \qquad \text{(IV.5)}$$

If the direct interactions are dominated by van der Waals forces and are factorizable as in Chapter III ($\chi^{d}{}_{ij} = -$ constant $\alpha_i \alpha_j$), then χ_{AB}, χ_{AS}, χ_{BS} are positive.

We are interested primarily in the case where $\chi_{AS} < 1/2$, $\chi_{BS} < 1/2$ (good solvent), and $\chi_{AB} > 0$ (trend toward segregation). There do not seem to exist many data on such a case, but there is an excellent study on ternary mixtures where the solvent is somewhat less good—i.e., where χ_{AS} and χ_{BS} are close to 0.5. The system[7] is as follows:

A = polystyrene B = polyisobutylene S = toluene

where $\chi_{AS} = 0.45$, and $\chi_{BS} = 0.48$ at room temperature.

IV.1.3. Specific features of polymer segregation

Precipitation of a polymer from a solvent of decreasing quality is a classical process.[8] Segregation effects in mixtures of two polymers are sometimes welcome—e.g., for the preparation of *composite materials* with special resistance to fracture—but they are often unwelcome because the resulting structures scatter light and result in a loss of transparency. The situation with two polymers in a good solvent is also of some practical importance. For example, high impact polystyrene is prepared from a solution of polybutadiene in styrene. When the polymerization of styrene is initiated, as soon as the polystyrene fraction reaches a (rather low) threshold, phase separation occurs. Again the result is a composite material, of complex texture, and interesting mechanical properties.

Segregation effects are also important for fundamental studies. For each demixing process there is a *critical point,* near which certain fluctuations of concentration in the mixture become anomalously large. Essentially all the work on the theory of these effects in polymer systems has been based on mean field ideas. However in many cases critical phenomena are now known to be qualitatively different from mean field predictions. One of our tasks in this chapter is to *classify* the critical points; some will be of the mean field type, others will be different.

From an experimental standpoint, polymer solutions and melts show segregation effects which are often very different from what we see in binary mixtures of small molecules:

(i) The spatial scales are enlarged because the building blocks are coils of size 100–500 Å.

(ii) All characteristic times are also enlarged. This enhancement is

particularly evident for polymer–polymer systems, where the chains are strongly entangled. On the one hand, this often prevents the use of certain powerful observation techniques (such as the photon beat method), but, on the other hand, it allows for relatively easy studies on quenching processes, where a homogeneous mixture is suddenly cooled to a temperature where it will segregate. There is a remarkable analogy here between studies on polymer melts and studies on metallic alloys; both types of systems have very slow diffusion processes.

(iii) A constant complication displayed by polymer systems is the existence of *polydispersity* in all practical samples—i.e., of a distribution of molecular weights (or of the polymerization index N). For our discussion of scaling laws in Chapters I, II, and III, polydispersity was not a serious complication. If the *shape* of the distribution is the same for all samples, a law such as $R_F \sim N^{3/5}$ remains valid for the average R_F as a function of the average N. However, when phase separations are involved, the effects of polydispersity can become much more dangerous. Our discussion in Section IV.1.1 suggests that the longer chains have a stronger trend toward segregation; two phases in equilibrium do not have the same weight distribution for their chains. These effects have been discussed in the Flory-Huggins framework.[9]

IV.2. Polymer–Polymer Systems

We start with segregation problems involving two polymers (rather than with the one solvent–one polymer problem) because polymer–polymer systems can be rather correctly described in terms of Flory-Huggins theory; no similar simplification exists for polymer–solvent systems.

In this section, we first reconstruct some classic results on thermodynamic properties[10] and then discuss spatial correlations, which are less well known.

IV.2.1. Thermodynamic principles

Our starting point is the Flory-Huggins free energy for a mixture of two polymer species (with degrees of polymerization N_A, N_B). This is a natural generalization of eq. (III.7)

$$\left.\frac{F}{T}\right|_{site} = \frac{\Phi_A}{N_A} \ln \Phi_A + \frac{\Phi_B}{N_B} \ln \Phi_B + \chi \Phi_A \Phi_B \qquad \text{(IV.6)}$$

where Φ_A, Φ_B are the volume fractions, related by $\Phi_A + \Phi_B = 1$. We shall often write $\Phi_A = \Phi$ and $\Phi_B = 1 - \Phi$. The first two terms in eq. (IV.6) represent an estimate of the entropy for arranging the A and B chains on the lattice. Surprisingly, they are not different from what one would have with two perfect gases of concentrations* Φ_A/N_A and Φ_B/N_B. The third term in eq. (IV.6) is the interaction term and is positive.

How do we discuss phase separation on the free energy $F(\Phi)$? The essential property is the curvature of $F(\Phi)$, as explained in Fig. IV.2.

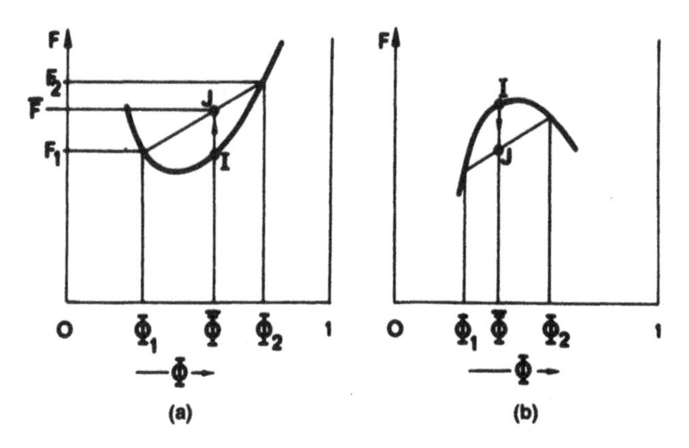

Figure IV.2.

Assume first that the sample is homogeneous (single phase), with a certain concentration $\overline{\Phi}$ (point I). Try then to decompose it into two phases, of concentrations Φ_1 and Φ_2. The relative weights of the two phases in the mixtures are f_1 and f_2. We then have

$$\overline{\Phi} = f_1 \Phi_1 + f_2 \Phi_2 \qquad (f_1, f_2 \text{ positive}) \tag{IV.7}$$

and we reach a free energy

$$\overline{F} = f_1 F_1 + f_2 F_2 \tag{IV.8}$$

This corresponds to point J in Fig. IV.2. The energy change is positive in case (a) and negative in case (b). Thus, case (b) imposes phase separation.

In case (a) near a concentration $\overline{\Phi}$ we have local stability, but we may still have an instability with respect to another branch of the phase diagram. This will become apparent in the plots of the Flory-Huggins free energy [eq. (IV.6)] (Fig. IV.3).

*In a perfect gas $F \sim Tc \ln c$; here if we set $c = \Phi/N$, we get $T\Phi/N\,(\ln \Phi - \ln N)$, and the second term (linear in Φ) can be dropped, as already explained.

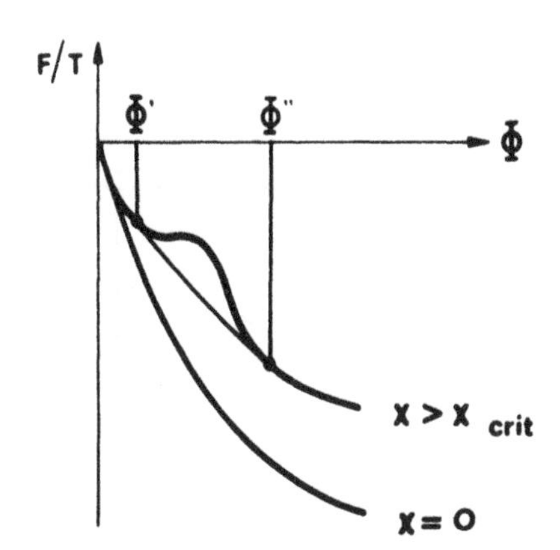

Figure IV.3.

For $\chi = 0$ (or χ finite but very small), the plot $F(\Phi)$ is convex everywhere; entropy effects are dominant, and they favor mixing. However, when χ becomes larger than a critical value χ_c (to be computed below), a region of negative curvature exists. Repeating the argument of Fig. IV.2 we see that a single phase exists only outside a certain interval (Φ' Φ''), while inside the interval the system breaks up into two phases, of concentrations Φ' and Φ''.

It is convenient to discuss these properties in terms of the quantity $\partial F/\partial\Phi = \mu$ where μ is what we call the *exchange* chemical potential (or more briefly the exchange potential); a change $\Phi \to \Phi + d\Phi$ represents an increase in the number of A monomers (equal to $d\Phi$ per site) but an equivalent decrease of B ($- d\Phi$ per site). The two coexisting phases at Φ' and Φ'' have equal μ values.

IV.2.2. The coexistence curve in the symmetrical case

We can now compute the coexistence curve—i.e., the plots of $\Phi'(\chi)$ and $\Phi''(\chi)$. We do this for only one specific example—the symmetrical *case* of $N_A = N_B = N$. Then the entire free energy diagram is symmetrical around $\Phi = 1/2$, and the equations are simplified. The exchange potential then vanishes at $\Phi = \Phi'$ or $\Phi = \Phi''$ as shown in Fig. IV.4. Returning to eq. (IV.6) and writing $\mu = 0$ gives the coexistence curve:

$$\frac{1}{N} \ln [\Phi/(1 - \Phi)] + \chi (1 - 2 \Phi) = 0 \qquad \text{(IV.9)}$$

This can be written as:

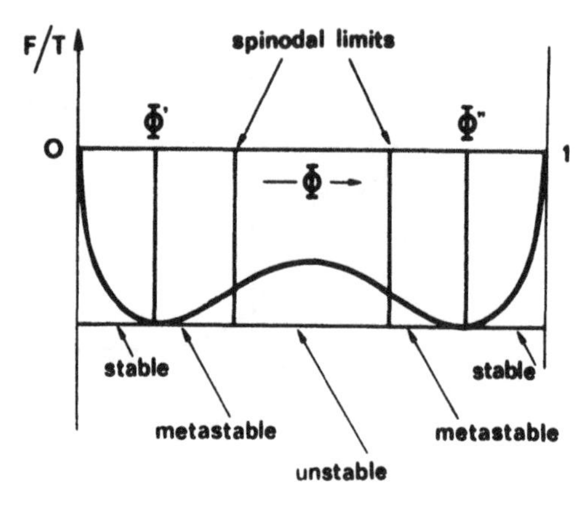

Figure IV.4.

$$\chi = \frac{1}{N}\frac{(-1)}{1 - 2\Phi} \ln [\Phi/(1 - \Phi)] \tag{IV.10}$$

The resulting coexistence curve is shown in Fig. IV.5. Phase separation occurs only when χ is larger than a threshold value:

$$\chi_c = \frac{2}{N} \tag{IV.11}$$

This very small value of χ_c is the essential reason for the strong incompatibility usually found between polymers. We need very special tricks (explained in Section IV.1) to realize values of χ below this threshold.

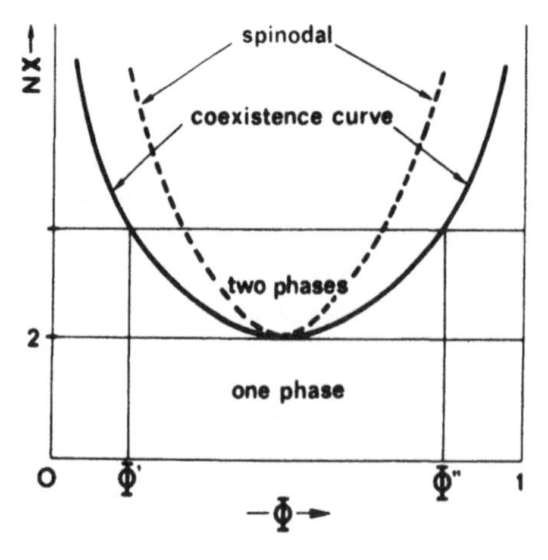

Figure IV.5.

IV.2.3. Metastable states and the spinodal curve

Equilibrium is reached only slowly in polymer melts of high molecular weight. This will facilitate certain quenching experiments, where the system is brought suddenly from the one-phase region to the two-phase region (e.g., by a change in temperature imposing a change in χ).

If we enter the two-phase region only slightly, demixing can take place only by nucleation of a droplet of one phase inside the other. This is a thermally activated process, implying the interfacial energy at the droplet surface, and is slow. However, if we go more deeply into the two-phase region, we reach a state where the interfacial energy vanishes (and changes sign). It is then favorable for the system to break up spontaneously into many small domains. This threshold defines what we call the *spinodal curve* in the χ, Φ plane.*

From the point of view of Fig. IV.2 the instability occurs whenever the $F(\Phi)$ plot is concave. The spinodal thus corresponds to the inflection points in Fig. IV.4, and is ruled by the equation

$$0 = \frac{\partial^2}{\partial\Phi^2}\left(\frac{F}{T}\right) = \frac{1}{N_A\Phi} + \frac{1}{N_B\,(1-\Phi)} - 2\chi \qquad \text{(IV.12)}$$

In eq. (IV.12) we have not set $N_A = N_B$ since the general formula remains quite simple. The plot of eq. (IV.12) for the symmetrical case is given in Fig. IV.5.

IV.2.4. The critical point

The critical point always corresponds to the minimum value of χ on the spinodal curve. From eq. (IV.12) this is obtained when $\Phi = \Phi_c$, where Φ_c satisfies

$$-\frac{1}{N_A\,\Phi_c^{\,2}} + \frac{1}{N_B(1-\Phi_c)^2} = 0 \qquad \text{(IV.13)}$$

$$\frac{\Phi_c}{1-\Phi_c} = \left(\frac{N_B}{N_A}\right)^{1/2}$$

$$\Phi_c = \frac{N_B^{\,1/2}}{N_A^{\,1/2} + N_B^{\,1/2}} \qquad \text{(IV.14)}$$

*It is not quite certain that the spinodal has a precise experimental meaning. As pointed out recently by K. Binder (unpublished), the nucleation processes have a low barrier when we get close to the nominal spinodal line, and the onset of observable instabilities is probably: (*i*) not sharp; (*ii*) dependent on the length of the observations. But for many practical purposes, the spinodal concept is helpful—particularly so at the mean field level which we discuss here.

In the symmetrical case $\Phi_c = 1/2$. On the other hand, when N_B becomes much smaller than N_A, the critical point shifts toward low concentrations of A. (Ultimately, when $N_B = 1$, we are led back to a polymer + solvent problem (discussed in the next section). From eq. (IV.12) the critical value of χ is:

$$\chi_c = (N_B^{1/2} + N_A^{1/2})^2/2\, N_A\, N_B \tag{IV.15}$$

When $N_B = N_A$, χ_c is very small and compatibility is exceptional. However, when the situation is unsymmetrical ($N_B \ll N_A$), then χ_c ($\cong 1/2\, N_B$) becomes somewhat larger, and compatibility is more frequent. We do not discuss the full coexistence curves near the critical point for the dissymetric case, but their qualitative aspect is shown in Fig. IV.6.

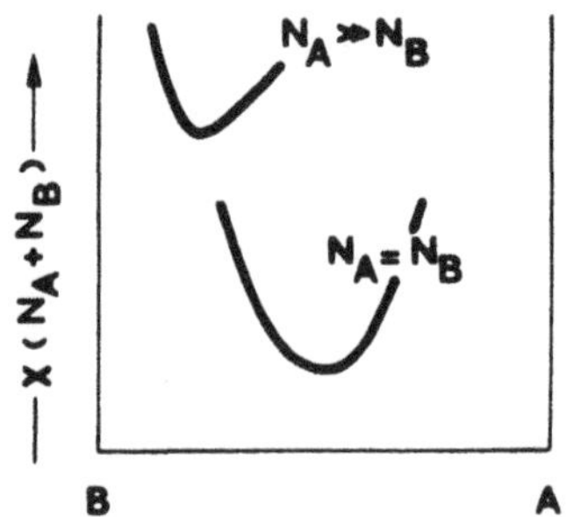

Figure IV.6.

IV.2.5. Critical fluctuations

Returning to a phase diagram such as that shown in Fig. IV.5 we now focus on the area of the one-phase region that is near the critical point. In this domain, the local concentration Φ has large fluctuations which can be detected by light scattering experiments. At present this kind of data is not very abundant for polymer–polymer systems (because of the long times required to reach equilibrium) but we hope that the situation will improve.

In a scattering experiment (using light, X-rays, or neutrons) the essential parameter is the scattering wave vector $\mathbf{q}$ (equal to $4\pi\lambda^{-1} \sin \theta/2$ where λ is the wavelength and θ is the scattering angle). What is measured is a correlation function between the concentration at two points

$$S_{ij}\,(\mathbf{r}_1 - \mathbf{r}_2) = \langle \Phi_i\,(\mathbf{r}_1)\, \Phi_j\,(\mathbf{r}_2) \rangle - \langle \Phi_i \rangle \langle \Phi_j \rangle \tag{IV.16}$$

where subscripts i and j represent the various species present (here i, j = A, B), and the brackets $\langle\ \rangle$ denote a thermal average. In our particular lattice model, because $\Phi_A + \Phi_B = 1$, there is only one independent correlation function

$$S_{AA} = S_{BB} = - S_{AB} = S \qquad (IV.17)$$

What is measured is the Fourier transform

$$S(\mathbf{q}) = a^{-3} \int d\mathbf{r} \exp(i\mathbf{q}\cdot\mathbf{r})\, S(\mathbf{r}) \qquad (IV.18)$$

where the factor a^{-3} is introduced to make $S(\mathbf{q})$ dimensionless. We may say that $S(\mathbf{q})$ is the scattering power (at a given $\mathbf{q}$) per site of the lattice. At first sight the calculation of the correlation $S(\mathbf{r})$ in a dense mixture of strongly interacting chains appears formidable. However, it is simple because *on the scale of one coil the chains remain nearly ideal* (very much like the one-component melts of Chapter II).

The complete calculation of correlations can then be performed by a "random phase" method described in Chapter IX. Here we quote the results. They can be expressed simply in terms of the Debye function $g_D(N, q)$ for the scattering by an ideal chain of N monomers (defined in Section I.1). Explicitly, one finds the simple formula

$$S^{-1}(q) = \frac{1}{\Phi g_D(N_A,q)} + \frac{1}{(1 - \Phi)\, g_D(N_B,q)} - 2\chi \qquad (IV.19)$$

We now discuss its consequence in detail

For $q = 0$ the Debye function $g_D\,(N, q = 0)$ is equal to N. Eq. (IV.19) is then identical to eq. (IV.12), and we have

$$TS^{-1}(0) = \frac{\partial^2 F}{\partial \Phi^2} \qquad (IV.20)$$

in agreement with a general thermodynamic theorem. Physically the low q limit corresponds to small-angle scattering. Eq. (IV.19) then tells us that the intensity diverges not only at the critical point but also at each point of the spinodal curve; light scattering is a good indicator of the vicinity of the spinodal.

For $q \to 0$ the inverse intensity may be written in a simple form. This is obtained by expansion of the Debye function at small q

$$g_D\,(N, q \to 0) = N\,(1 - \frac{1}{3} q^2 R_G^2) \quad (qR_G < 1) \qquad (IV.21)$$

where $R_G^2 = Na^2/6$ is the gyration radius of the coil. Eq. (IV.21) corresponds to a general theorem due to A. Guinier.[11] Inserting this into eq. (IV.19) we get:

$$S^{-1}(q) = 2(\chi_s(\Phi) - \chi) + \frac{q^2\, a^2}{18} \frac{1}{\Phi\,(1 - \Phi)} \qquad \text{(IV.22)}$$

where χ_s is the value of χ on the spinodal as defined in eq. (IV.12). It is convenient to rewrite this equation in the standard form

$$S(q) = \frac{S_{AA}\,(0)}{1 + q^2 \xi_s^{\,2}} \qquad \text{(IV.23)}$$

where ξ_s is a certain correlation length* defined explicitly by

$$\xi_s = \frac{a}{6}\, [\Phi(1 - \Phi)\,(\chi_s(\Phi) - \chi)]^{-1/2} \qquad \text{(IV.24)}$$

Note that ξ_s diverges near the spinodal with a square-root singularity.

For $qR_G > 1$ the g_D functions are strongly reduced, the factors $1/g_D$ increase, and the last term (-2χ) becomes negligible. In this limit we have

$$g_D(N,q) \rightarrow \frac{12}{q^2\, a^2} \quad \text{(independent of } N\text{)}$$

and the scattering intensity differs from this only by a normalization factor

$$S(q) \rightarrow \frac{\Phi\,(1 - \Phi)\, 12}{q^2 a^2} \qquad (qR_G > 1)$$

It is also of interest to rewrite these results in terms of the spatial correlation function $S(\mathbf{r})$, shown in Fig. IV.7. At relatively small distances

$$S(\mathbf{r}) = \frac{3}{\pi}\, \Phi(1 - \Phi) \frac{a}{r} \qquad (r < R_G)$$

while at large distances we expect an "Ornstein-Zernike form":[12]

$$S(\mathbf{r}) = \frac{a^3 S_{AA}(q = 0)}{4\pi r \xi_s^{\,2}} \exp\,(-r/\xi_s)$$
$$= \frac{9}{2\pi}\, \Phi(1 - \Phi) \frac{a}{r} \exp\,(-r/\xi_s)$$

*Subscript *s* stands for segregation and is used to distinguish this length from the "size of the transient network" discussed in Chapter III.

Let us end this section by a remark on the complications induced by *polydispersity*. Fortunately, for the properties discussed above, in the one-

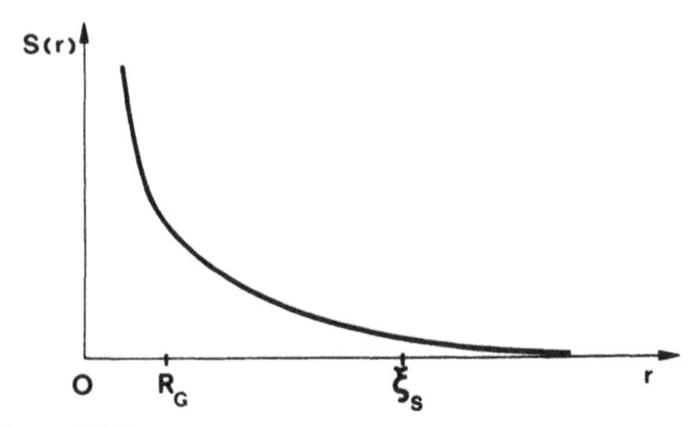

Figure IV.7.

Concentration-concentration correlation function for a mixture of two polymers near the consolute point. ξ_S is the correlation length for this critical point and is much larger than the size R_G of one chain. For all $r < \xi_S$ the correlation decays essentially like $1/r$. For $r > \xi_S$ it drops exponentially.

phase region, polydispersity does not alter the picture too seriously. In the spinodal equation [eq. (IV.12)] the polymerization numbers N_A and N_B are replaced by their weight averages:[13]

$$N_w = \sum_{\substack{\text{all chains}\\ \text{of type } i}} N_i^2 \Big/ \sum_{\substack{\text{all chains}\\ \text{of type } i}} N_i$$

Also $S^{-1}(q)$ remains a linear function of the interaction parameter χ as in eq. (IV.19)[14]

$$S^{-1}(q) = \frac{1}{N_{Aw}\,\Phi} + \frac{1}{N_{Bw}(1-\Phi)} - 2\chi + \frac{q^2a^2}{3}\left[\frac{\overline{N_A^2}}{N_{Aw}^2\,\Phi} + \frac{\overline{N_B^2}}{N_{Bw}^2\,(1-\Phi)}\right] \qquad \text{(IV.25)}$$

where N_{Aw} is a weight average

$$N_{Aw} = \sum_{A\ chains} N_A^2 \Big/ \sum_{A\ chains} N_A$$

and $\overline{N_A^2}$ is a higher average

$$\overline{N_A^2} = \sum_{A\ chains} N_A^3 \Big/ \sum_{A\ chains} N_A \qquad \text{(IV.26)}$$

The essential conclusion is that polydispersity effects renormalize certain coefficients but do not seriously affect the behavior in the region where a single phase is present. This is encouraging for future experiments.

IV.2.6. Absence of anomalous exponents

Most critical phenomena do not follow simple mean field laws. For example, with small molecules ($N_A = N_B = 1$) our eq. (IV.24) for the correlation length ξ_s is *wrong*. In this case the correct ξ_s diverges as

$$\xi_s \sim (\chi \sim \chi_c)^{-\nu_s} \qquad (N_A = N_B = 1)$$

and the exponent ν_s is of order 2/3 (in three dimensions) instead of being equal to 1/2.[12]

Fortunately these complications do not exist when both N_A and N_B are large; the mean field (Flory-Huggins) theory is *qualitatively correct for polymer mixtures without solvent*. This can be shown from a detailed study of fluctuation effects.[15] We can summarize the results by the following statement, which is used often in this chapter. If we focus our attention on the species (A) with the longest chains ($N_A > N_B$), each chain has a size $R_{oA} \cong aN_A^{1/2}$ and spans a volume $R_{oA}^3 \sim a^3 N_A^{3/2}$. We now define a parameter P equal to the average number of other chains of the same type (A) occupying this volume. Then there are two cases: 1) if P is of order unity, fluctuations effects are dangerously large and the mean field picture breaks down, and 2) if $P \gg 1$ all fluctuations effects are greatly reduced—i.e., each chain essentially experiences an average field due to all others, and the Flory-Huggins theory applies.

The scaling form of P is simple. The number of A chains per unit volume is $\Phi/N_A a^3$, and thus we have

$$P \sim \frac{\Phi}{N_A a^3} R_{oA}^3 \cong \Phi N_A^{1/2} \tag{IV.27}$$

Since we are concerned primarily with the area around the critical point, we substitute the value of eq. (IV.14) for Φ and find

$$P \sim \frac{(N_B N_A)^{1/2}}{N_B^{1/2} + N_A^{1/2}} \tag{IV.28}$$

Therefore,

(*i*) In the symmetrical case ($N_A = N_B = N$) we see that $P \sim N^{1/2}/2$ is very large and that the mean field approach is correct, as stated.

(ii) When N_B becomes much smaller than N_A, we have

$$P \to N_B^{1/2} \quad \text{(IV.29)}$$

and the mean field remains correct whenever N_B is still much larger than unity.

(iii) When $N_B = 1$ (polymer plus solvent), P is automatically of order unity, and the mean field description of the critical point is unacceptable.

IV.3. Polymer Plus Poor Solvent

IV.3.1. Regions in the phase diagram

We now consider a set of chains A (degree of polymerization N_A) in a solvent of small molecules B ($N_B = 1$). The chain–solvent interactions are always characterized by a Flory parameter χ [eq. (IV.6)] which depends on temperature T. In most (but not all) systems $\chi(T)$ is a decreasing function of temperature.* Experimental phase diagrams are always given in terms of a concentration c and a temperature T. For our general discussion it is more convenient to use more the fundamental quantities $\Phi = ca^3$ (volume fraction) and $\chi(T)$. The phase diagram then has an universal structure, shown in Fig. IV.8.

The following points are essential in connection with this phase diagram.

(i) The particular temperature $T = \Theta$ at which $\chi = 1/2$ corresponds to an exact cancellation between steric repulsion and van der Waals attraction between monomers. (The excluded volume parameter v of eq. (III.10) vanishes at $T = \Theta$.) Thus at $T = \Theta$, dilute chains are nearly ideal.

(ii) At lower χ values, steric repulsion dominates: the chains tend to swell. We enter the good solvent regime when we cross a certain line L. Beyond L the excluded volume parameter $v = (1 - 2\chi)a^3$ dominates the interactions. Returning to the free energy expression, eq. (III.8)

$$\left.\frac{F}{T}\right|_{site} = \frac{\Phi}{N}\ln \Phi + \frac{1}{2}(1 - 2\chi)\,\Phi^2 + \frac{1}{6}\Phi^3 + \dots \quad \text{(IV.30)}$$

we see that the binary interaction term(Φ^2) dominates the three-body term (Φ^3) as soon as

$$\Phi < 3\,(1 - 2\chi) \quad \text{(IV.31)}$$

*Whenever the interactions AA, AB, and BB are not very sensitive to T, the parameter $\chi \sim$ (interaction)/T varies roughly as $1/T$.

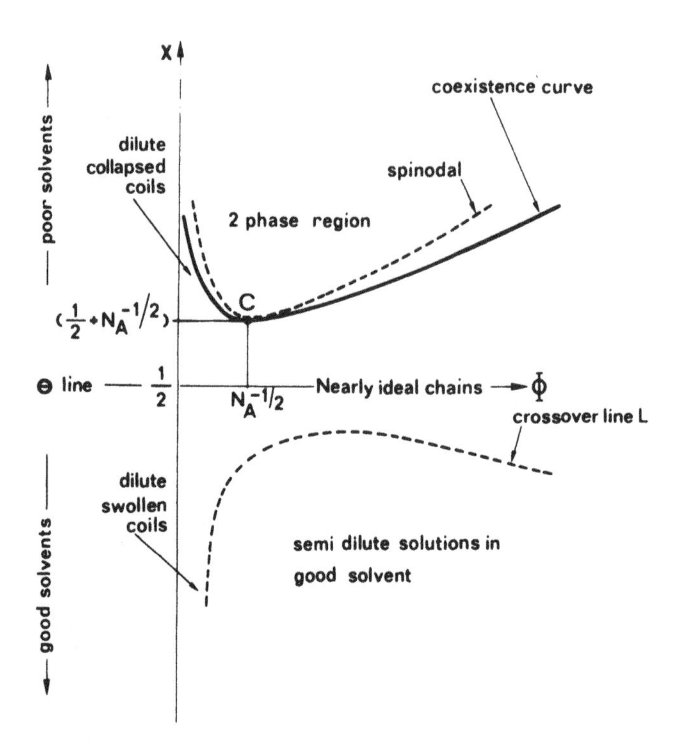

Figure IV.8.

Phase diagram for a polymer–solvent system. χ is the Flory interaction parameter, and Φ is the volume fraction occupied by the polymer. The condition $\chi = 1/2$ defines the Flory Θ temperature. In usual cases such as polystyrene–cyclohexane, χ is a decreasing function of the temperature T; high temperatures correspond to the lower part of the diagram.

This defines approximately the position of line L. Of course, L is *not* a sharp boundary; it defines a region of crossover between ideal and swollen chains.

(iii) When $\chi > 1/2$, we are dealing with a poor solvent, and we notice the appearance of a two-phase region. There is a critical point C. The Flory-Huggins predictions for C are (from eqs. (IV.14, IV.15), setting $N_B = 1$)

$$\Phi_c = N_A^{-1/2} \tag{IV.32}$$

$$\chi_c - \frac{1}{2} = N_A^{-1/2} \tag{IV.33}$$

Thus, the critical point occurs at very low concentrations. The laws [eqs. (IV.32, IV.33)] are rather well confirmed both by experiment (and by more refined theoretical analysis of the fluctuations), apart from corrections in the coefficients. Thus the Flory-Huggins theory locates C correctly.* However, the Flory-Huggins theory cannot predict the structure of the coexistence curve near C (as explained in the preceding paragraph).

Note that at point C, the different coils are essentially *closely packed;* since χ is very near 1/2, the coils are nearly ideal, with size $R_0 = N_A^{1/2}a$; the close packing density is

$$N/R_0^3 \cong N^{-1/2}a^{-3} \quad \text{(IV.34)}$$

and is thus comparable with $a^{-3}\Phi_c$.

At the low limit Φ in the figure, for $\chi > 1/2$, we find separate coils which tend to become more compact than ideal chains because of the trend toward segregation. This collapsed structure occurs only in very dilute systems and is difficult to obtain experimentally, but neutron data have been collected recently on one example.[16]

IV.3.2. A single coil near $T = \Theta$

We now focus on the line $\chi = 1/2$ (Θ line). The most striking feature is that in the free energy expansion of eq. (III.9) or eq. (IV.30) the Φ^2 term disappears. There remain, however, higher order terms and mainly (for not too large a Φ) the Φ^3 term. This situation is depicted in Fig. IV.9, where the solid lines represent polymer chains; the dotted line represents a pair interaction (between two monomers which are very close) and is associated with the Φ^2 term. The dot with three arms represents a three-body interaction (i.e. the Φ^3 term).

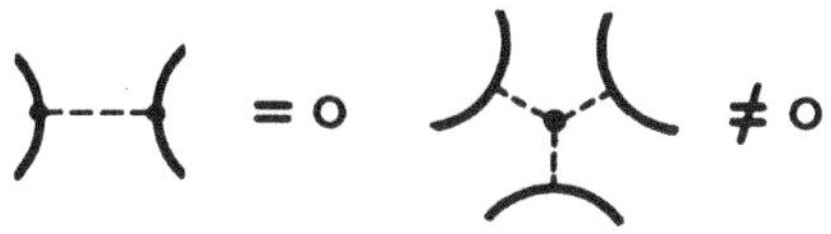

$= 0$ $\neq 0$

Figure IV.9.

At $T = \Theta$ the two-body term vanishes by cancellation between repulsions and attractions. Clearly this tends to make polymer solutions more ideal; however, the three-body term is still important in some effects. We delineate these effects and discuss first the single coil problem.

SHIFT OF THE THETA POINT

Looking at eq. (III.9) we can say that the three-body interactions w^2

*There are, however, some weak (logarithmic) corrections to eqs. (IV.32, IV.33). The origin of these corrections is explained in Chapter XI.

represent a concentration-dependent correction to the two-body terms. For example, in the chemical potential $\partial F/\partial c = \mu$, we may write

$$c^{-1}\frac{\mu}{T} = v + \frac{w^2}{2}c = \tilde{v} \tag{IV.35}$$

introducing an effective two-body coupling constant $\tilde{v}$. However, it would be a mistake to assume that the concentration, c, to be used in the w^2c correction, is the average concentration inside one coil

$$\bar{c} \cong N/R_0^3 \sim N^{-1/2} \quad \text{(ideal coil)} \tag{IV.36}$$

It is in fact much larger than $\bar{c}$, for the following reason.

In the region where we measure $\tilde{v}$, we know that at least one monomer is already present. Thus, the local concentration, c, to be used in $\tilde{v}$, is a pair correlation function $g(\mathbf{rr}')$ taken for distances $\mathbf{r} - \mathbf{r}'$ comparable with the range of the three-body interactions ($|\mathbf{r} - \mathbf{r}'| \sim a$).

Thus the shift of the two-body coefficient is

$$\tilde{v} - v \cong w^2 g(a) \cong a^3 \tag{IV.37}$$

The real Flory point $\tilde{\Theta}$ is defined by the vanishing of $\tilde{v}$ and is thus shifted significantly.

This renormalization of Θ is conceptually important but does not lead to very interesting experimental properties. All practical measurements determine $\tilde{v}$ and not v. Thus, in the following paragraph we omit the distinction between $\tilde{\Theta}$ and Θ.

DEVIATIONS FROM IDEALITY AT THE COMPENSATION POINT

When our coil is exactly at the Flory point, we say that it has a *quasi-ideal* behavior. The prefix quasi is used to recall that some interactions are still present because the three-body term has some residual effects (apart from the renormalization of v). Thus, some subtle correlations remain at $T = \Theta$. Since they are probably too small to be observed, we shall not insist very much on their properties. Mathematically, however, they are associated with unusual *logarithmic factors*. The origin of such factors can be understood from the following crude argument.

Assume that one monomer is located at the origin O. Then since the chain is nearly ideal, the density of surrounding monomers is [as shown in (eq. I.17)] in three dimensions

$$g(r) \sim \frac{1}{a^2 r} \quad \text{(IV.38)}$$

Let us compute the energy F_3 caused by the c^3 interaction near the origin, using $c = g(r)$ as the local concentration. We have

$$F_3 \cong Tw^2 \int g^3(r) d\mathbf{r} \quad \text{(IV.39)}$$

$$\cong Tw^2 a^{-6} \int \frac{1}{r^3} 4\pi r^2 \, dr \cong Tw^2 a^{-6} \ln \left(\frac{r_{max}}{r_{min}} \right) \quad \text{(IV.40)}$$

$$\cong Tw^2 a^{-6} \ln \frac{r_{max}}{r_{min}} \quad \text{(IV.41)}$$

where r_{min} and r_{max} are two physical cutoffs. If we are just at the Flory point, the upper limit will be the coil size

$$r_{max} \cong R_0 = N^{1/2} a \quad \text{(IV.42)}$$

while the lower limit is the monomer size $r_{min} = a$. This gives

$$\ln \frac{r_{max}}{r_{min}} = \frac{1}{2} \ln N + \text{constant} \quad \text{(IV.43)}$$

Thus, the interaction energy, computed with ideal chain correlations, shows certain logarithmic anomalies. The above discussion is oversimplified, but it gives us a qualitative feeling for the intricacies at the compensation point. We discuss some deeper aspects of these logarithmic singularities in Chapter XI.

IV.3.3. Semi-dilute solutions at $T = \Theta$

Let us start with single coils at the Θ point, with a radius $R_0 = N_A{}^{1/2} a$, and an internal concentration Φ_c [eq. (IV.14)]; let us then increase the concentration Φ. When Φ becomes higher than Φ_c (but still smaller than 1), the coils overlap, and we reach a well-defined semi-dilute regime, which differs strongly from the semi-dilute regime in good solvents discussed in Chapter III. Neutron data on the system polystyrene–cyclohexane near $T =$

Θ have been taken by the Saclay group,[17] and the related scaling laws have been constructed theoretically by Jannink and Daoud.[18] We summarize the results briefly.

(i) The correlation function for Φ decays according to a simple Ornstein-Zernike law

$$\frac{1}{\Phi}[\langle\Phi(0)\Phi(r)\rangle - \Phi^2] \cong \frac{a}{r}\exp(-r/\xi) \qquad \text{(IV.44)}$$

At short distances this coincides with the pair correlation for a single, ideal chain. At larger distances $r > \xi$ it is reduced. The correlation length ξ depends on concentration but is independent of the polymerization index. Scaling suggests

$$\xi \cong R_{0A}\left(\frac{\Phi_c}{\Phi}\right)^{m_\xi} \qquad \text{(IV.45)}$$

where m_ξ must be equal to unity so that the powers of N_A cancel. The resulting law is simply:

$$\xi \cong a\Phi^{-1} \qquad \text{(IV.46)}$$

(ii) It is important to realize that in this case ξ does *not* give us the mesh size of the transient network. There are many contacts and entanglements which are important to define the network but which do not show up in the correlations because the pair interaction vanishes.

(iii) The osmotic pressure Π is not very different from the Flory-Huggins prediction. From the free energy [eq. (IV.6)] (with $N_B = 1$ and $\chi = 1/2$) one derives

$$\Pi = a^{-3}\,\Phi^2\,\frac{\partial}{\partial\Phi}\left(\frac{F_{site}}{\Phi}\right) = \frac{T}{a^3}\left[\frac{\Phi}{N_A} + \frac{1}{3}\Phi^3 + \ldots\right] \cong \frac{T}{3a^3}\,\Phi^3 \qquad \text{(IV.47)}$$

Note that the relation

$$\Pi \cong \frac{T}{\xi^3} \qquad \text{(IV.48)}$$

is maintained, ξ being given by eq. (IV.46).

(iv) On the theoretical side, all the above formulas are based on mean field ideas, which are rather good at the Θ point, but they ignore logarithmic corrections similar to eq. (IV.41). Corrections of this sort are probably present in the osmotic pressure, and so forth, but they are difficult to detect experimentally.

IV.3.4. Semi-dilute solutions: crossover between good and poor solvent

This has also been studied experimentally and theoretically by the Saclay group.[17] Let us start from the good solvent side, at a fixed concentration $c = \Phi a^{-3}$, and progressively reduce the excluded volume parameter $v = a^3 (1 - 2\chi)$. When $v \sim a^3$ (athermal solvent), the scaling formulas of Chapter III hold. When v becomes much smaller than a^3, the single chain radius R decreases and the overlap concentration $c^\star = N/R^3$ increases.* To make this more precise, we use the Flory formula for R as a function of v [eq. (I.38)]

$$R \cong v^{1/5} a^{2/5} N^{3/5} \tag{IV.49}$$

obtaining

$$c^\star = N^{-4/5} v^{-3/5} a^{-6/5} \tag{IV.50}$$

Let us now repeat the arguments of Chapter III to find the correlation length ξ in the semi-dilute regime. We again write

$$\xi \cong R \left(\frac{c^\star}{c}\right)^{3/4} \tag{IV.51}$$

the exponent being chosen to eliminate N as usual. Inserting eqs. (IV.49, IV.50) we get

$$\xi \cong a \left(\frac{a^3}{v}\right)^{1/4} \Phi^{-3/4} \quad \text{(good solvent)} \tag{IV.52}$$

Eq. (IV.52) agrees rather well with the neutron data when the temperature T (controlling $v(T)$) is varied at constant Φ.[17] Let us now see when the

*In this section we drop the subscript A on the polymerization index ($N_A \rightarrow N$) to make comparisons with Chapter III easier.

formula [eq. (IV.52)] for good solvents merges with the formula [eq. (IV.46)] corresponding to theta solvents. Comparing the two, we find that crossover takes place when

$$\Phi \cong \frac{v}{a^3} = 1 - 2\chi$$

a condition which agrees qualitatively with our earlier definition of the crossover line L [eq. (IV.31)].

The "nearly good solvent" regimes described by eq. (IV.52) can be described more physically in terms of spatial correlations. The basic idea is that for small v, a short portion of a chain, with a number p of monomers must be nearly ideal. We see this from the perturbation expansion [eq. (I.42)] where we find no effect of v if $v/a^3\ (p^{1/2}) < 1$. There is a certain value of $p[(p = g_B \sim (a^3/v)^2]$, beyond which excluded volume effects become important. A single chain will appear ideal at scales $r < r_B$ where

$$r_B \cong a(g_B)^{1/2} \cong a\frac{a^3}{v} \cong \frac{a}{1 - 2\chi} \qquad \text{(IV.53)}$$

while at scales $r > r_B$ it will show excluded volume effects.*

How does this extend to a semi-dilute solution? The answer is that, on the good solvent side of line L in the phase diagram, the length r_B is smaller than the correlation length ξ. Thus we distinguish three types of spatial scales:

$a < r < r_B$ ideal (plus logarithmic corrections)

$r_B < r < \xi$ excluded volume type

$\xi < r < R_o$ ideal

The existence of a low r ideal region, and of a crossover at r_B, has been seen in one neutron experiment.[17]

It is of interest to see if ξ is larger than r_B as predicted. Using eqs. (IV.52, IV.53) we see that

$$\frac{\xi}{r_B} \cong \left(\frac{1 - 2\chi}{\Phi}\right)^{3/4} \qquad \text{(IV.54)}$$

*Eq. (IV.53) ignores all the remaining interaction effects in the quasi-ideal state (the logarithmic corrections). For the present quality of experiments this is probably quite sufficient.

and the right side is indeed larger than 1 when we are on the good solvent side of the crossover line (L) in the phase diagram (Fig. IV.8).

IV.3.5. Vicinity of the coexistence curve

Returning to the phase diagram (Fig. IV.8) we can distinguish three parts in the coexistence curve as discussed below. *Near the critical point* the coexistence curve is not described correctly by mean field theory. For example, the difference between the coexisting concentrations Φ' and Φ'' at a given temperature (or χ) near critical point is predicted in mean field theory to behave as:

$$\Phi' - \Phi'' = (\chi - \chi_c)^{1/2} N^{-1/4} \qquad \text{(IV.55)}$$

but the correct exponent is expected to be different and to coincide with what has been found in liquid–gas critical points:

$$\Phi' - \Phi'' \sim (\chi - \chi_c)^{\beta} f(N) \qquad \text{(IV.56)}$$

where the exponent β is close to 1/3.[12]* Experiments measuring various nonmean field exponents near the critical point have been reported.[19] The main practical difficulty arises from polydispersity effects, for which there exists no theory outside of mean field theory.

The semi-dilute side is much simpler. In this region each coil is interacting with a large number (P) of other coils ($P \sim \Phi/\Phi^\star$). The discussion in Section IV.2.6 then shows that the mean field description becomes adequate. The coexistence curve can then be derived from the Flory-Huggins free energy [eq. (IV.6)]. The values of Φ', Φ'' for a given χ can be obtained by imposing two conditions:

(i) Equality of the osmotic pressures, $\Pi(\Phi') = \Pi(\Phi'')$.

(ii) Equality of the exchange chemical potential $\mu(\Phi') = \mu(\Phi'')$.

For the discussion of the semi-dilute branch (i.e., of the root Φ'' which is much larger than Φ_c) the first condition is enough. This is because the osmotic pressure of the dilute phase $\Pi \cong T/N\Phi'$ is entirely negligible, and thus the condition defining Φ'' is simply $\Pi(\Phi'') = 0$ or explicitly

$$T\left[\frac{\Phi''}{N} + \frac{1}{2}(\Phi'')^2 (1 - 2\chi) + \frac{1}{3}(\Phi'')^3\right] = 0$$

*The structure of the N-dependent factor $f(N)$ can be predicted from scaling requirements: $f(N) \sim N^{-1/2+\beta/2}$.

Again, neglecting a perfect-gas term Φ''/N (which is small compared with the others when $\Phi'' > \Phi_c$) we get:

$$\Phi'' = \frac{3}{2}(2\chi - 1) \qquad (\chi - \chi_c \gg N^{-1/2}) \qquad \text{(IV.57)}$$

Similarly, the spinodal in the semi-dilute limit can be obtained directly from the mean field formula [eq. (IV.12)] with $N_B = 1$. Always dropping terms of order Φ/N, we arrive at

$$\Phi_{spinodal} = 1 - \frac{1}{2\chi} \qquad (\chi - \chi_c \gg N^{-1/2}) \qquad \text{(IV.58)}$$

Light scattering studies on this branch have been carried out by a british group,[20] and they seem to confirm the validity of the mean field description for $\Phi \gg \Phi_c$. (The region $\Phi \cong \Phi_c$ is still under dispute).

On the dilute side, in a poor solvent, we have a few coils which tend to be *more compact* than ideal coils. This collapsed structure has been studied by many theorists, mainly by self-consistent field methods.[21] To a first approximation we may simply say that (at a given χ) the coils build up an internal concentration Φ_{in} which is controlled by the interactions and is independent of chain length. This means that:

$$\Phi_{in} = \Phi'' = 3\,(\chi - 1/2) \qquad \text{(IV.59)}$$

The radius of the coil R is then such that

$$a^{-3}\,\Phi_{in}\,\frac{4\pi}{3}R^3 = N \qquad \text{(IV.60)}$$

Thus in this collapsed regime

$$R \sim aN^{1/3}\,(\chi - 1/2)^{-1/3} \qquad \text{(IV.61)}$$

In a recent neutron experiment by Nierlich, Cotton, and Farnoux[16] this regime was studied on *short* ($M = 29{,}000$) polystyrene chains in cyclohexane. Short chains were chosen because of the structure of the partition law for a polymer between two phases [eq. (IV.1)]. If the polymer is long, there are essentially no chains in the dilute phase ($\Phi' \rightarrow 0$), and the experiment becomes impossible. However, with these short chains ($N \sim 300$)

the Saclay group could follow R as a function of temperature (i.e., of χ). They find for R $(\chi - 1/2)$ an experimental exponent of -0.32 (± 0.05) in very good agreement with the collapse law [eq. (IV.61)].

An interesting feature related to the coexistence curve is the structure of the interface between the dilute (Φ') and the semi-dilute (Φ'') phase. No details are known about this structure, but scaling gives us two predictions:

(i) The thickness of the interface scales like the correlation length ξ of eq. (IV.46)

$$\xi \sim a\Phi''^{-1} \sim a\,(\chi - 1/2)^{-1} \qquad (\Phi'' \gg \Phi_c) \qquad \text{(IV.62)}$$

(ii) The interfacial energy scales like

$$\frac{T}{\xi^2} = \frac{T}{a^2}(\Phi'')^2 = \frac{T}{a^2}\frac{1}{N}\left(\frac{\Phi''}{\Phi_c}\right)^2 \qquad \text{(IV.63)}$$

In eq. (IV.63) the weakness of this surface tension is striking; T/a^2 is a natural unit of surface energy and is of order $\sim$ 100 ergs/cm^2. However, the $1/N$ factor reduces it enormously. We conclude that polymer–solvent systems in the semi-dilute regime have diffuse interfaces and low interfacial energies; the latter feature is well known experimentally.

SUMMARY

Flexible polymers in poor solvent show a quasi-ideal behavior at a certain compensation point Θ. At a slightly lower solvent quality, phase separation occurs, and there is an equilibrium between a nearly pure solvent phase (containing a few chains, each being severely contracted) and a polymer-rich phase. The latter can be described by the Flory-Huggins theory. At the onset of phase separation (at the critical point) each polymer coil behaves like one individual argon atom at the liquid–gas transition of argon, and the Flory-Huggins approximation is not valid.

IV.4.
Polymer Plus Polymer Plus Solvent

The case of two incompatible polymers (A, B) dissolved in a common solvent (S) is interesting but is less well known. We discuss it only briefly.

IV.4.1. Good solvent and strong segregation factor

The Flory-Huggins interactions for this problem are given in eq. (IV.4). We assume that $\chi_{AS} < 1/2$ and $\chi_{BS} < 1/2$ so that the solvent S is good for both A and B. The third parameter χ_{AB} will be called the segregation factor. We assume that it is positive and not too small (but it need not be as large as 0.5). Then the AB pair tends to segregate strongly.

To present things in simple terms it is convenient to select a "symmetrical" case where $N_A = N_B$ and where the interaction parameters χ_{AS} and χ_{BS} are also equal. We assume these conditions in what follows. Then the phase diagram (represented at one given temperature as a function of the two independent concentrations Φ_A and Φ_B) is symmetric around the first bisector, and is as shown in Fig. IV.10.

There is a critical point C with a certain total concentration Φ_c. The mean field prediction for the critical point, based on a free energy

$$\frac{1}{T}F\bigg|_{site} = \frac{\Phi_A}{N}\ln\Phi_A + \frac{\Phi_B}{N}\ln\Phi_B + \Phi_S\ln\Phi_S + \chi_{AS}(\Phi_A+\Phi_B)\Phi_S + \chi_{AB}\Phi_A\Phi_B \tag{IV.64}$$

is of the form[7]

$$\Phi_c = N^{-1}\chi_{AB}{}^{-1} \cong N^{-1} \quad \text{(mean field)} \tag{IV.65}$$

However, eq. (IV.65) is *wrong*, as shown by the following argument. Let us start from the origin in Fig. (IV.10) and progressively increase the

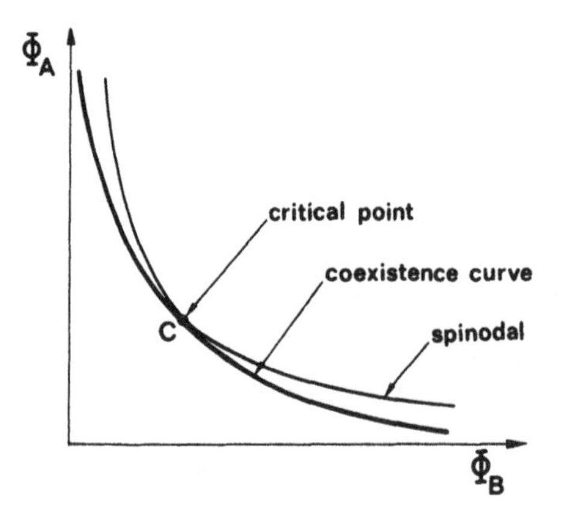

Figure IV.10.

solute concentrations (keeping $\Phi_A = \Phi_B = \Phi/2$). At the beginning we have separate, swollen coils of size $R_F \cong aN^{3/5}$. These coils behave like hard spheres; they cannot interpenetrate, and thus the differences between A and B are not seen. This state persists to the overlap concentration $\Phi^* \sim a^3 N/R_F^3 \sim N^{-4/5}$. Then, when Φ exceeds Φ^*, the coils do interpenetrate, and if χ_{AB} is strong, demixing will soon take place. Thus the correct estimate for Φ_c is

$$\Phi_c \cong \Phi^* \cong N^{-4/5} \quad \text{(strong segregation)} \qquad \text{(IV.66)}$$

The next question is related to the critical behavior near $\Phi = \Phi_c$. Since the coils are just beginning to overlap in the critical region, we see that the parameter P of Section IV.2.6. (giving the number of coils interacting with one of them) is of order unity. Then we expect critical exponents which are not of the mean field type but rather are related to those of the liquid-gas transition.[12]

IV.4.2. Good solvent and weak segregation factor

We now discuss the case where χ_{AB} is positive but small and start again from the dilute end. When we reach the overlap concentration Φ^*, the factor χ_{AB} is not strong enough to induce segregation. We may still increase the concentration, keeping a single phase, which is a *semi-dilute mixture of A and B chains*. The interaction χ_{AB} is then a weak perturbation superimposed on a familiar problem—the problem of Chapter III. As seen in this chapter, the probability of contact between two monomers is much smaller than predicted by mean field theory; the reduction amounts to a factor of $\Phi^{1/4}$, where Φ is the total concentration of monomers ($\Phi = \Phi_A + \Phi_B$). This applies in particular to the AB contacts. Their number is reduced by $\Phi^{1/4}$. Thus we are led to use the mean field theory with a *renormalized segregation factor*[22]

$$\tilde{\chi}_{AB} = \chi_{AB}\, \Phi^{1/4} \qquad \text{(IV.67)}$$

We can now obtain the critical point from eq. (IV.65) by replacing χ_{AB} by $\tilde{\chi}_{AB}$. The result is

$$\Phi_c = N^{-1}\, \Phi_c^{-1/4}\, \chi_{AB}^{-1}$$

$$\Phi_c = (N\, \chi_{AB})^{-4/5} \qquad (\chi_{AB} \ll 1) \qquad \text{(IV.68)}$$

We do not yet know of any real example corresponding to this case, but it should be looked for. The critical point described by eq. (IV.68) is unusual. It is easy to see that in this case mean field exponents must hold; the parameter P is now much larger than unity.

IV.4.3. Theta solvents[22]

Consider the case $\chi_{AS} = \chi_{BS} = 1/2$ where the dilute A and B chains are nearly ideal. Here the critical overlap concentration Φ^* becomes $Na^3 R_0^{-3} \sim N^{-1/2}$. Below Φ^* the chains still repel each other strongly (because of the three-body interaction of Fig. IV.9), and segregation cannot take place. However, as soon as $\Phi > \Phi^*$, we have full overlap and phase separation occurs. Thus,

$$\Phi_c \sim N^{-1/2} \qquad (\Theta \text{ solvent, strong } \chi_{AB}) \qquad \text{(IV.69)}$$

The case studied by the Utrecht group[7] and mentioned in Section I.1, seems to correspond closely to this situation (see the numerical values after eq. (IV.5)). The phase diagram measured for various molecular weights (but keeping N_A/N_B constant) do seem to give a variation of Φ_c with N which is close to the inverse square-root law of eq. (IV.69).

REFERENCES

1. P. G. de Gennes, *The Physics of Liquid Crystals,* Oxford University Press, London, 1976.
2. B. Kronberg, Ph.D. Thesis, McGill University, Montreal, 1977.
3. A. Dubault, M. Casagrande, M. Veyssie, *Mol. Cryst. Lett.* in press.
4. S. Krause, *J. Macromol. Sci. Rev. Macromol. Chem.* **C7,** 251 (1972).
5. T. Kwei, T. Nishi, R. Roberts, *Macromolecules* **7,** 667 (1974).
6. *Molecular Forces,* Pontifical Academy of Science, Ed., North-Holland, Amsterdam, 1967.
7. M. Van Esker, A. Vrij, *J. Polym. Sci.* **14,** 1943 (1976). M. Van Esker, J. Laven, A. Broeckman, A. Vrij, *J. Polym. Sci.* **14,** 1953 (1976). M. Van Esker, A. Vrij, *J. Polym. Sci.* **14,** 1967 (1976).
8. P. Flory, *Principles of Polymer Chemistry,* Cornell University Press, Ithaca, New York, 1971.
9. R. Koningsveld, *Discuss. Faraday Soc.* **49** (1970).S. Koningsveld, L. Kleintjens, in *Macromolecular Chemistry,* Vol. 8, p. 197, Butterworths, London, 1973.
10. R. L. Scott, *J. Chem. Phys.* **17,** 279 (1949). H. Tompa, *Trans. Faraday Soc.* **45,** 1142 (1949).
11. A. Guinier, G. Fournet, *Small Angle Scattering of X-Rays,* John Wiley and Sons, New York, 1955.

12. H. E. Stanley, *Introduction to Phase Transitions,* Oxford University Press, London, 1971.
13. W. Stockmayer, *J. Chem. Phys.* **17,** 588 (1949).
14. J. F. Joanny, *C. R. Acad. Sci. Paris* **286B,** 89 (1978).
15. P. G. de Gennes, *J. Phys. (Paris) Lett.* **38L,** 441 (1977). J. F. Joanny, Ph.D. Collège de France, 1978.
16. M. Nierlich, J. P. Cotton, B. Farnoux, *J. Chem. Phys.* **69,** 1379 (1978).
17. J. P. Cotton *et al., J. Chem. Phys.* **65,** 1101 (1976). B. Farnoux *et al., J. Phys. (Paris)* **39,** 77 (1978).
18. M. Daoud, G. Jannink, *J. Phys. (Paris)* **37,** 973 (1976).
19. N. Kuwahara, J. Kojima, M. Kaneko, B. Chu, *Polym. Sci.* **11,** 2307 (1973).
20. K. Derham, J. Goldsbrough, M. Gordon, *Pure Appl. Chem.* **38,** 97 (1974).
21. S. F. Edwards, *J. Non-Cryst. Solids* **4,** 417 (1970). I. M. Lifshitz, I. M. Grosberg, *Soviet Phys. JETP* **38,** 1198 (1974).
22. P. G. de Gennes, *J. Polym. Sci. (Phys.)* **16,** 1883 (1978).

V

Polymer Gels

V.1.
Preparation of Gels

A polymer gel is a network of flexible chains, with the general structure shown in Fig. V.1. Structures of this type can be obtained by chemical or physical processes. Since the final gel properties are sensitive

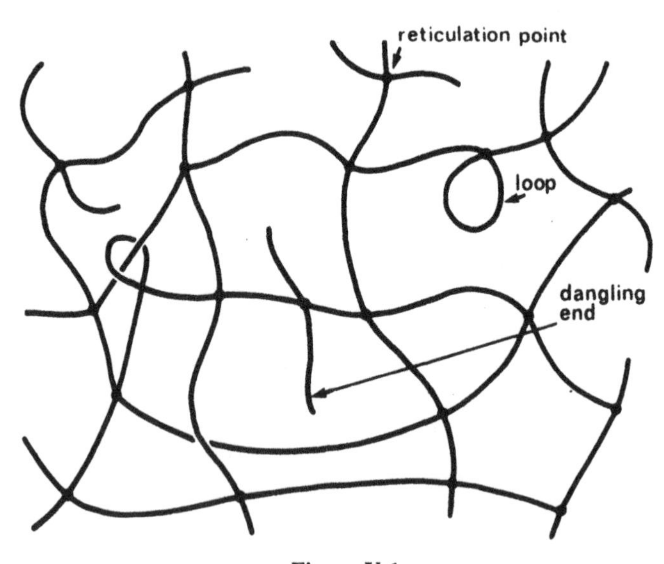

Figure V.1.

to the preparation methods, we give a brief list of them. For more details see Refs. 1, 2, and 3.

V.1.1. Chemical pathways

A first and conceptually simple method of gel preparation is based on *condensation* of polyfunctional units. A typical example would be:[4]

trialcohol + diisocyanate

The alcohol functions (o) and the isocyanate functions (o) condense according to the reaction:

$$\underset{\text{alcohol}}{ROH} + \underset{\text{isocyanate}}{ONCR'} \rightarrow \underset{\text{urethane}}{R{-}O{-}CO{-}NH{-}R'}$$

The reaction, when proceeding for long enough times, will lead to branched objects, each trialcohol becoming a branch point when its three functions are reacted.

A second approach uses *additive* polymerization. For example, if we start with a vinylic monomer

$$R{-}CH{=}CH{-}R'$$

and open the double bond by a free radical reaction, we generate (mainly) linear chains:

$$\left[\begin{array}{cc} CH{-} & CH \\ | & | \\ R & R' \end{array}\right]_N$$

However, if we add a fraction of the divinyl derivative, to the mixture

$$CH_2{=}CH{-}R_2{-}CH{=}CH_2$$

the two double bonds will participate in the construction of two distinct chains, and $-R_2-$ will become a crosslinking bridge in the structure.[5]

A third approach amounts to start from preexisting chains carrying suitable chemical functions, and to attach different chains as shown in Fig. V.2.

For some research purposes it is interesting to prepare gels where the length of the chains between crosslink is controlled and is the same for most chains in the gel. This type of program has been pursued actively at Strasbourg.[6] A typical case has the following two stages:

(i) By anionic polymerization, prepare polystyrene chains of well-defined molecular weight and with reactive ends.

(ii) React the ends with a polyfunctional unit (such as divinylbenzene) to create the crosslinks.

This gives rise to gels with a rather well-defined mesh size; however, there are some complications related either to the polyfunctional units (which can polymerize in "nodules" larger than a single unit) or to accidental "mistakes" such as closed loops. Such a loop is shown in Fig. V.1.

The above list is not exhaustive. For example, an important class of gels is made by polycondensation of silicates[7] or aluminates; the physics of these gels is complex and is not discussed here.

V.1.2. Unorthodox gelation processes

Here we present some special reactions which lead (or should lead) to unusual gel structures. Most of what is described is conjecture but may stimulate future research.

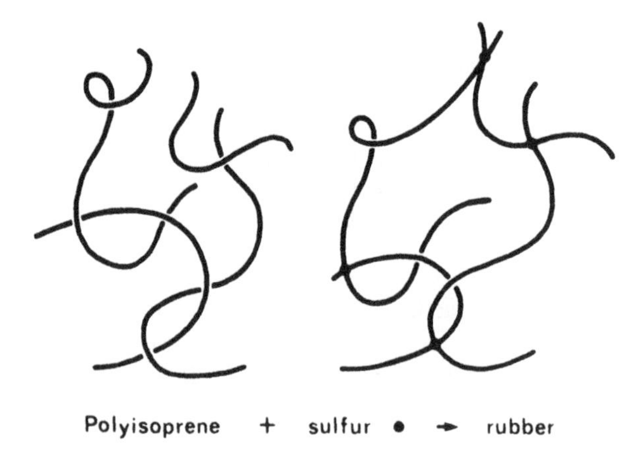

Figure V.2.

CROSSLINKING IN SPECIAL MATRIX SYSTEMS

It should be possible to incorporate polymer chains in various liquid crystalline systems which impose certain conformations on them. For example, if we have a lamellar phase of lipid + water, it may be possible to incorporate a hydrophilic polymer into the water layer, obtaining the structures in Fig. V.3. Assuming that the ternary system can be observed in certain phase diagrams, one might then crosslink the chains and wash out the lipid with suitable solvents. An unusual (and anisotropic) gel should result.[8]

Another example (which up to now seems very difficult to achieve) is based on chains dissolved in a *cholesteric* phase. This is a liquid where the molecules locally have one direction of alignment but where this direction has a helical twist in space.[9] If we start with chains which are not optically active, crosslink them by an optically inactive agent, and then wash out the cholesteric solvent (replacing it by an achiral solvent), we should obtain a gel which has an *optical rotatory power* (a memory of its preparative state) although all its components do not distinguish right from left.[8]

Many other proposals of this kind could be made. Generally, the notion of preparing gels inside a pre-existing, organized, structure may become important in the future.

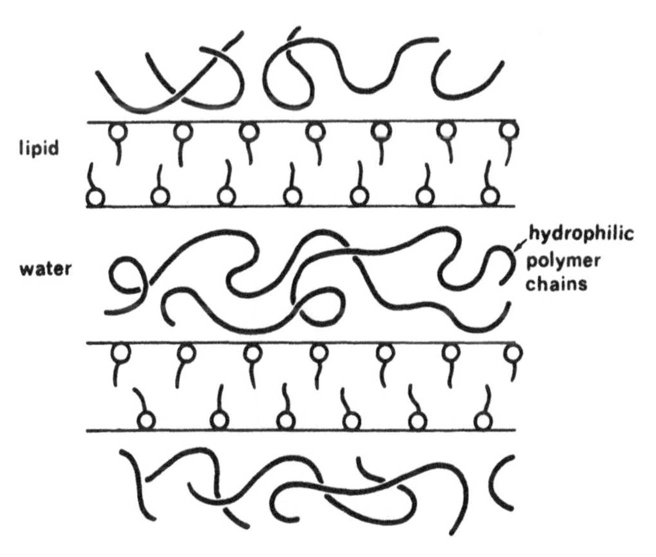

Figure V.3.

KNOTTED STRUCTURES

If we start with a swollen gel A that contains certain free chains B, we can, in a later stage, crosslink the B chains among themselves, with no attachment between A and B. This leads to a system of two mutually interpenetrating gels[10] that are strongly knotted.

It is also possible to imagine a gel which would be made without any chemical bonding between the constituent chains, all the attachment being realized by suitable knots. We call this an "olympic gel" in analogy with the Olympic rings. The geometry would be of the type shown in Fig. V.4. It is not possible to prepare such a gel directly, because cyclization competes with linear addition of chains.

Figure V.4.

A possible method is based on *two steps:*

(i) With a low concentration ($c \sim c^\star$) of chains which are reactive at both ends, perform a cyclization (Fig. V.5). It is difficult to show that the reaction has occurred (the change in chain size from cyclization could be monitored by light scattering), but it is not essential to demonstrate cyclization at this stage.

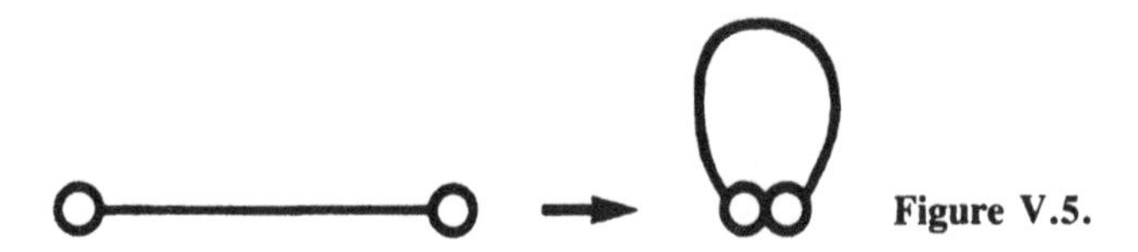

Figure V.5.

(ii) The product is then concentrated, leading to $c_2 \gg c^\star$ (semi-dilute solution of rings). To this solution, one adds a small fraction ($\delta c \sim c^\star$) of the original linear chains and starts a new cyclization. On the average, each new ring will be knotted with a number $P \sim c_2/N\ (R^3)$ of pre-existing

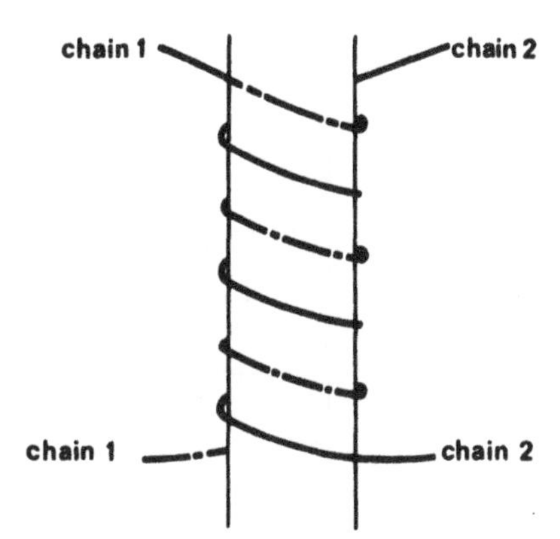

Figure V.6.

rings (each of size R). The number P is large if $c_2 \gg c^\star$.* At this point, the whole population of rings should reach the olympic state.†

V.1.3. Physical gelation

In the network structure of Fig. V.1 the crosslinks need not be produced by chemical reaction. Any physical process which favors association between certain (but not all) points on different chains may also lead to gels. Many examples of this are found with biological molecules, such as gelatin[11] or certain polysaccharides.[12] In many of these systems the association process is still disputed. There are three main possibilities:

(*i*) Formation of *helical structures* with two (or more) strands (Fig. V.6).[13]

(*ii*) Formation of a *microcrystal,* for which a highly idealized picture is shown in Fig. V.7. Independently of the details of the association, we must remember one essential point: the structures (Figs. V.6, V.7) must not be able to enlarge very much. If they could, they would progressively invade the whole system of chains and we would be left with an array of microcrystallites. One important factor blocking the growth of the association regions is *tacticity.*[1] If the chains are not stereoregular, they cannot crystallize (or build up helixes) over long lengths; then gel formation is natural. Many cases of type (*i*) or (*ii*) show some irreversibility. For example, a 1.5% solution of gelatin in water is, at high temperatures, a simple

*More precisely, if we are in a good solvent, the formulas of Chapter III tell us that $P \sim (c_2/c^\star)^{5/8}$.

†However, the numerical coefficient in P is unknown and may be small: the experiment may require very large ratios $c/c^\star$.

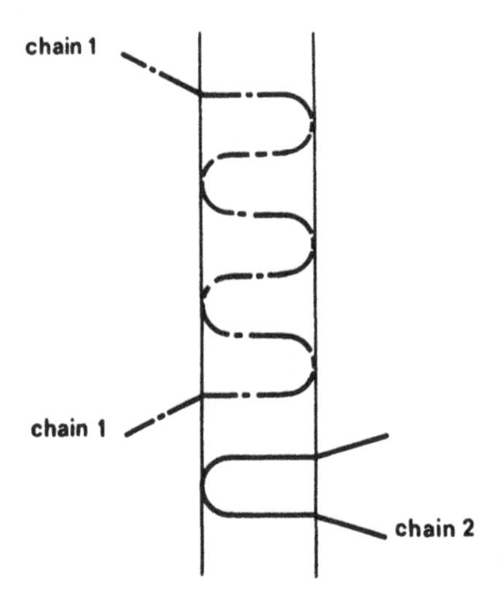

Figure V.7.

solution of chains, or a *sol*. If we cool it, we get a gel. If we raise the temperature again, we recover the sol. However, the transition temperature measured at increasing temperatures is often somewhat different (higher) than the transition point measured on cooling. The existence of these hysteresis effects is not surprising when we think of the complexity of the attachment procedures shown in Fig. V.6 and V.7.

(*iii*) Formation of *nodules* with block copolymers. If we have chains made of three blocks BAB in a solvent which is good for A and poor for B, the B portions will tend to coalesce into *nodules* (or alternatively in sheets or in rods). Depending on the temperature and other similar variables, the B monomers inside the nodules may be either in a solid state (crystalline or glass) or in a fluid state (micelles).

The first case (solid nodules) is not very different conceptually from (*i*) and (*ii*) above. The second case (fluid nodules) is more interesting because it may lead to a much more reversible sol–gel transition.

V.1.4. Strong gelation versus weak gelation

When we start with a sol, we have molecules (chains or smaller partners) which are independent, and the system is a conventional *liquid*. When we go to the gel phase, a finite fraction of the chains belong to an infinite network, and the system can resist stresses. It deforms elastically like an *isotropic solid* and can be characterized by two elastic moduli (or Lamé coefficients). How do we commute from one form to the other?

Let us return first to a case of chemical crosslinking. If we increase the fraction p of crosslinks in the sol phase, we begin to build up branched molecules which become longer and larger. At a certain critical value of p ($p = p_c$), a huge molecule appears, which is present in all parts of the reaction vessel. Beyond p_c, this molecule becomes more and more crosslinked and branched. (There are still some other finite molecules, but their number decreases rapidly when p increases). In this case there is a well-defined gelation threshold, $p = p_c$.

However, this is not true for some other gelation processes. Consider a gel of chains which associate by a physical process and where the crosslinks are not strong. Under any weak (but finite) stress, the crosslinks will eventually split, and the long-time behavior of the material will always be liquid-like. There is no strict gel point in such a system. There is, of course, a certain crossover region near a temperature, T_g, where the system switches from viscous behavior to elastic behavior at the frequency, ω, used for the experiment, but the value of T_g depends on ω. Thus, in this case, gelation is conceptually similar to a *glass transition*. It is not an equilibrium process, but it corresponds to the progressive freezing of a certain number of degrees of freedom.[14]

We distinguish between these two types of behavior as follows:

(i) When the crosslinks, once made, are completely stable (for the stresses and the time scales involved in the experiments), we say that we have a *strong gelation* process, and we expect a sharp threshold.

(ii) When the crosslinks are not completely stable but are associated with a reaction (bonding $\leftrightarrow$ nonbonding) that can proceed in both directions, we speak of a *weak gelation* process, and we expect to find some of the intricacies of glass transitions.

In most cases chemical crosslinking leads to strong gelation, while physical crosslinking may lead to either strong or weak gelation, depending on the case at hand.

The criterion for strong gelation may be formulated as follows. We start from the sol side and increase p, finding larger and longer molecules (or "clusters" as they are often called in the theoretical literature). If it is possible to stop the reaction at a given p and to subject the clusters to various treatments (dilution, change of solvents, shear flows, and so forth) without cutting them into pieces, we say that we are in the strong gelation regime. This regime is interesting because it is universal. The scaling laws involved are discussed in the next section.

V.1.5. Relationship between preparation and properties of gels

A gel is a frozen system. To understand it, we need two kinds of statistical information.

(i) What was the situation at the moment of preparation? Were the chains dilute or semi-dilute? Was the solvent good or poor? What is the level at which the chemical reaction was stopped?

(ii) What is the situation at the moment of study (solvent, temperature, etc.). Thus, gels (very much like glasses) must be described in terms of two ensembles, the "preparative ensemble" and the "final ensemble." This is much more complex than the usual equilibrium systems, where a single ensemble (ruled by Boltzmann exponentials) is required. Preparation fixes a number of constraints which are not easy to specify. The first general discussion on these two situations came from Edwards.[15] From a practical point of view, the sensitivity of gel properties to preparation is an interesting feature but is far from being under control. Let us take some simple examples.

Gels can be "dry" (without solvent) or "swollen" (with a good solvent). A dry gel which was prepared in the same state is often called *normal*. A dry gel which was prepared in a swollen state is often called *supercoiled* and is very different.

In many cases the implantation of crosslinks in a system of chains tends to induce a *segregation* between chains and solvent. Each crosslink forces two chains to come in close contact and thus promotes an effective *attraction* between chains. If the original interchain repulsion was not high enough (i.e., if we had a poor solvent), some segregation may occur. However, because solvent expulsion is slow in the gel phase, it will often happen that phase separation does not take place in a macroscopic sense. What we have rather is the formation of very small "pockets" which have high chain concentrations and others which are rich in solvent. Many observations (by light scattering, electron microscopy,* and so forth) show strange heterogeneties in gels. Sometimes we see ribbons, fibrils, nodules, and so forth, with sizes in the range 200 to 1,000 Å. These effects (decorated with the majestic name "microsyneresis") have often discouraged the experimentalists because the corresponding gels are irregular. However, the situation is not that bad. Gels prepared in good solvents under suitable conditions can be quite homogeneous and reproducible, even if they are not made with calibrated chains as in Section V.2.

In this chapter, the emphasis is on swollen gels in good solvents (assuming that they have been prepared under similar conditions). However, we also describe briefly the effects of a decrease in the quality of the solvent and the resulting microsyneresis.

*Of course, in electron microscopy, the sample treatments alter the gel significantly and often favor segregation.

V.2. The Sol–Gel Transition

V.2.1. The classical picture

We now restrict our attention to strong gelation processes and discuss the area of the threshold. The classical picture for this transition[16,17] is based on a "tree approximation," where the growing clusters are represented in Fig. V.8. The essential simplification underlying this approach is

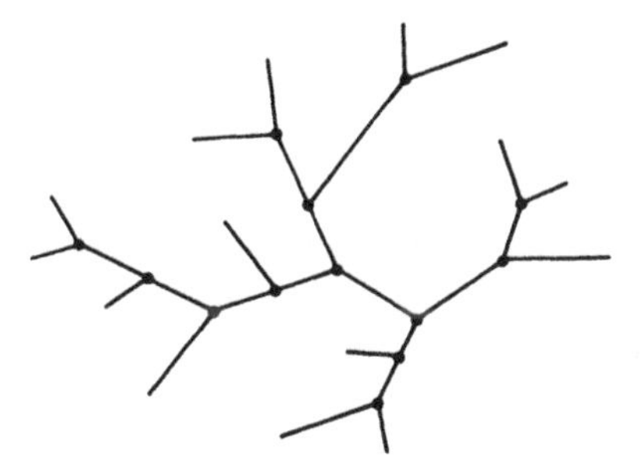

Figure V.8.

that one assumes *no closed cycles* and *no steric hinderances*. The chain is assumed to branch off freely, the branches never being limited in their growth by the existence of other branches in the same cluster: the trees are taken to be ideal. Detailed calculations along these lines are summarized in Section V.2.4, but we do not want to insist on them because the tree approximation is clearly a gross oversimplification. Excluded volume effects in a branched molecule are expected to be even stronger than in linear chains.

However, the tree approximation has been accepted by polymer scientists because it gives good values for the threshold (i.e., for the fraction of reacted bonds p_c which is reached when the sol-gel transition takes place). It has been pointed out only recently (by D. Stauffer[18] and by the present author[19]) that despite this agreement on p_c, the behavior near the threshold ($p \rightarrow p_c$) must differ widely from the predictions of classical theory. In what follows, we insist on these aspects, which are not appreciated enough.

V.2.2. Gelation without solvent: the percolation model

A simple model for gelation without solvent is again a lattice model, where each lattice site (with z neighbors) represents one polyfunctional unit with z reacting arms. Two neighboring units can react, and we represent such a reacted bond by a heavy line, as in Fig. V.9. This

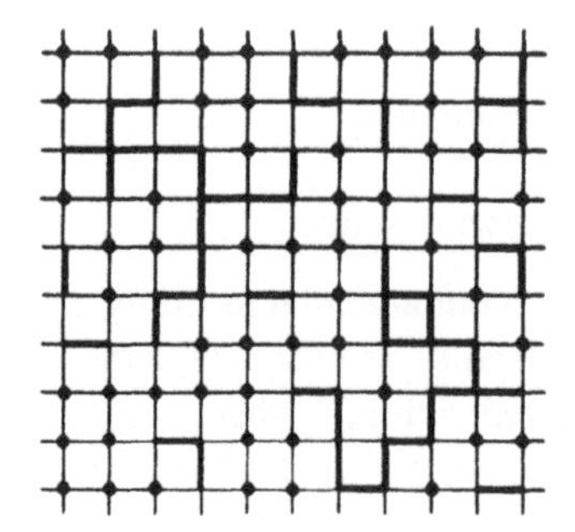

Figure V.9.

gives certain "molecules." We assume that these bonds appear at random. At one moment the fraction of reacted bonds is called p. Clearly at small p we have only small molecules (or clusters), but when p exceeds a well-defined value (p_c), we get an infinite cluster.

This type of problem was first introduced by Hammersley, and he coined the word *percolation* to describe it. Since there are many reviews of percolation theory,[20,21] here we give only selected results concerning the behavior near the threshold.

V.2.3. Large clusters below the gelation threshold

We consider the regime where p is slightly smaller than p_c ($p = p_c - \Delta p$). We then have clusters with a broad distribution of sizes and shapes; for the percolation model, the properties of these clusters *in the reaction bath* are well known.[20,21]

The weight average polymerization index N_w diverges (for $\Delta p \to 0$) according to the law

$$N_w \cong \Delta p^{-\gamma} \tag{V.1}$$

where γ is of order 1.8 in three dimensions. Note that eq. (V.1) is universal; it holds independently of the lattice chosen, of the detailed shape of the monomers, etc., provided only that: 1) we are close enough to the threshold, so that the clusters are indeed large, and 2) the building units (the monomers) are small (the counter example of vulcanization is discussed separately).

All the equations discussed in this section have the same level of universality.

Eq. (V.1) could be compared with some light scattering data.[22] Usually the latter have been interpreted in terms of the tree approximation value,[23,24] which is much lower ($\gamma = 1$). This should not disturb us: in the field of magnetic phase transitions, where critical exponents are also rather different from the mean field values, it took more than 30 years to convince the experimentalists that mean field theory was wrong.

Another property of great interest is the *distribution of molecular weights* of the clusters, which has been constructed by Stauffer.[18] Let us call c_n the concentration of monomers belonging the clusters of n units (note that our notation differs from that of Ref. 18). This is normalized by:

$$\sum_{n=1}^{\infty} c_n = c(= a^{-3}) \quad \text{total concentration} \tag{V.2}$$

$$\sum_{n=1}^{\infty} n c_n = c N_w \tag{V.3}$$

(*i*) If we are just at the critical point ($p = p_c$), c_n decreases like a power law

$$c_n \cong a^{-3}\, n^{-[(\gamma+2\beta)/(\gamma+\beta)]} \tag{V.4}$$

where β is another characteristic exponent ($\beta = 0.39$). Thus, the weight distribution decreases as $n^{-1.2}$ at the threshold.

(*ii*) If we are slightly below the threshold, the distribution is modified, mainly by the introduction of a cutoff at a polymerization index[16]

$$N_l \cong \Delta p^{-\gamma-\beta} \cong \Delta p^{-2.1} \tag{V.5}$$

where $\Delta p = |p - p_c|$. In many physical measurements, it is this cutoff value (rather than N_w) which is essential. For example, if we look at an average size of clusters, we have to define it with great care. The root-mean square size for clusters of n monomers increases with n like a power law

$$R(n) \cong a n^{\nu/\gamma+\beta} \tag{V.6}$$

where ν is still another exponent ($\nu = 0.85$).

The size at the cutoff ($n = N_l$) is called the *correlation* length $\xi(\Delta p)$, and scales like

$$\xi(\Delta p) \equiv R(N_l) \cong a \Delta p^{-\nu} \tag{V.7}$$

However, it is not ξ which is measured in most studies on the clusters size:

(i) For certain purposes, we need the weight average of the radius of gyration squared for all clusters

$$R_w^2 = \frac{1}{c} \sum_n c_n \, R^2(n) \tag{V.8}$$

Inserting eq. (V.4) plus a cutoff at N_l, the result is

$$\begin{aligned} R_w^2 &\cong a^2 N_l^{2\nu-\beta/\gamma+\beta} \cong a^2 \Delta p^{-2\nu+\beta} \\ &\cong \xi^2 \Delta p^{\beta} \end{aligned} \tag{V.9}$$

(ii) Light scattering experiments can be performed on the clusters. However, these experiments should *not* be done directly on the reacting mixture, where signals are dominated by interference between different clusters and are weak. What must be done is to quench the reaction at a certain level (a certain p) and then to dilute the system. One can then determine an average molecular weight and an average size for the diluted clusters. The scaling law for the molecular weight is given in eq. (V.1). On the other hand, the gyration radius $\tilde{R}(n)$ is much more difficult to interpret. It is not certain that $\tilde{R}(n)$ scales like the cluster radius in the reaction bath $R(n)$ because the excluded volume effects in the reaction bath (where each cluster is near many others) are very different from those in a dilute regime:

$$\tilde{R}(n) > R(n) \tag{V.10}$$

V.2.4. Gel properties just above threshold

We now take $p = p_c + \Delta p$ with Δp positive and small. Here we have an infinite cluster plus some finite clusters. The essential properties are:

(i) The gel fraction S_∞ (the fraction of monomers belonging to the infinite cluster) increases rapidly with $\Delta p = p - p_c$

$$S_\infty \cong (\Delta p)^{\beta} \tag{V.11}$$

where β has been given above.

(ii) The Lamé coefficients of the gel (the elastic moduli E) increase much more slowly

$$E \cong \Delta p^{t} \tag{V.12}$$

In particular, for $d = 3$, $t \sim 1.7$ to 1.9.* Elastic measurements by M.

*For theoretical discussions of the exponent t, see Ref. 19.

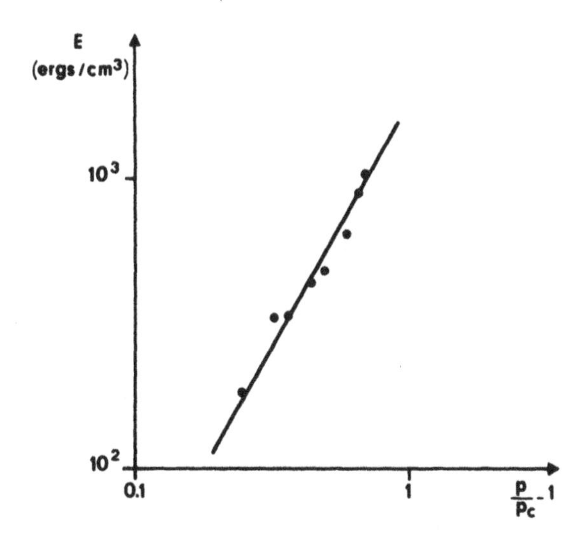

Figure V.10.
Elastic modules E of an aqueous solution of gelatin (weight concentration 5.7%, molecular weight 105,000) above the sol–gel transition. This particular experiment suggests $E \sim (p-p_c)^{1.69}$. Courtesy of M. Adam. After Peniche-Covacs *et al.*, "Gel and gelling processes," *Discuss. Faraday Soc.* **57**, 165 (1974).

Gordon and co-workers[25,26] on gelatin just above the threshold give an exponent $t = 1.7$ which is not too far from this prediction (Fig. V.10).*

Why is t much larger than β? At first we might have thought that the elastic modulus E is proportional to the gel fraction S_∞ (giving $t = \beta$). The difference stems from *dangling chains* in the infinite cluster; examples of dangling ends are given in Fig. V.1. The dangling chains contribute to the gel fraction but do not contribute to the elastic modulus. This explains why $E \ll S_\infty$ or why $t > \beta$. This point was originally noticed by polymer physicists (Flory, James and Guth, etc.) in connection with rubber elasticity. Later the same remark was made by Thouless[27] on a different but related problem—namely, the *electrical conductance of a random conducting network.* Returning to the percolation lattice of Fig. V.9, let us replace the "reacted bonds" by electric conductors, all nonreacted bonds being nonconducting. Then we obtain an electric network which has a certain macroscopic conductance (Σ) when we are above the threshold. It turns out that Σ is the exact analog of the elastic modulus. The argument for this is simple:[19]

*I am indebted to M. Adam for this observation and for the plot of Fig. V.10.

(i) For an elastic deformation with displacements x_i on the ith monomer, a natural form of the elastic energy is

$$E_{el} = \frac{1}{2} \sum_{ij} K_{ij} (x_i - x_j)^2 \qquad (V.13)$$

where the spring constant K_{ij} exists only for first-neighbor sites which are *connected*. Then $K_{ij} = K$. Writing that E_{el} is a minimum, leads to the balance of forces at each monomer

$$\sum_j K_{ij} (x_j - x_i) = 0 \qquad \text{for all } i \qquad (V.14)$$

(*ii*) In the electric network problem K_{ij} is the local conductance and x_i is the voltage on node (i). Eq. (V.14) is then the Kirchoff equation, expressing that the sum of all currents flowing towards side i vanishes. The quantity $2E_{el}$ then represents the Joule dissipation in the network; minimization of eq. (V.13) corresponds to the principle of minimum entropy. On a macroscopic scale, we have a linear relationship between current density J and electric field $(-\nabla x)$

$$J = -\Sigma \nabla x$$

In the gel problem, the analog of $-J$ is the mechanical stress, and ∇x is the strain, $-J = E \nabla x$ where E is an elastic modulus. Hence *E and Σ scale in the same way*. This remark is of practical use because a vast literature has been collected on random electric networks.[27,28]

V.2.5. A quick glance at the classical theory

All the scaling laws derived from the percolation model and quoted above are very different from the predictions of classical theory (the tree approximation):[16,17]

$$\nu_{classical} = 1/2 \qquad \nu_{d=3} \cong 0.85$$

$$\beta_{classical} = 1 \qquad \beta_{d=3} = 0.39$$

$$t_{classical} = 3^* \qquad t_{d=3} \cong 1.8$$

Thus, we might be tempted to skip a description of the classical theory. However, we cannot avoid a brief sketch of the calculations, which are

*The calculation of t in the classical theory is nontrivial and is due to M. Gordon, *Proc. Int. Rubber Conf.*, Moscow, 1969.

simple and have some instructive features.* We consider one case, where each monomer can link to z others. Starting from a given monomer, which we call the *root,* we look for the probability $w_n(p) = c_n/c$ that it belongs to a cluster of n monomers. It is convenient to obtain the w values though a generating function $F_0(\theta)$ such that

$$F_0(\theta) = \sum_n w_n(p)\, \theta^n \qquad \text{(V.15)}$$

The variable θ is introduced solely for mathematical purposes; whenever we add one monomer to a molecule, we must insert an extra power of θ.

The construction of F_0 proceeds in two steps. First we look at all possible linkages of our first monomer; they are shown in Fig. V.11 (for $z = 3$). For each lateral tree there is another generating function $F_1(\theta)$. In terms of F_1 we may translate Fig. V.11 into the equation:

$$\begin{aligned} F_0(\theta) &= (1-p)^z + zp\theta F_1\,(1-p)^{z-1} + z\frac{(z-1)}{2}\,(p\theta F_1)^2\,(1-p)^{z-2} \\ &\quad + \ldots + (p\theta F_1)^z \\ &= (1 - p + p\theta F_1)^z \end{aligned} \qquad \text{(V.16)}$$

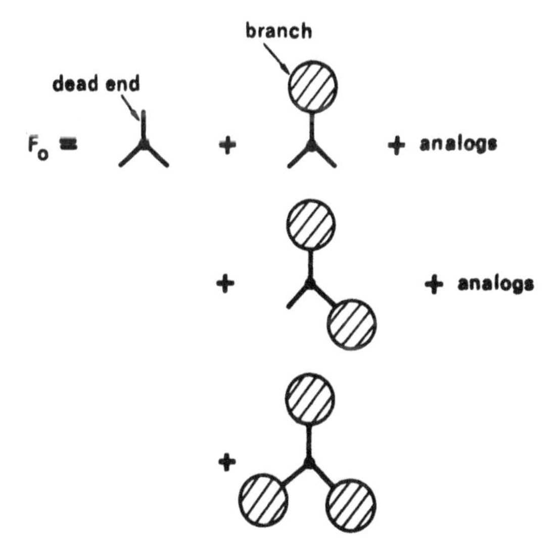

Figure V.11.

*Our presentation follows the line of M. Gordon, *Proc. Roy. Soc. (London),* **268A,** 240 (1962).

For example, the second term in the counting corresponds to one single reacted bond, starting from the root. The bond (factor p) brings one new monomer (factor θ) and possibly one lateral tree (factor F_1); the other $(z - 1)$ available functions from the root are not reacted (factor $(1 - p)^{z-1}$). Finally there are z choices for this single reacted bond (factor z). Eq. (V.16) totals all possibilities with 0, 1, 2, . . . , z reacted bonds on the root, all different possibilities being exclusive and independent. We now write a similar equation for F_1. The difference is that when we deal with a second monomer, one chemical function on the monomer is known to have reacted. Thus Fig. V.11 is replaced by Fig. V.12. A nice feature here is that the new "branches" involve the *same* function F_1. Fig. V.12 gives us a *closed* equation for F_1

$$F_1 = (1-p + \theta p F_1)^{z-1} \qquad \text{(V.17)}$$

From this law we can derive $F_1(\theta)$, and thus $F_0(\theta)$, giving all statistical properties. For instance consider

$$F_0(\theta = 1) = \sum_n w_n\,(p)$$

When we are below the threshold, we expect to have only finite clusters. The sum of the probabilities w_n for all cluster sizes must total 1, but when $p > p_c$, we have another probability—namely, the starting monomer can belong to the infinite cluster, and this event has a probability $S_\infty(p)$ (the gel fraction). Then

$$F_0(\theta = 1) = 1 - S_\infty(p) \qquad \text{(V.18)}$$

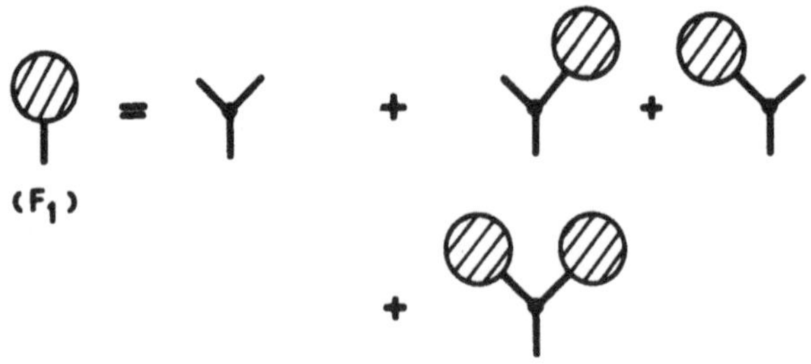

(F_1)

Figure V.12.

Fixing $\theta = 1$ we can solve for F_1 in eq. (V.17) or more conveniently solve for p as a function of F_1. There are two roots

$$\begin{cases} p = \dfrac{1 - F_1^{1/(z-1)}}{1 - F_1} \\ p = \text{indeterminate if } F_1 = 1 \end{cases}$$

The complete plot of $F_1(\theta = 1)$ as functions of p is shown in Fig. V.13. The two branches intersect at the critical value of p:$p_c = 1/(z - 1)$. The plot of F_0 as a function of p is also very similar to Fig. V.13. From it one obtains the gel fraction $S_\infty = 1 - F_0$. This starts linearly at the threshold, i.e., the exponent β of eq. (V.8) is 1 in this theory. By similar arguments we find that $\gamma = 1$ and $\nu = 1/2$.

The tree approximation discussed here has the significance and limitations of a *mean field* theory. Basically we have let the chains grow, ignoring any possible distortions in their shape and in their probability of occurrence which result from excluded volume effects and cyclization effects. This is the source of the difference between the real exponents and their mean field value.

V.2.6. The classical theory works in six dimensions

In Chapter I, we saw that *linear* chains become ideal at dimensionalities (d) equal to or larger than 4. For the *branched* chains of interest here, we can expect a similar property. However, branched chains are more compact than linear chains, and it will require a higher dimensionality to reach structures which are "open" enough to be ideal. We discuss this here through a qualitative method, based on an idea of Ginsburg for phase transitions, and recently transposed to the sol–gel transition.[32]

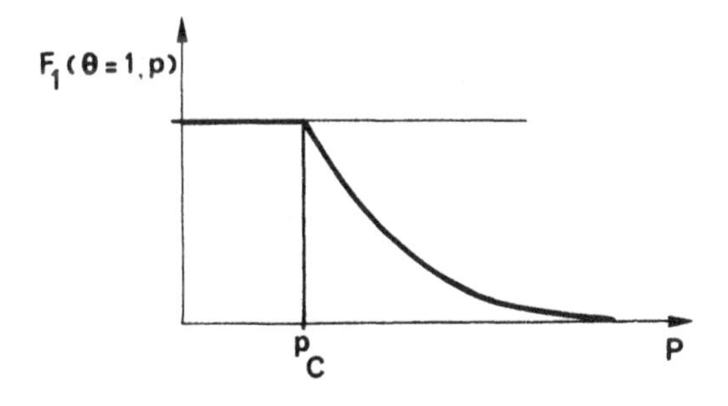

Figure V.13.

The deviations from the mean field (classical) exponents are due to the fluctuations in the gel fraction inside one correlation volume $\Omega = \xi^d$. We shall estimate these fluctuations, assuming that the classical exponents hold, and see if they are large or small when compared to the average gel fraction. We shall see that they are indeed small whenever the dimensionality d is larger than 6: in this case it was thus correct to assume classical behavior.

In a volume Ω the average number of sites belonging to the infinite cluster is

$$\bar{\nu}_{gel} = S_{\infty}(p)\left(\frac{\xi}{a}\right)^d \sim \Delta p^{\beta-\nu d} \qquad \text{(V.19)}$$

We now look at the fluctuations of ν_{gel} (which we call $\delta\nu_{gel}$) or at the fluctuations ($\delta\nu_c$) of the number of sites ν_c which belong to *finite* clusters. The two are linked, since the sum $\nu_g + \nu_c$ is the total number of sites in the volume Ω and is constant. Thus $\delta\nu_g = -\delta\nu_c$. The average square $\delta\nu_c^2$ is easy to compute: since each site is (on the average) connected with N_w other sites, we have

$$\delta\nu_c^2 \cong \nu_c N_w \qquad \text{(V.20)}$$

Near threshold the gel fraction is small and ν_c is essentially equal to the total number of sites $(\xi/a)^d$. Thus

$$\delta\nu_g^2 = (\delta\nu_c)^2 = \left(\frac{\xi}{a}\right)^d N_w \cong \left(\frac{\xi}{a}\right)^d \Delta p^{-\gamma} \qquad \text{(V.21)}$$

We can now compare the fluctuations to the average, and see whether they are really dangerous. We define:

$$x = \delta\nu_g^2 \,/\, (\nu_g)^2 \cong \Delta p^{-\gamma-2\beta+\nu d} \qquad \text{(V.22)}$$

Now we insert in this formula the classical exponents ($\gamma = 1$, $\beta = 1$, $\nu = 1/2$) and find:

$$x = \Delta p^{-3+d/2} \qquad \text{(V.23)}$$

We conclude that, when $d > 6$, x is small near the critical point, the fluctuations are not dangerous, and the mean field approach makes sense.[29,30] But there is a critical dimensionality $d_c = 6$ below which the mean field idea is not self-consistent. The fact that our world has a value of $d (d = 3)$

which is much smaller than d_c, explains why the tree approximation is so poor in practice.

V.2.7. The special case of vulcanization

There is one sol–gel transition for which the tree approximation is expected to give the correct critical behavior near the threshold. This is the case where we start from a *dense* system of linear chains (degree of polymerization $N \gg 1$) and where we crosslink them. A typical practical example is the vulcanization of rubbers.[1]

The cascade theory of Section V.2.5. can be applied directly to this situation. The only special feature is that the functionality z becomes very large; if each monomer can be a partner in a crosslink, we have $z = N$. Then the threshold is low $p_c = 1/(z-1) \cong N^{-1}$. Near threshold we expect classical exponents. A detailed argument for this is given in Ref. 32. To understand it on simple physical grounds, it is again convenient to discuss the number (P) of chains which are likely to crosslink directly with one given chain (in the melt). The region of space spanned by one particular chain C is of volume $R_0^3 = N^{3/2} a^3$, and the number of chains per unit volume is $1/Na^3$. Any chain C′ which has a good overlap with the volume R_0^3 is certainly in direct contact at some points with C, the number of CC′ contacts being of order $N^2a^3/R_0^3 \sim N^{1/2}$.

Thus the total number of chains C′ likely to be attached to C is of the order

$$P \cong \frac{1}{Na^3} R_0^3 \cong N^{1/2} \tag{V.24}$$

We see that P is much larger than unity. This ensures that all deviations from a mean field picture are weak. The sol–gel transition is thus correctly described by the tree approximation.

Remark. This cross-over from percolation exponents to mean field exponents may occur in situations other than vulcanization. Consider for instance a condensation reaction between a difunctional unit AA and a trifunctional unit BB′B, in a case where one of the groups (B′) is much less reactive than the two others (BB).* Then the early polymerization products are linear chains:

$$\ldots \text{AA—BB}'\text{B—AA} \ldots$$

*I am indebted to R. Audebert and M. Adam for pointing out this possibility.

and it is only at a late stage that branching takes place through the functions B′, leading to structures such as:

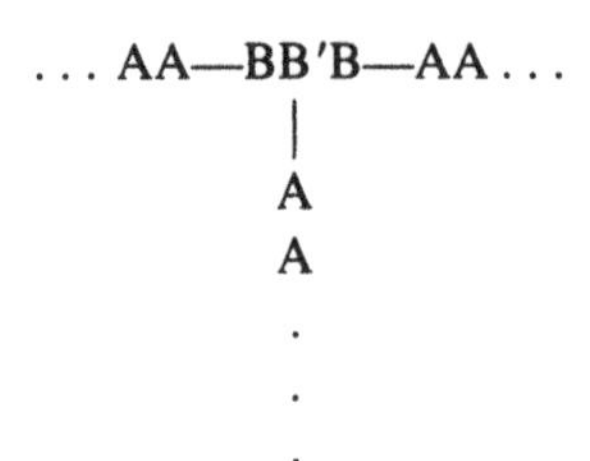

On the whole, near the sol → gel transition, we are dealing with a cross-linking of preexisting long chains: if these chains are rather concentrated, the above discussion may apply, and we expect a trend towards mean field behavior. Thus many practical cases may be intermediate between percolation and mean field: this may explain (at least partly) the large discrepancies between critical exponents measured in different systems.

V.2.8. Dilution effects: competition between gelation and precipitation

It is clear that the percolation model is a very crude representation of any gelation processes. We shall now discuss two possible criticisms: (*i*) the monomers are not on a lattice, but are *disordered;* (*ii*) in many practical cases, the monomers are mixed with a *solvent,* and this feature is absent in the percolation model.

The first effect is probably not essential from our point of view: when we drop the lattice picture, and replace it by a random, dense set of z functional monomers, there may be some corrections to the threshold value p_c; but we do not expect changes in the critical exponents.*

The second effect (dilution) is much more important. As we shall see, when the monomers become attached together by chemical links, they tend to set up clusters in the solvent, that is, to precipitate. Thus there is a competition between gelation and precipitation, which is essential for many practical applications.

In the percolation model, the monomers are closely packed. The model describes polyfunctional polymerization in a melt. What happens if we dilute the reactants? When we fix a reaction level p, we impose certain conditions on the monomers. They cannot be spread at random on the lattice because this would not give the correct p value. Thus, dilution leads

*There are some solid-state phase transitions where the introduction of disorder in the crystal lattice changes the exponents, but it may be shown that percolation does not belong to this class.[33]

to a problem of *percolation between correlated objects*. Numerical data on this situation have recently been collected.[34] We shall present the problem here in the case of a very good (athermal) solvent, using again a lattice model. Each lattice site (i) must be occupied either by one monomer (occupation number $n_i = 1$) or by one solvent molecule ($n_i = 0$).

We assume that whenever two adjacent sites (i,j) are occupied by two monomers ($n_i n_j = 1$) a chemical bond is instantly established between the two monomers. With this model, we want to investigate the situation in the reaction bath for a given concentration $\phi = \langle n_i \rangle$ and for a given number of reacted bonds:

$$R = \sum_{i>j}{}' \langle n_i \, n_j \rangle$$

where the sum Σ' is restricted to nearest neighbor pairs. The relation between R and the fraction of reacted bonds p is:

$$p = R/R_{max}$$

where the maximum value R_{max} is equal to $N_s \phi z/2$ (N_s being the number of sites on the lattice). To understand this, note that the number of monomers is $N_s \phi$, and that maximum linkage is reached where all monomers coalesce in one fraction of the total volume: then each monomer is linked to z others, and the number of pairs is $N_s \phi z/2$ (omitting surface corrections).

Our model for gelation in a good solvent can be described in terms of a sum of states, or partition function

$$Z = \sum_{(n_k)} \delta \left(\sum_j n_j - N_s \phi \right) \delta \left(\sum_{ij} n_i n_j - R \right)$$

As usual, it is convenient to remove the constraints described by ϕ and R, and to calculate not Z, but a related grand partition function Ξ

$$\Xi = \sum_{(n_k)} exp \left[\alpha \sum_i n_i + \beta \sum_{ij}{}' n_i n_j \right]$$

where e^{α} is a fugacity for monomers, and e^{β} a fugacity for bonds. Their relations to ϕ and R have the standard form

$$N_s \phi = \frac{\partial \log \Xi}{\partial \alpha}$$

$$R = \frac{\partial \log \Xi}{\partial \beta}$$

In practice R is an increasing function of β. For $\beta = 0$ (no correlations between monomers) we have

$$\langle n_i n_j \rangle = \langle n_i \rangle \langle n_j \rangle = \phi^2$$
$$R_{\beta=0} \equiv R_0 = N_s z \phi^2/2$$

This corresponds to the situation immediately after starting the reaction. At later times, R increases beyond R_0, and β is *positive*.

At this point it is useful to observe that Ξ has exactly the form of a partition function for the so called "lattice gas model." In this model each n_i describes the occupancy of a lattice site by an atom, and neighboring atoms have an attractive interaction energy $-J$. The temperature of the lattice gas is J/β. Because of the attraction J the model gives isotherms with a liquid/vapor transition.

Similar features hold for our gelation problem: in the reaction bath, even if the solvent is very good, the branched polymers which are generated will tend to segregate if the reaction is near completion. *Chemical bonding is equivalent to an attractive interaction!*

This point is essential, and explains many of the difficulties encountered in the preparation of gels. To make it more precise, consider Fig. V.14, which gives the phase diagram of our system in terms of the variables ϕ (concentration) and β^{-1} (temperature of the corresponding lattice gas). The diagram shows a one-phase region (corresponding to spatially homogeneous systems) limited by a coexistence curve Γ. The one-phase region itself is divided into two parts by a sol–gel transition line Δ with the following features:

(*i*) For high equivalent temperatures β^{-1}, the Δ line reaches an asymptote $\phi = \phi_{cs}$. This regime (with randomly distributed monomers) corresponds to what is called "site percolation." Numerical values of ϕ_{cs} for various lattices are known. The critical exponents β, γ, ν, . . . for site percolation are identical to the corresponding exponents for bond percolation.

(*ii*) At lower β^{-1}, for a fixed ϕ, the correlations increase, and the gel phase becomes more prevalent. Mean field calculations of this effect have been performed; see R. Kikuchi, *J. Chem. Phys.* **53,** 2713 (1970), and A. Coniglio, *Phys. Rev. Lett.* **B13,** 2194 (1976).

Thus the Δ line is displaced towards the low ϕ side of Fig. V.14.

(*iii*) Finally, the Δ line hits the coexistence line at a certain point P_3. Note that in general P_3 is quite different from the critical point C of the coexistence curve (P_3 occurs at much lower concentrations).

Having established the qualitative structure of our phase diagram, we can now consider some typical reaction paths. In all cases we start with some

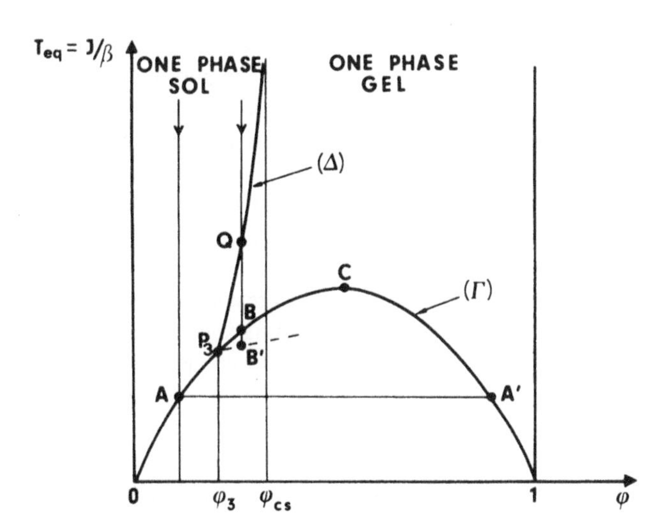

Figure V.14.
Equivalent phase diagram for a gelating system in semi-dilute solutions with a good solvent. ϕ is the monomer concentration (and each monomer is polyfunctional). T_{eq} is an equivalent temperature, which decreases from infinity to low values when the chemical reaction progresses. The particular model used in the text assumes instant reaction between monomers which are in contact. But the qualitative features of the diagram are more general.

fixed value of ϕ, and let the reaction proceed, R increasing from R_0 towards R_{max}. This corresponds to a progressive decrease in the equivalent temperature β^{-1} (from $+\infty$ to zero). From Fig. V.14 we find three situations—depending on the concentration.

(i) If $\phi < \phi_3$ we remain in a sol up to point A, at which some gel begins to precipitate in the sol (the gel concentration being described by A′).

(ii) At intermediate concentrations ($\phi_3 < \phi < \phi_{cs}$) we start again with a sol, and reach a homogeneous gel at point Q. The critical exponents for this sol-gel transition are still of the percolation type, for the following reason: at point Q the range ξ_c of the correlations imposed on the *monomers* by chemical bonding is finite (because Q does not coincide with the critical point C). Then, by suitable redefinition of the polyfunctional units (making them larger than ξ_c) we can always return to a simple percolation problem. What happens if we go beyond the sol-gel threshold? At high enough reaction levels we reach point B, and naively we might think that some segregation occurs at this moment. This is not true, however, because we are dealing with a gel: crosslinking prevents a macroscopic segregation. On a local scale (sizes comparable to the distance between crosslinks) segregation may take place, but this is expected to require higher reaction rates, and should occur at a point B′ below B. These "microphase separations" are very important in practical fabrications of gels, but are still poorly understood.

(*iii*) At high concentrations $\phi > \phi_{cs}$ our model gives instant gelation: clearly in this regime, a more detailed model allowing for contact without reaction is required.

To summarize: a gelation process in the presence of solvent always brings in a trend towards segregation of the gelating species. However, by a suitable choice of the concentration in the reaction bath, one can still observe a well-defined sol–gel transition. The critical exponents observed in this case may still be of the percolation type. The latter statement has been proven more formally in one case = quasi equilibrium with mixtures of linear chains and crosslinking agents in a athermal solvent.*

V.3. Gels in Good Solvents

We now focus on gels which are well beyond the gelation threshold. We assume that they have been prepared in good solvents, and that, at the moment of study, they are also in good solvents. This is the best situation if we wish to avoid segregation effects and the resulting heterogeneities. Also, for simplicity, we focus our attention on *calibrated* gels, where the number N of monomers between adjacent crosslinks is well defined. The classic picture for these gels is from Flory[1] and is very successful. We present it here in different language.

V.3.1. The c^* theorem

Let us start with a solution of chains (polymerization index N) in a good solvent (excluded volume parameter $v = a^3 (1 - 2\chi) > 0$). The chains repel each other, and this is reflected in the existence of a positive osmotic pressure Π.

We now begin to attach the chains together, for example by reaction of the chain ends with certain z-functional molecules (z being equal to 3, 4, etc.), and we let them choose their density. They would like to separate from each other as much as possible; however, each coil must remain in contact with its neighbors because of the crosslinks. The net result is shown in Fig. V.15.

*T. C. Lubensky and J. Isaacson, *Phys. Rev. Lett.* **41**, 829 (1978); *Phys. Rev. Lett.* **42**, 410 (E) (1979); *Phys. Rev.* **A20**, 2130 (1979).

Figure V.15.

What we have is a set of closely packed coils sealed together by the crosslinks. The situation is reminiscent of the overlap threshold in semidilute solutions (Chapter III). Thus, the gel *automatically maintains a concentration c proportional to c⋆.*

A detailed formula for $c^\star$ at arbitrary $\chi < 1/2$ was worked out in Chapter IV [eq. (IV.50)]. This gives

$$c = k(z)\, c^\star = k(z)\, N^{-4/5}\, v^{-3/5}\, a^{-6/5} \qquad \text{(V.25)}$$

where $k(z)$ is a constant number, of order unity, depending on the functionality z of the crosslinks and on the preparation conditions.

Eq. (V.25) summarizes the *Flory theory of gels.*[1] Changing the chemical nature of the solvent amounts to changing the excluded volume parameter v; if v increases (better solvent), $c^\star$ decreases (swelling). Eq. (V.25) has been confirmed by macroscopic measurements on many gel systems. Experimentally, it is important to *wait* long enough to choose a correct equilibration of the solvent. Since we cannot stir the system, concentrations are equalized only slowly by diffusion processes. Equilibration times are of order L^2/D where L is a sample size and D is a diffusion coefficient. Typical values of D are in the range 10^{-6} to 10^{-7} cm²/sec, and the resulting times are around one day.

V.3.2. Pair correlations in the gel

In his original derivation of equations similar to eq. (V.25) Flory assumed gaussian statistics for the chains plus a mean field estimate for the repulsive energies.[1] His theory is successful; the scientific community has naturally concluded that the chains in a swollen gel are gaussian to a very good approximation. This is entirely wrong for the following reason.

Flory's calculation is quite similar to his discussion of a single chain in a good solvent, which we analyzed in Section I.3. In this case we saw that an excellent result came from a cancellation between two serious approximations—one related to the use of gaussian statistics, and one due to the neglect of correlations between chains. The same cancellation occurs for swollen gels, and the success of the theory does not tell us that the chains are gaussian.

The correct structure of pair correlations in the gel can be read from Fig. V.16. At short distances, the correlation function

$$g(\mathbf{r}) = \frac{1}{c}[\langle c(o)\, c(\mathbf{r})\rangle - c^2]$$

is dominated entirely by correlations inside one chain and follows the Edwards law [eq. (I.31)]. It is only when $g(\mathbf{r})$ goes down to values of the order $c = kc^\star$ that the existence of a gel phase affects the correlation. This crossover point corresponds to r values comparable with the single coil size R_F (given by eq. (IV.49)). At larger distances density fluctuations are limited by the macroscopic rigidity of the gel, and $g(\mathbf{r})$ decays rapidly in space.

The latter statement can be made more precise from a generalized form of elastic theory at long wavelengths, corresponding to the free energy (per cm^3)

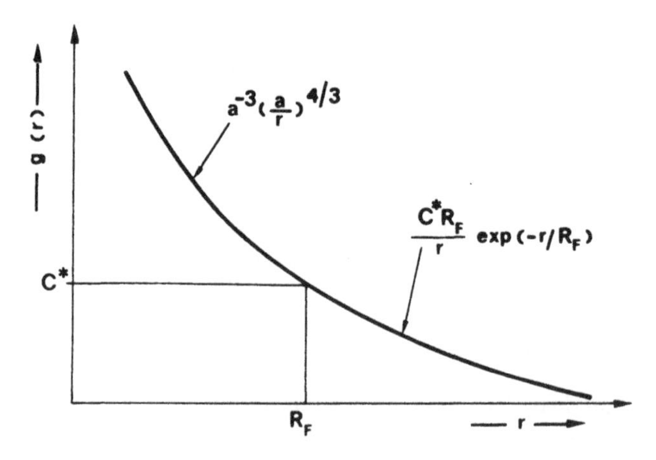

Figure V.16.

$$F = \frac{1}{2} E \left(\frac{\delta c}{c}\right)^2 + \frac{1}{2} L \left[\nabla \left(\frac{\delta c}{c}\right)\right]^2 + \text{higher gradient terms} \quad \text{(V.26)}$$

where δc is the local change in concentration, E is the bulk rigidity, and L represents a higher order correction (usually ignored in continuous elasticity). Since the only characteristic length available is R_F, scaling means that

$$L \cong E\, R_F^2 \quad \text{(V.27)}$$

Going to Fourier transforms, eq. (V.26) gives a sum of terms for different wave vectors q

$$F = \frac{1}{2} \sum_q \left(\frac{\delta c_q}{c}\right)^2 (E + Lq^2)$$

Applying the equipartition theorem to each mode q, we get

$$\frac{1}{c^2} \langle (\delta c_q)^2 \rangle = \frac{T}{E + Lq^2} \qquad (qR_F < 1) \quad \text{(V.28)}$$

Note that eq. (V.28) applies only for $qR_F \ll 1$ because eq. (V.26) assumes slow spatial variations. Returning to real space, we can transform eq. (V.28) into

$$g(\mathbf{r}) = \frac{c\,T}{4\pi L} \frac{1}{r} \exp -\left[r \left(\frac{E}{L}\right)^{1/2}\right] \qquad (r > R_F) \quad \text{(V.29)}$$

$$\cong \frac{N}{R_F^2} \frac{1}{r} \exp - \left(\text{constant} \frac{r}{R_F}\right) \quad \text{(V.30)}$$

The scaling form of the coefficient in eq. (V.30) has been obtained separately from the requirement that at $r = R_F$, the correlation function $g(r)$ must be comparable with the average concentration c ($\sim c^\star$). Thus a byproduct of our discussion is to give (by comparison between eqs. V.30 and V.29) the scaling form of the elastic moduli:

$$E \cong \frac{L}{R_F^2} \cong \frac{c^\star}{N} T \cong \frac{cT}{N} \quad \text{(V.31)}$$

The whole picture subtended by Fig. V.15 may be simply stated in terms of suitable "blobs." To each chain (N monomers) we associate a blob of size R_F. The blobs are essentially closely packed (the exact packing depending on the functionality of the gel and on the conditions of preparation). Inside one blob, the *correlations are of the excluded volume type*—i.e., the blobs are not gaussian. Neighboring blobs are coupled by elastic forces; using eq. (I.45) to predict the spring constant of one blob, we may easily rederive eq. (V.31).

This picture can be tested in principle by various scattering measurements. Neutron data on gels have been taken by the Strasbourg group.[35] However, the main emphasis has been on a different type of measurement where a certain part of the gel structure is *labeled;* in particular it is comparatively easy to deuterate the crosslinks and to measure the correlation between them. The resulting diffraction pattern is very similar to diffraction by an *amorphous solid;* the crosslinks maintain a certain average distance (or order R_F), and this gives rise to a diffuse peak (at $q \sim 1/R_F$) in the scattering pattern. Two crosslinks cannot come very close together; this would imply a large overlap between two neighboring blobs and an energy which would then become quite large.

Under the stimulation of H. Benoît, similar experiments with labeled centers were also made with *solutions* of star-shaped polymers which have the same geometry and the same concentration[35] (Fig. V.17). The scattering patterns for both situations are of the same type. This is not surprising since both systems are at $c = c^\star$ and are very similar to a dense fluid of hard spheres.

V.3.3. Elasticity of swollen gels

We have seen in eq. (V.31) that the bulk modulus E of the gel should scale like $(c/N)T$. A similar scaling law should also hold for the shear modulus, which is more easily accessible to experiment. (In what follows, since we are interested only in scaling properties, we use the same symbol E for both.) It is possible to test eq. (V.31) by varying either the quality of the solvent (i.e., v) or the length of the chain (i.e., N). Recall that c, v, and N are always linked by the $c^\star$ theorem [eq. (V.25)].

For a given solvent (fixed v) it is often convenient to eliminate N between eqs. (V.31) and (V.25), obtaining

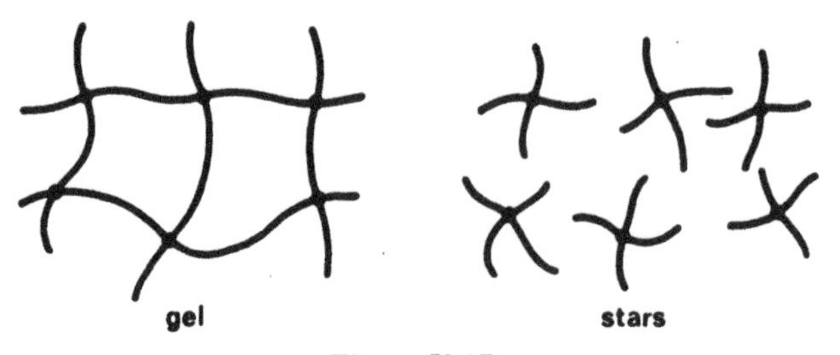

Figure V.17.

$$E \cong Tc^{2.25} \, (v^{3/4} \, a^{3/2}) \tag{V.32}$$

Thus, the elastic modulus should scale like the power 2.25 of concentration. Recent elastic data of Belkebir-Mrani[36] have been reanalyzed along these lines.[38] They do show exponents which are close to 2.25 if the comparison is made at *fixed functionality* z (the numerical coefficient in eq. V.32 depends on z).

For a fixed N and a solvent of variable quality, the elastic modulus should scale linearly with concentration ($E \cong cT/N$). This is also well confirmed.

The above discussion was restricted to linear elasticity—i.e., to the regime where the relationship between stress (σ) and deformation (λ) is of the form

$$\sigma = E(\lambda - 1) \tag{V.33}$$

(We define λ for longitudinal deformations as the ratio of the extended length of the sample to the length at rest.) This is obtained when $\sigma \ll E$. The opposite limit $\sigma \gg E$ would be of great interest. Unfortunately, gels usually break at low σ values, and these strong deformations are difficult to study. However, they are important because the stress is then sensitive to the nongaussian character of the individual chains. Section I.4 showed that swollen chains have a nonlinear relationship between force and elongation; this should show up in $\sigma(\lambda)$. The prediction is[37]

$$\sigma = E \, \lambda^{5/2} \quad \text{(real gel; } \sigma \gg E) \tag{V.34}$$

for longitudinal extension at constant c.* Compare this with the law for gaussian chains

$$\sigma = E\lambda^2 \quad \text{(gaussian; } \sigma \gg E) \tag{V.35}$$

Some readers may be surprised by the occurence of a quadratic law [eq. (V.35)] for gaussian chains which have a linear spring behavior. The reason is simple. When we extend our sample very much, its lateral dimensions decrease, and σ (which is a force per unit area of cross-section) increases by one extra power of λ. The really interesting feature is the difference between eq. (V.34) and eq. (V.35), which reflects the Fisher-Pincus scaling law for swollen chains in strong extension [eq.

*Since the experiment must be done relatively fast, the gel cannot change its solvent content during elongation.

(I.47)]. We hope that rapid elongation studies (before fracture) on swollen gels will become feasible soon.

V.3.4. Spinodal decomposition

We now extend our discussion to a slightly more complex situation. Starting with a swollen gel in a good solvent, we decrease the quality of the solvent slightly. This can often be done simply by lowering the temperature. What happens then? The answer depends on the speed of the cooling process.

(i) If cooling is very slow, we have a smooth contraction of the gel and when the solvent becomes really poor, we may reach a "collapse transition," where the gel expels most of the solvent.

(ii) If the cooling is somewhat faster (minutes rather than a day) the elimination of solvent cannot be performed, and we work effectively at fixed concentration. However, even in this swollen gel, an instability may occur:[39] we find small regions which are alternately dense and dilute.

These effects can be well described within the Flory theory for gels. Always using a lattice model for the chains, this corresponds to a free energy of the form

$$\frac{1}{T}F\bigg|_{site} = (1 - \Phi)\ln(1 - \Phi) + \chi\Phi(1 - \Phi) + Q\frac{3R^2}{Na^2}\frac{c}{N} \quad (V.36)$$

Notice the difference with solutions [eq. (III.7)]: 1) the translational entropy of the chains ($\Phi/N \ln \Phi$) is not present since the chains are attached, and 2) there is an elastic energy term, which is taken to be of the ideal chain form. R is the size of one chain and is related to $\Phi = ca^3$ through

$$\frac{c}{N}R^3 = 1 \quad (V.37)$$

Finally Q is a numerical coefficient that depends on functionality and on the conditions of preparation and which is poorly known. (The original Flory theory contained further terms, aiming at a more precise description of the elastic behavior at small R, but these terms are not well justified, and they do not play much role in case *(ii)*; we ignore them systematically.)

The equilibrium condition corresponds to a minimum of energy per monomer, F_{site}/Φ. The minimum is to be taken with respect to Φ, but we must remember that Φ and R are linked through eq. (V.37). Minimization gives a relationship between size R and temperature (the latter coming in though χ or $v = a^3 (1 - 2\chi)$) which is qualitatively shown in Fig. V.18. For $v > 0$ the curve is smooth—the gel contracts progressively. Ultimately at $v < 0$, the solvent is completely expelled.

Let us now turn to the more interesting process where v is decreased but the concentration Φ does not have enough time to relax. There is then a trend toward segregation—fluctuations in concentration are enhanced. This is equivalent to saying that the bulk modulus E becomes weaker and finally vanishes at a certain temperature, $T_s(\Phi)$. Ignoring all complications due to the tensorial nature of the stresses, we can write E in terms of a derivative of the osmotic pressure Π

$$E = \Phi \frac{\partial \Pi}{\partial \Phi} \tag{V.38}$$

where Π is defined as usual by eq. (III.12)

$$a^3 \Pi = \Phi^2 \frac{\partial}{\partial \Phi} \left(\frac{F_{site}}{\Phi} \right) \tag{V.39}$$

When eqs. (V.38, V.39) are used with the form of the free energy in eq. (V.36), one finds that E reaches 0 on a certain spinodal curve. If we start at certain Φ values (or certain R) corresponding to point A in Fig. V.18, we

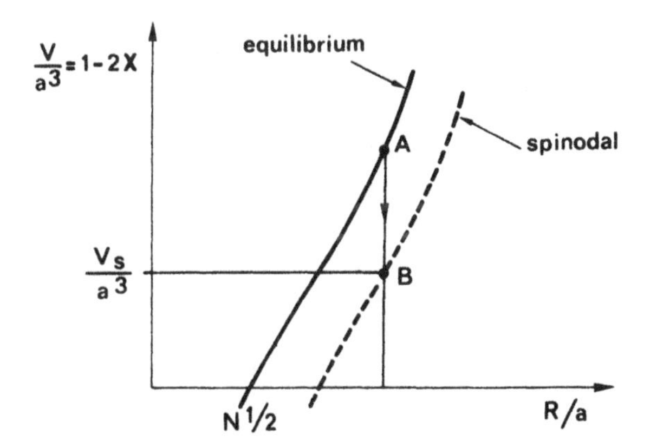

Figure V.18.

can decrease v to point B ($v = v_s$), keeping the gel stable. If we go beyond point B, microsyneresis occurs.

Near point B the fluctuations are large and the intensity of light scattering at small angles I tends to diverge. In a mean field theory of the Flory type the divergence is of the form $I \sim (v - v_s)^{-1}$. This is compatible with Tanaka's data.[39] However, in this problem, since each blob is interacting only with a restricted number of other blobs ($P \sim 1$), there is no reason to believe that the exponents are of a mean field type.

In connection with this spinodal decomposition, if we start with a good solvent, the original radius R is very swollen. Then the final value v_s of the excluded volume parameter is still largely positive; χ_s is significantly smaller than 1/2. Thus *we do not need a poor solvent to have syneresis;* all that is required is a slight lowering of solvent quality. This is another reason for the frequent heterogeneities in gels.

Very little is known about the region below the spinodal. Depending on the detailed conditions, we expect domains which can be either lamellar, rod-like, or spherical. If we wait long enough, some domains can coalesce, following the trend toward a general expulsion of solvents. The problem here is reminiscent of certain situations in metallurgy, but physical techniques available for an *in situ* study of the domains are not numerous. In Chapter VII we return to certain *dynamic* features of the spinodal transition, which have been probed by the M.I.T. group.

V.3.5. Summary

Swollen gels obey simple scaling laws, which are independent of preparation conditions. The gel can be visualized as a collection of adjacent blobs, each blob being associated with one chain and having properties very similar to those of a single chain. The blobs are not gaussian. However, a Flory model based on gaussian statistics gives a good description of the gel properties—thanks to a remarkable compensation of errors. A relatively slight lowering of solvent quality is enough to induce spinodal decomposition in a gel.

REFERENCES

1. P. Flory, *Principles of Polymer Chemistry,* Cornell University Press, Ithaca, N.Y., 1971. K. Dusek, W. Prins, *Adv. Polym. Sci.* **6,** 1, (1969).
2. *Polymer Networks,* A. Chompff and S. Newman, Eds., Plenum Press, New York, 1971.
3. "Gel and gelling processes," *Discuss. Faraday Soc.* **57** (1974).
4. G. Allen, P. Egerton, D. Walsh, *Polymer* **17,** 65 (1975).

5. P. Flory, *J. Am. Chem. Soc.* **63,** 3096 (1941).
6. G. Beinert, A. Belkebir-Mrani, J. Herz, G. Hild, P. Rempp, *Discuss. Faraday Soc.* **57,** 27 (1974).
7. J. G. Vail, *Soluble Silicates,* Van Nostrand-Reinhold, New York, 1952.
8. P. G. de Gennes, *Phys. Lett.* **28A,** 725 (1969).
9. P. G. de Gennes, *The Physics of Liquid Crystals,* Oxford University Press, London, 1976.
10. H. L. Frisch, D. Krempner, K. Frisch, T. Kwei, *Polymer Networks,* A. Chompff and S. Newman, Eds., p. 451, Plenum Press, New York, 1971.
11. G. Stainby, *Recent Advances in Gelatin and Glue Research,* Oxford University Press, London, 1958. P. Flory, R. Garret, *J. Am. Chem. Soc.* **80,** 4836 (1958). P. Flory, E. Weaver, *J. Am. Chem. Soc.* **82,** 451 (1960).
12. T. Bryce *et al., Discuss. Faraday Soc.* **57,** 221 (1974).
13. For an introduction to the physics of helical macromolecules see T. Birshtein, O. Ptitsyn, *Conformations of Macromolecules,* Interscience Publishers, New York, 1966.
14. L. Rogovina, G. Slonimskii, *Russian Chem. Rev.* **43,** 503 (1974).
15. S. F. Edwards, *Polymer Networks,* A. Chompff and S. Newman, Eds., p. 83, Plenum Press, New York, 1971.
16. W. Stockmayer, *J. Chem. Phys.* **11,** 45 (1943). W. Stockmayer, *J. Chem. Phys.* **12,** 125 (1944). B. Zimm, W. Stockmayer, *J. Chem. Phys.* **17,** 1301 (1949).
17. G. R. Dobson, M. Gordon, *J. Chem. Phys.* **43,** 705 (1965). M. Gordon, G. Scantlebury, *J. Chem. Soc.* **1** (1967). M. Gordon, G. Scantlebury, *J. Polym. Sci.* **C16,** 3933 (1968). M. Gordon, T. Ward, R. Whitney, *Polymer Networks,* A. Chompff and S. Newman, Eds., p. 1, Plenum Press, New York, 1971.
18. D. Stauffer, *J. C. S., Faraday Trans. II* **72,** 1354 (1976).
19. P. G. de Gennes, *J. Phys. (Paris) Lett.* **37L,** 1 (1976).
20. S. Kirkpatrick, *Rev. Mod. Phys.* **45,** 574 (1973).
21. J. W. Essam, *Phase Transitions and Critical Phenomena,* C. Domb and M. Green, Eds., Vol. 2, p. 197, Academic Press, New York, 1972.
22. W. Burchard, K. Kajiwara, J. Kalal, J. Kennedy, *Macromolecules* **6,** 642 (1973).
23. K. Kajiwara, W. Burchard, M. Gordon, *Br. Polym. J.* **2,** 110 (1970).
24. K. Kajiwara, M. Gordon, *J. Chem. Phys.* **59,** 3623 (1973).
25. C. Peniche-Covacs, S. Dev, M. Gordon, M. Judd, K. Kajiwara, "Gel and gelling processes," *Discuss. Faraday Soc.* **57,** 165 (1974).
26. D. Paul, *J. Appl. Polym. Sci.* **11,** 439 (1967).
27. B. Last, D. Thouless, *Phys. Rev. Lett.* **27,** 1719 (1971).
28. R. Fisch, A. B. Harris, *Phys. Rev.* **18B,** 416 (1978).
29. G. Toulouse, P. Pfeuty, *Introduction aux Groupes de Renormalisation,* Presses Universitaires de Grenoble, France, 1975.
30. A. B. Harris, *et al., Phys. Rev. Lett.* **35,** 327 (1975).
31. P. G. de Gennes, *Biophysics* **6,** 715 (1968).
32. P. G. de Gennes, *J. Phys. (Paris) Lett.* **38L,** 355 (1977).

33. A. B. Harris, T. Lubensky, *Phys. Rev. Lett.* **33,** 1540 (1974).
34. E. Stoll, C. Domb, T. Schneider, *Israel Acad. Sci.* **13** (2), 303 (1978).
35. H. Benoît *et al., J. Polym. Sci. A2* **14,** 2119 (1976). See also Refs. 36, 37.
36. A. Belkebir-Mrani, Ph.D. Thesis, Strasbourg, 1976. See also Ref. 6.
37. S. Daoudi, *J. Phys. (Paris)* **38,** 1301 (1977).
38. J. P. Munch *et al., J. Phys. (Paris)* **38,** 971 (1977).
39. T. Tanaka, S. Ishiwata, C. Ishimoto, *Phys. Rev. Lett.* **38,** 771 (1977).

Part B

DYNAMICS

VI

Dynamics of a Single Chain

VI.1. Historical Background

The random motions of a flexible chain floating in a solvent are fascinatingly complex. Some years ago, thanks to a series of careful mechanical experiments[1] together with some elegant theoretical work[2,3,4] these motions appeared to be well understood. However, there are some serious flaws in the classical picture.[5] Thus, we do not review it in detail (it is lucidly described in Ref. 6) but present only basic ideas, show their limitations, and proceed directly to more general scaling concepts.

VI.1.1. The Rouse model

The classical picture is based on the notion of *relaxation modes* for one chain. It first appeared in a 1953 paper by P. E. Rouse[3] and was based on the following model:

(*i*) *Ideal chain.* Rouse described the chain as a succession of "beads" $\mathbf{r}_1 \ldots \mathbf{r}_n \ldots \mathbf{r}_{n+1}$ separated by "springs" along the vectors $\mathbf{a}_1 \ldots \mathbf{a}_N$ (Fig. VI.1). Physically, a spring can be thought of as a sequence of monomers (or subchain) which is long enough to obey gaussian statistics. The elastic energy for the subchain is then given by the analog of eq. (I.8)

$$F_{n,n+1} = \frac{3T}{2} \frac{(\mathbf{r}_{n+1} - \mathbf{r}_n)^2}{\langle(\mathbf{r}_{n+1} - \mathbf{r}_n)^2\rangle} = \frac{3T}{2} \frac{\mathbf{a}_n^2}{a^2} \qquad \text{(VI.1)}$$

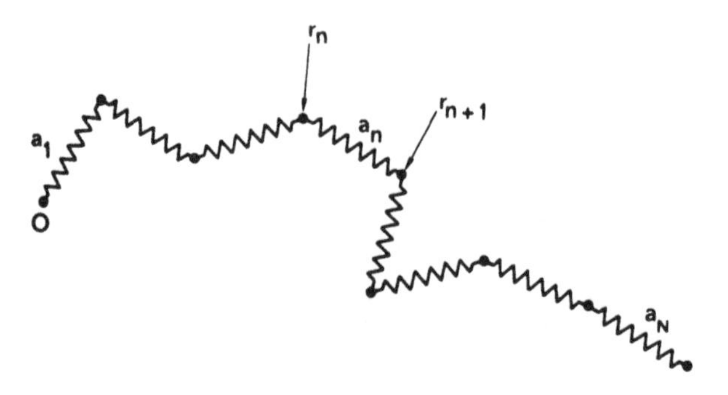

Figure VI.1.

where a^2 is the mean size of one subchain. The total elastic energy is the sum

$$F_{el} = \sum_{o}^{N-1} F_{n,n+1} \tag{VI.2}$$

(ii) Phantom chain. A physical chain cannot cross itself. Processes such as that shown in Fig. VI.2 (where CD goes from "above" AB to "below" AB), are forbidden. Rouse ignored this complication. Following tradition, we say that the Rouse model corresponds to "phantom chains."

(iii) Locality of response. Each bead experiences a force φ_n from its two neighbors

$$\varphi_n = -\frac{\partial F_{el}}{\partial \mathbf{r}_n} = \frac{3T}{a^2}[(\mathbf{r}_{n+1} - \mathbf{r}_n) + (\mathbf{r}_{n-1} - \mathbf{r}_n)] \tag{VI.3}$$

We then assume that the velocity of bead **n** is a linear function of the forces applied to n (and its neighbors)

$$\frac{\partial \mathbf{r}_n}{\partial t} = \sum_m \mu_{nm} \varphi_m \tag{VI.4}$$

where μ_{nm} has the dimension of a *mobility* and is nonzero only for n close to m (locality assumption). Then by a suitable redefinition of the sub-

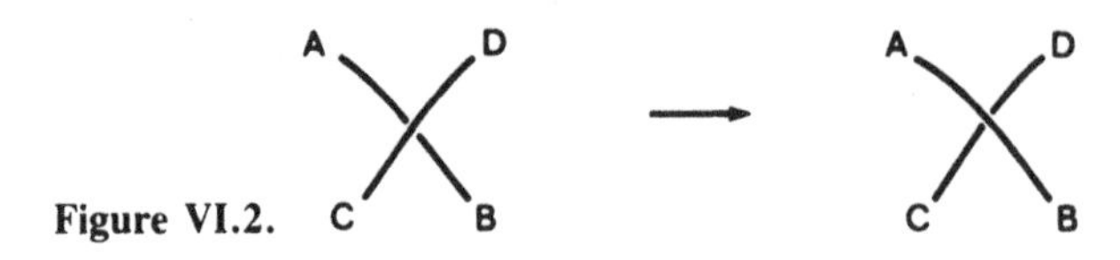

Figure VI.2.

chains, it is always possible to arrange that $\mu_{nm} = 0$ for $n \neq m$, and to keep only one mobility constant $\mu_{nn} = \mu$. Thus eq. (VI.4) reduces to:

$$\frac{\partial \mathbf{r}_n}{\partial t} = \mu \varphi_n = \frac{3\,T\mu}{a^2}(\mathbf{r}_{n+1} - 2\,\mathbf{r}_n + \mathbf{r}_{n-1})$$

$$\cong \frac{3T\,\mu}{a^2}\frac{\partial^2 \mathbf{r}}{\partial n^2} \qquad \text{(VI.5)}$$

where we have gone to the continuous limit $\mathbf{r}_{n+1} - \mathbf{r}_n \rightarrow \partial \mathbf{r}/\partial n$. In this last form eq. (VI.5) must be supplemented by boundary conditions at both ends of the chain. They are*

$$\left.\frac{\partial \mathbf{r}_n}{\partial n}\right|_0 = \left.\frac{\partial \mathbf{r}_n}{\partial n}\right|_N = 0 \qquad \text{(VI.6)}$$

On the whole, the three assumptions lead us to a very simple "Rouse equation" (VI.5) describing the relaxation of an elongated state of the chain. Because it is a *linear* equation, the solutions can be analyzed in terms of eigenmodes

$$\mathbf{r}_{np}(t) = \cos\frac{\pi p n}{N}\exp(-t/\tau_p)\boldsymbol{\alpha}_p \qquad \text{(VI.7)}$$

where $p = 1, 2, \ldots$ is an integer, and the cosine is chosen to match the boundary conditions. $\boldsymbol{\alpha}_p$ is the mode amplitude. The time τ_p is the relaxation time of mode p, and is given by

$$\frac{1}{\tau_p} = 3\pi^2\frac{T\mu}{a^2}\left(\frac{p}{N}\right)^2 = W\left(\frac{p}{N}\right)^2 \qquad \text{(VI.8)}$$

This is a quadratic dispersion relationship $(1/\tau_p \sim p^2)$. Note that the longest relaxation time $(p = 1)$ scales like N^2 in the Rouse model.

VI.1.2. Weakness of internal friction effects

What is the microscopic origin of the "bead mobility" μ? In a *liquid* (as considered originally by Rouse), μ would naturally be associated with the

*To understand eq. (VI.6), notice that the last bead $(\mathbf{r}_{N+1})$ sees only one spring and thus $\partial \mathbf{r}_{N+1}/\partial t = 3T\mu/a^2\,(\mathbf{r}_N - \mathbf{r}_{N+1})$. This may be lumped into eq. (VI.5) if we add one extra spring $(N+1, N+2)$ and impose $(\mathbf{r}_{N+2} - \mathbf{r}_{N+1}) \equiv 0$. This condition becomes equivalent to eq. (VI.6) in the continuous limit.

hydrodynamic friction of one bead on the solvent. If the bead behaves as a sphere of hydrodynamic radius a_H, it will have a friction coefficient

$$\mu^{-1} = 6\pi\eta_o a_H \tag{VI.9}$$

It is also conceivable (at least theoretically) that one chain is inserted into a solid-state matrix and moves (through vacancy diffusion in the solid) by random processes such as that shown in Fig. VI.3. Two types of barriers oppose this process. One is associated with the vacancies; the other is associated with the changes in chain conformation in going from the first state to the second: "internal barriers." Detailed calculations[7] and Monte Carlo analogs[8] for this "solid-state process" have been given. For large N and finite p, they always end up with an equation of the Rouse form;[9] the resulting μ depends strongly on conformational barriers.

Returning to the chain in a solvent, we may ask whether or not μ can contain any effect of internal barriers also for this case? The answer is no. If we are dealing with long chains ($N \to \infty$), *internal barriers are not seen in the first modes*. This is a delicate point, proved first (in a different language) by Kuhn.[10] We approach it by the following two methods.

THE LIMIT OF UNIFORM TRANSLATION

Let us consider a Rouse chain to which we apply certain external forces ($\mathbf{f}_{en}$ on bead n). This would correspond physically to sedimentation or to electrophoresis. Eq. (VI.5) is modified:

$$\frac{\partial \mathbf{r}_n}{\partial t} = \mu\left[\mathbf{f}_{en} + \frac{3T}{a^2}\frac{\partial^2 \mathbf{r}}{\partial n^2}\right] \tag{VI.10}$$

Consider the case of equal forces on all beads, $\mathbf{f}_{en} = \mathbf{f}_e = \mathbf{f}_{tot} N^{-1}$ (where $\mathbf{f}_{tot} = \sum_n \mathbf{f}_{en}$ is the total force). Then eq. (VI.10) has a solution with uniform velocity and vanishing internal forces ($\partial^2\mathbf{r}/\partial n^2 = 0$).

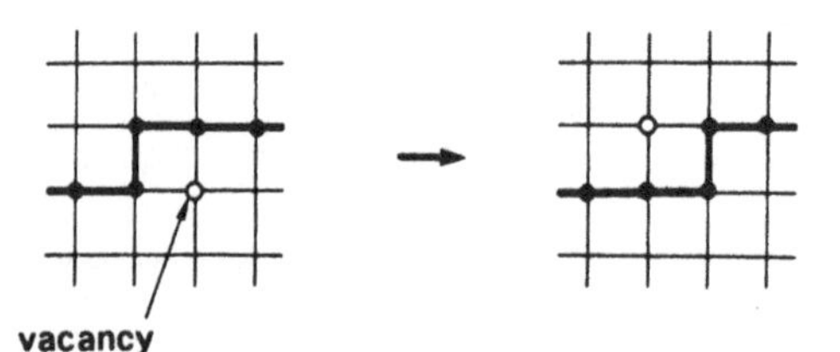

Figure VI.3.

$$\frac{\partial \mathbf{r}_n}{\partial t} = V = \mu N^{-1} \mathbf{f}_{tot} \tag{VI.11}$$

Thus, in the Rouse model, μN^{-1} is the overall mobility of the chain. Consider now a chain (floating in a solvent) with very high conformational barriers. Even if it behaves as a completely frozen shape (a twisted piece of rigid wire), it will show a mobility in the solvent, and this mobility will depend only on the solvent viscosity, not on internal barriers. Thus, in a liquid matrix, internal barriers are completely irrelevant for the uniform translation properties. This remains true for the first Rouse modes if N is large enough. The point is that in the first mode ($\mathbf{r}_n \sim \cos(\pi n/N)$ any fraction of the chain (from n to m, with $|n - m| \ll N$) moves essentially as if it were in uniform translation.

UNIFORM DEFORMATION

We now present a different (and more quantitative) argument for the Kuhn theorem. Starting with an ideal chain, we attach one end at the origin and pull the other end, imposing a certain velocity $\mathbf{V}_e$ (subscript e is extremity) (Fig. VI.4):

$$\frac{\partial \mathbf{r}_{N+1}}{\partial t} = \mathbf{V}_e$$

What are the friction forces involved (in a Rouse model)? The solvent forces are simple. The velocity for the n-th monomer is

$$\frac{\partial \mathbf{r}_n}{\partial t} = \frac{n}{N+1}\mathbf{V}_e \cong \frac{n}{N}\mathbf{V}_e$$

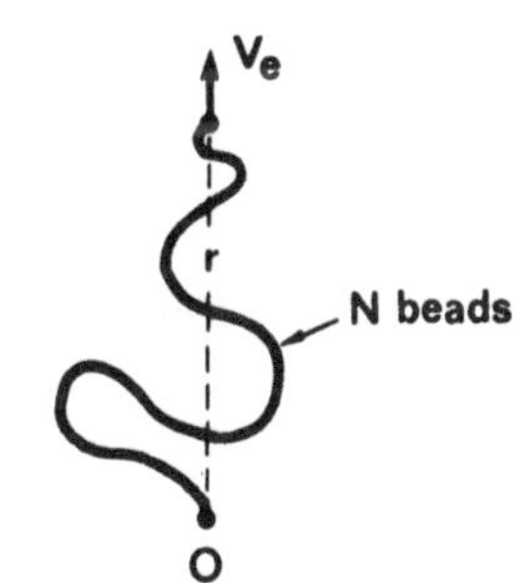

Figure VI.4.

and the total solvent friction force is (in terms of $\zeta_s = 1/\mu$: friction coefficient per bead)

$$\mathbf{f}_s = \sum_n \zeta_s \frac{\partial \mathbf{r}_n}{\partial t} = \frac{N}{2} \zeta_s \mathbf{V}_e \qquad \text{(VI.12)}$$

Let us estimate the other friction force $\mathbf{f}_i$ arising from internal barriers (omitting the solvent). It is convenient to think in terms of a given $\mathbf{f}_i$ and to look at what would be the resultant $\mathbf{V}_e$. If one chain contained only two monomers, which can be in two conformations (Fig. VI.5), we would expect

$$\mathbf{V}_e \cong \zeta_i^{-1} \mathbf{f}_i \quad (N \text{ small}) \qquad \text{(VI.13)}$$

where ζ_i contains an activation energy associated with the barriers between conformations (i) and (ii). Let us now turn to a chain with many units (N large) and impose a given $\mathbf{f}_i$ at the end. This tension is transmitted all along the chain. All monomers in a favorable (kink) position have a chance to stretch; thus, the overall stretching $\mathbf{V}_e$ is the sum of N contributions [eq. (VI.13)]

$$\mathbf{V}_e = N \zeta_i^{-1} f_i \quad (N \text{ large}) \qquad \text{(VI.14)}$$

Solving for the force, we have

$$\mathbf{f}_i = N^{-1} \zeta_i \mathbf{V}_e \qquad \text{(VI.15)}$$

We can now compare the two types of friction [eqs. (VI.12, VI.15)]. The ratio is

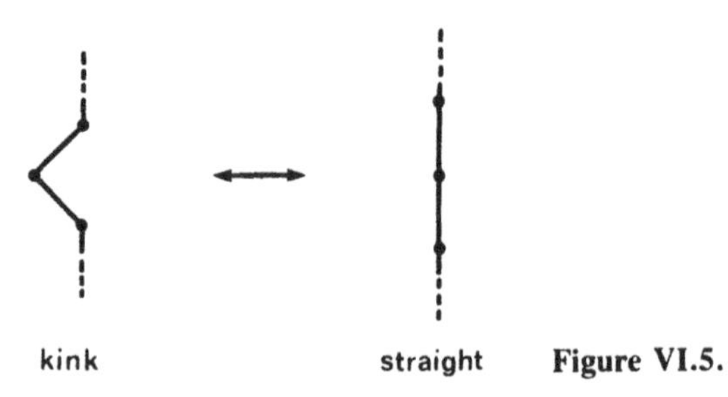

Figure VI.5.

$$\frac{f_i}{f_e} \cong \frac{\zeta_i}{\zeta_s} \frac{1}{N^2} \qquad \text{(VI.16)}$$

This is the precise form of the Kuhn theorem. Whenever N is larger than $(\zeta_i/\zeta)^{1/2}$, internal friction is negligible. [The detailed form of eq. (VI.16) is specific for the Rouse model, but similar results (with slightly different power laws) occur in more realistic situations.]

VI.1.3. Critique of the mode concept

The three assumptions (listed in Section VI.1.1) that define the Rouse model are often unacceptable. The assumption of *localized responses* is not correct because of "backflow" effects. Whenever we apply a force $\mathbf{f}_n$ to one monomer in a fluid, the result is a distorted velocity field in the whole fluid. This "backflow" decreases only slowly with distance (like $|\mathbf{r} - \mathbf{r}_n|^{-1}$. It drives other monomers into motion. The net result is that the mobility matrix μ_{nm}, introduced in eq. (VI.4), now becomes a very slowly decreasing function of the chemical interval $|n - m|$; this effect profoundly modifies the mode structure.[4]

The assumption of *ideal chain elasticity* is incorrect in a good solvent. As discussed in eq. (I.45), the spring constant of a swollen chain is much smaller than the spring constant of an ideal chain. The resulting corrections have been incorporated only recently into the theory.[5,11]

The assumption of *phantom chain* behavior amounts to neglect of any effect of knots along the chain. This may be shown to be correct in good solvents, but it may become more dangerous in a Θ solvent. In the latter case the coils are much more compact, and they tend to be more knotted.

For a long time it was considered that for Θ solvents the theory was in good shape (after incorporation of the backflow corrections). However, when we think of knots, this is less obvious. Single chain dynamics in a Θ solvent may be very complex[12] and are not discussed here.

On the whole, we are not even sure that the very concept of *modes* retains a fundamental validity when the three corrections are included. On the original Rouse equation [eq. (VI.5)], modes emerged naturally because the equation was *linear*. However, if we incorporate the backflow corrections properly, the mobility matrix μ_{nm} becomes a function of the distance $|\mathbf{r}_n - \mathbf{r}_m|$; the equation is then nonlinear, and all modes get mixed. The result is seen more easily on what experts in mechanics call a spectrum of relaxation times (or rates).[13] This is shown in Fig. VI.6.

The sharp peaks correspond to a mode picture. The actual spectrum for

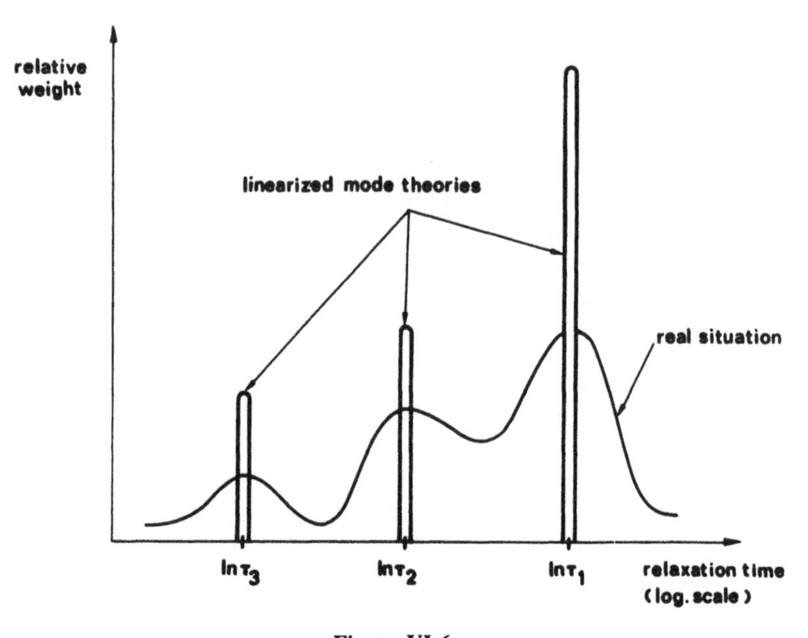

Figure VI.6.

dilute chains is not well known experimentally (because of the weak signals involved) but is probably much more like the smooth curve. It is not even certain that any bumps remain on the curve. Many nonlinearities can be responsible for the broadening—they may come not only from the backflow terms but also from the excluded volume interactions and possibly from knot formation.

This discussion may seem to be negative and discouraging. However, the situation is not completely desperate; scaling laws come to our help. One important aspect of scaling laws is that they ignore many unimportant details, such as the precise shape of the relaxation spectrum in Fig. VI.6, and they give predictions which hold independently of a specific model. In the following sections we present an overview of the "dynamical scaling laws" at the present level of understanding. We emphasize, of course, the limitations. In all branches of statistical physics, dynamical scaling is more complex and less universal than static scaling.[14] Macromolecules are no exception. However, the experimental situation has progressed significantly in the past few years, thanks to the "photon beat method" which probes chain motions in a convenient frequency range ($1-10^{26}$ cycles). Here we insist more on these data than on conventional measurements, which have been fully reviewed by Ferry.[13]

VI.2. Dynamic Scaling in Good Solvents

VI.2.1. The Kirkwood approximation for chain mobility

The mobility μ_{chain} of one chain is measured in experiments (such as sedimentation, electrophoresis, etc.) where a given external force $\mathbf{f}_{tot}$ is applied to the chain, and the resulting drift velocity $\mathbf{V}$ is measured. At low forces (such that the chain shapes are unperturbed)

$$\mathbf{V} = \mu_{chain}\, \mathbf{f}_{tot} \tag{VI.17}$$

Another, related, parameter, is the diffusion constant D of the chain; it is measured by direct monitoring of concentration profiles or more conveniently by inelastic scattering of laser light. Diffusion is related to mobility through the Einstein formula

$$D = \mu_{chain}\, T \tag{VI.18}$$

A good, rigorous, starting point for the theoretical discussion of D is the formula relating D to the spontaneous fluctuations of the velocity V (due to Brownian motion). In three dimensions

$$D = \frac{1}{3}\int_{t_1}^{\infty} \langle\, \mathbf{V}(t_1)\cdot \mathbf{V}(t_2)\rangle\, dt_2 \tag{VI.19}$$

where $\langle\ \rangle$ represents a thermal average, $\mathbf{V}(t)$ is the velocity of the center of gravity of the chain and can be rewritten as an average over all monomers

$$\begin{aligned} \mathbf{V}(t) &= \frac{1}{N}\sum_n \mathbf{v}_n \\ &= \frac{1}{N}\int c(\mathbf{r},t)\mathbf{v}(\mathbf{r}t)d\mathbf{r} \end{aligned} \tag{VI.20}$$

where c is the local monomer concentration and $\mathbf{v}$ is the *local* velocity ($\mathbf{v}$ is the monomer velocity, but we can also interpret it as the solvent velocity at the same point*). Inserting eq. (VI.20) into eq. (VI.9), we arrive at

*If we think of the monomer as a small sphere floating in a continuous liquid, the velocity of the sphere is exactly equal to that of the liquid on its surface.

$$D = \frac{1}{3} N^{-2} \int_{t_1}^{\infty} \langle c(\mathbf{r}_1 t_1)\ c(\mathbf{r}_2 t_2)\ \mathbf{v}(\mathbf{r}_1 t_1)\ \mathbf{v}(\mathbf{r}_2 t_2) \rangle\ d\mathbf{r}_1\ d\mathbf{r}_2\ dt_2 \qquad \text{(VI.21)}$$

At this point we introduce the first approximation. We split the correlation function $\langle cc\ vv \rangle$ into a concentration part and a velocity part. In modern jargon this corresponds to a "mode-mode coupling theory," which has been applied with great success to binary mixtures near their critical point by Kawasaki[14] and Ferrell.[15] However, for the present problem, the idea (in a different language) goes back to Kirkwood and Risemann.[16]

We make a further (minor) simplification. We assume that the essential time dependence is contained in the $\langle vv \rangle$ part of the correlation, while the $\langle cc \rangle$ part may be taken at equal times. This may be justified by a detailed study of the $\langle cc \rangle$ dynamics, along the lines of Section VI.2.2.

$$\langle c\ (\mathbf{r}_1 t_1)\ c\ (\mathbf{r}_2 t_2) \rangle \rightarrow \langle c\ (\mathbf{r}_1 t_1)\ c\ (\mathbf{r}_2 t_1) \rangle = \frac{N}{\Omega} g(\mathbf{r}_1 - \mathbf{r}_2) \qquad \text{(VI.22)}$$

Here $g(r)$ is the static pair correlation discussed in Chapter I and normalized always by $\int g(\mathbf{r}) d\mathbf{r} = N$. Ω is the total volume allowed for the chain, and the factor N/Ω ensures that if we integrate eq. (VI.22) over $\mathbf{r}_1$ and $\mathbf{r}_2$, we get N^2 as desired.

The velocity correlation is now calculated for a *pure* solvent—i.e., we omit any effect of the polymer on $\langle vv \rangle$. This is the central approximation. For an incompressible, viscous fluid of viscosity η_s, a Fourier component v_q of the velocity is ruled by the equations

$$\frac{\partial}{\partial t}(\rho \mathbf{v}_q) + \eta_s q^2 \mathbf{v}_q = 0 \qquad \text{(VI.23)}$$

$$\mathbf{q} \cdot \mathbf{v}_q = 0 \qquad \text{(VI.24)}$$

where ρ is the density. Eq. (VI.23) is the Navier-Stokes equation.* Eq. (VI.24) says that the velocity is transverse (div $\mathbf{v} = 0$). From eq. (VI.23) we get for each transverse component

$$\langle v_{-q}(0)\ v_q(t) \rangle = \langle v_{-q}(0)\ v_q(0) \rangle \exp -(\eta_s q^2 t/\rho) \qquad \text{(VI.25)}$$

and the equal times average $\langle v_{-q} v_q \rangle$ is derived from the equipartition theorem for kinetic energy

*There is no pressure term p in eq. (VI.23) because we are interested only in transverse velocities ($q v_q = 0$) while ∇p is longitudinal.

$$\frac{1}{2}\,\rho\,\langle v_{-q} v_q\rangle = \frac{1}{2}\,T \tag{VI.26}$$

Finally, integrating eq. (VI.25) over time, using eq. (VI.26) and transforming to real space, one arrives at

$$\frac{1}{T}\int_0^\infty \langle v_\alpha(00)\; v_\beta(\mathbf{r}t)\rangle\, dt = \frac{1}{8\pi\,\eta_s r}\left[\delta_{\alpha\beta} + \frac{r_\alpha\, r_\beta}{r^2}\right] = \mathfrak{T}_{\alpha\beta}(\mathbf{r}) \tag{VI.27}$$

Here α and β are component indices: α, $\beta = x,y,z$.

The right side of eq. (VI.27) is often called the Oseen tensor $\mathfrak{T}_{\alpha\beta}$. The Oseen tensor gives the velocity response of a fluid at point r when a weak force is applied at the origin. As usual this response function can be expressed in terms of correlation functions; hence the equality, eq. (VI.27).

Having discussed the $\langle vv\rangle$ correlation, we can now return to the basic formula for D, integrate over r_1 (obtaining one factor Ω), and we arrive at

$$\begin{aligned} D &= \frac{1}{3N}\int d\mathbf{r}\; g\,(\mathbf{r})\; T\sum_\alpha \mathfrak{T}_{\alpha\alpha}\,(r) \\ &= N^{-1}\int d\mathbf{r}\; g\,(\mathbf{r})\frac{T}{6\pi\eta_s r} \end{aligned} \tag{VI.28}$$

Eq. (VI.28) expresses a dynamic quantity in terms of static correlations. This is the great achievement of these mode–mode coupling methods. We can apply eq. (VI.28) for good or for poor solvents, using the corresponding discussion for $g(\mathbf{r})$ in Chapter I. The essential points are:

(i) The function, g, diverges only mildly at the origin. For example, in a good solvent $g \sim r^{-4/3}$. Thus the integrand has the form

$$\int 4\pi r^2\, dr\, \frac{1}{r^{4/3}}\frac{1}{r}$$

and there is no dangerous singularity near $r = 0$. At large r the function g drops rapidly, and the convergence is also good.

(ii) The pair correlation g obeys a scaling law of the form

$$g(r) = \frac{N}{R^3}\,\tilde{g}\left(\frac{r}{R}\right) \tag{VI.29}$$

which is equivalent to the forms quoted in Chapter I. Here $R = R_0$ in a Θ solvent, and $R = R_F$ in a good solvent.

Then, changing variables from r to $\mathbf{x} = \mathbf{r}/R$ in eq. (VI.28) we arrive at

$$D = \frac{T}{6\pi\eta_s R}\int dx\tilde{g}(x)\,x \tag{VI.30}$$

$$= \text{constant}\,\frac{T}{6\pi\eta_s R} \tag{VI.31}$$

This may be expressed in terms of a chain friction coefficient (the inverse of the mobility), which is:

$$\zeta_{tot} = \mu_{tot}^{-1} = \frac{T}{D} \cong 6\pi\eta_s R \tag{VI.32}$$

very much like for a solid sphere of radius R. We conclude that in the Kirkwood approximation the chain has a hydrodynamic radius which is proportional to its radius of gyration.

Experimentally, one finds that the hydrodynamic radius does scale like $N^{1/2}$ in Θ solvents, but in good solvents the situation is less clear. Diffusion or sedimentation experiments on polystyrene in toluene[17] give $D \sim N^{-\nu_{app}}$ where the apparent exponent ν_{app} is of order 0.53–0.54. More recent photon beat experiments[18] give $\nu_{app} = 0.55 \pm 0.02$ for polystyrene in benzene.

Two types of explanations have been proposed to explain the difference between ν_{app} and ν:

(i) A fundamental flaw in the Kirkwood approximation. For the analog problems in phase transitions, the Kawasaki-Ferrell exponents are only approximate because the viscosity of the fluid itself is renormalized. To include this possibility, some authors have written[19]

$$\mu_{tot} \sim N^{(z-2)\nu}$$

where z is a "dynamic scaling exponent";[14] the reason for this notation will become apparent later [eq. (VI.49)]. The data which we quoted on polystyrene would suggest $z \sim 2.9$ (i.e., $z < 3$). However, this is surprising; one would always expect the effective viscosity to be increased by the presence of polymer and thus the mobility to be below the Kirkwood result. This corresponds to $z > 3$. A theoretical inequality obtained under rather general assumptions by des Cloizeaux[20] also gives $z > 3$. Thus, this first explanation is doubtful.

(ii) Technical complications. Many "good solvents" of polystyrene are in fact moderately good, with values of the Flory parameter χ not much lower than 1/2, or equivalently with $u = v/a^3 = 1 - 2\chi$ much smaller than unity. We know from Chapter III that in such a case, the chain is still ideal at short distances ($r < au^{-1}$) and is swollen only at large scales ($r > au^{-1}$). Des Cloizeaux and Weill[21] have suggested that, in such a case, the average of $1/r$ coming into eq. (VI.28) may require very large values of N to reach its asymptotic form (proportional to $1/R_F \sim N^{-\nu}$). On the other hand the radius of gyration (which is measured in static experiments) is based on an average of r^2, which is much more sensitive to the behavior at large scales where the scaling laws for swollen chains hold. This may well explain the discrepancy between static and dynamic measurements of the coil size $R(N)$. However, to prove the point in detail will require careful experiments at variable χ and high N. In the following discussion we remain with the simple Kirkwood approximation.

VI.2.2. Inelastic scattering of light

The principle of these experiments is simple. A strictly monochromatic light beam (wavelength λ, frequency ω_0) is scattered by the polymer solution. The scattering angle is θ, and the scattering wave vector is $4\pi/\lambda$ (sin $\theta/2) = q$. Because of the motions in the scattering system, the outgoing beam contains all frequencies. We measure the intensity at one outgoing frequency $\omega_0 + \omega$ and call it $S(q, \omega)$. In the photon beat method the frequency shift ω may range from 1 to 10^6 cycles—a very convenient domain. What is measured is a time-dependent correlation function $\langle cc \rangle$ first introduced by Van Hove:[22]

$$S(\mathbf{q}\omega) = \frac{1}{2\pi} \int_{-\infty}^{\infty} dt \exp (i\omega t) \int d\mathbf{r} \exp (i\mathbf{q}\cdot\mathbf{r}) \langle c\ (00)\ c\ (\mathbf{r},t) \rangle \quad \text{(VI.33)}$$

When we fix $\mathbf{q}$, we define the spatial scales in which we are interested. They are of order q^{-1}. Then, at fixed $\mathbf{q}$, the plot of $S(\mathbf{q}\omega)$ versus ω gives us the power spectrum of one Fourier component $c_q = \int c \exp (i\mathbf{q}\cdot\mathbf{r})\, d\mathbf{r}$. Physically we may say that we probe the characteristic frequencies for fluctuations of size q^{-1}.

Turning now to the dilute chain problem in a good solvent, we distinguish between two regimes. The usual situation corresponds to $qR_F < 1$.

In this case the coils behave like point scatterers, and what we see is their overall Brownian motion, controlled by the diffusion coefficient D. A Fourier component c_q decays in time as $\exp(-Dq^2t)$, and this gives a Lorentzian form for S as a function of ω

$$S(q\omega) = \text{constant}\ \frac{\Gamma_q}{\Gamma_q^2 + \omega^2} \qquad \text{(VI.34)}$$

where the half-width at half maximum is

$$\Gamma_q = Dq^2 \qquad (qR_F < 1) \qquad \text{(VI.35)}$$

Using eq. (VI.31) we see that

$$\Gamma_q \cong \frac{T}{6\pi\eta_s R_F} q^2 \qquad \text{(VI.36)}$$

Let us now go to larger q values ($qR_F > 1$). This situation is difficult to achieve. It requires very long chains (for which R_F is larger than 1,000 Å). Such long chains—with typical molecular weights in the range of 10^7—are difficult to prepare and difficult to maintain without breaking. However, experiments have been carried out (on polystyrene in various solvents) with values qR_F of order 10 or more. What happens in that limit? The frequency plot (the plot of $S(q\omega)$ versus ω at fixed q) shows a broader peak, which is no longer Lorentzian and which has a certain half-width Γ_q.

The dynamic scaling assumption for Γ_q amounts to:

$$\Gamma_q = \frac{T}{6\pi\eta_s R_F} q^2\ \varphi_\Gamma\ (qR_F) \qquad \text{(VI.37)}$$

where $\varphi_\Gamma\ (x)$ is a dimensionless function of $x = qR_F$. At low x we must have $\varphi_\Gamma\ (x \to 0) \cong 1$ to comply with eq. (VI.36).

At large x, we probe portions of the chain (with a size q^{-1}) which are much smaller than the overall coil size. Then we expect to measure a characteristic frequency Γ_q which becomes independent of N (i.e., independent of end effects on the chain). This requires that

$$\varphi_\Gamma\ (x \to \infty) \cong x \qquad \text{(VI.38)}$$

Thus at high q we expect a width

$$\Gamma_q = \text{constant}\ \frac{Tq^3}{\eta_s} \quad \text{(Kirkwood approximation)} \qquad \text{(VI.39)}$$

A result of this form was first found theoretically on a specific model (ideal chains plus backflow),[23] but its range of validity is much larger than was suspected at the time. Experimentally, for polystyrene in good solvents, the plots of Γ_q versus q indicate[18] (Fig. VI.7)

$$\Gamma_q \sim q^{2.85 \pm 0.05} \qquad \text{(VI.40)}$$

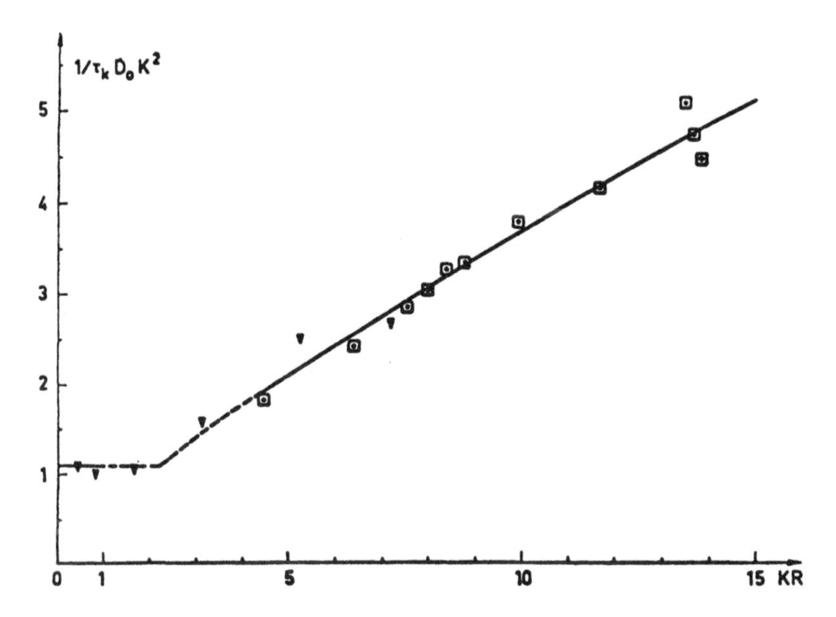

Figure VI.7.
Characteristic width of the photon beat spectrum $1/\tau_K$ of polystyrene solutions as a function of the wave vector K. The width is normalized by the diffusion width $D_0 K^2$, and the wave vector is expressed through KR when R is the coil radius. The squares and the triangles correspond to two very different molecular weights, suggesting that there is indeed a universal scaling form. From M. Delsanti, Ph.D. Thesis, Orsay, 1978.

VI.2.3. The fundamental relaxation time

In the Kirkwood approximation a single chain has a hydrodynamic radius R which scales like its geometric radius:

$$\left.\begin{aligned} R &\sim N^{1/2}\ (\Theta \text{ solvent}, v = 0) \\ R &\sim N^{3/5}\ \left(\text{good solvent}, \frac{v}{a^3} \sim 1\right) \end{aligned}\right\} \qquad \text{(VI.41)}$$

To this radius is associated a diffusion coefficient $D = T/6\pi\eta_s R$. From D and R we can construct a characteristic time

$$\tau = \frac{R^2}{D} \cong \frac{\eta_s R^3}{T} \quad \text{(Kirkwood approximation)} \qquad \text{(VI.42)}$$

The scaling law for the inelastic scattering of light at a fixed wave vector q [eq. (VI.37)] is (within the Kirkwood approximation)

$$\Gamma_q \tau = (qR)^2 \varphi_\Gamma (qR) \qquad \text{(VI.43)}$$

where $\varphi_\Gamma(x) \rightarrow 1$ for $x \rightarrow 0$ and $\varphi_\Gamma(x) \rightarrow$ constant x at large x.

Now we want to obtain a more physical feeling for this fundamental time τ, which is the analog of the first relaxation time in the mode picture. A very useful qualitative model for understanding the meaning of τ is the dumbbell model introduced by Kuhn.[10,24] Here one does not look at all the variables $\mathbf{r}_1 \ldots \mathbf{r}_{n+1}$ giving the position of all beads on the chain, but one concentrates on a single variable—the total elongation $\mathbf{r} \equiv \mathbf{r}_{n+1} - \mathbf{r}_1$.

One then pictures the whole chain as a spring, with a certain elastic energy

$$F_{el} = \frac{1}{2} K r^2 \qquad \text{(VI.44)}$$

and with a certain friction constant ζ_{tot}; from a scaling point of view, ζ_{tot} and the inverse translational mobility μ_{tot}^{-1} have the same properties; they both deal with friction for long wavelength deformations of the chain. The friction force on the spring is $\zeta_{tot}\, \partial r/\partial t$, and this balances the elastic force

$$\zeta_{tot} \frac{\partial \mathbf{r}}{\partial t} = -K\mathbf{r} \qquad \text{(VI.45)}$$

This leads to a characteristic relaxation time for chain deformation

$$\tau = \zeta_{tot}/K \qquad \text{(VI.46)}$$

The various models of chain dynamics correspond to different assumptions of the spring constant and the friction constant.

If we choose $K \cong T/R_o^2$ (corresponding to an ideal, phantom, chain; see eq. (I.8)) and $\zeta_{tot} \sim N$ (additivity of individual frictions), we get the Rouse model and $\tau \sim N^2$.

If we incorporate backflow effects at the Kirkwood level, the friction is modified $\zeta_{tot} \sim 6\ \pi\eta_s\ R$. If the chain is still an ideal phantom chain ($R = R_o$), we get:

$$\tau \cong \frac{\eta_s R_o^3}{T} \tag{VI.47}$$

which we might call the Zimm formula since Zimm first computed the detailed structure of modes for this case.[4]

If we consider a real chain in good solvent, we must use $\zeta_{tot} = 6\pi\eta_s R_F$ within the Kirkwood approximation, but the spring constant is also changed.[11] Eq. (VI.44) tells us that $K \sim T/R_F^2$. Then we get

$$\tau \cong \frac{\eta_s R_F^3}{T} \tag{VI.48}$$

Note that eqs. (VI.47, VI.48) agree with the general scaling formula, eq. (VI.42).

If we want to go beyond the Kirkwood approximation and incorporate a new dynamical exponent z (introduced on p. 170), we must alter the friction term and use the last eq. on page 176: $\zeta_{tot} \sim R_F^{z-2}$. This gives

$$\tau \sim R_F^z \tag{VI.49}$$

But, as pointed out before, there is no convincing argument for a z value different from 3.

We end this discussion of the fundamental relaxation time τ by remarks on the structure of eq. (VI.42):

(i) Apart from coefficients, τ coincides with the relaxation time associated with rotations of a solid Brownian sphere of radius R.[25] Thus rotation and deformation have the same characteristic time in solution. This simplification is the basis of the simple scaling laws that we found.

(ii) The orders of magnitude of τ are interesting. Typically for $R \sim 500$ Å and $\eta_s \sim 1$ poise, the time is around 10^{-5} sec and is suitable for mechanical studies.

(*iii*) The N dependence of τ (in a good solvent, within the Kirkwood approximation) is $\tau \sim N^{9/5}$. This is not very far from the Rouse prediction ($\tau \sim N^2$). Thus the corrections due to the backflow and to the excluded volume effects cancel to a large extent. This may explain some unexpected successes of the Rouse model in interpreting certain properties of dilute solutions.

(*iv*) An interesting determination of τ uses dielectric properties in one special case—namely, when the chain carries longitudinal dipoles along the backbone, as shown in Fig. VI.8. Usually each unit will also have a

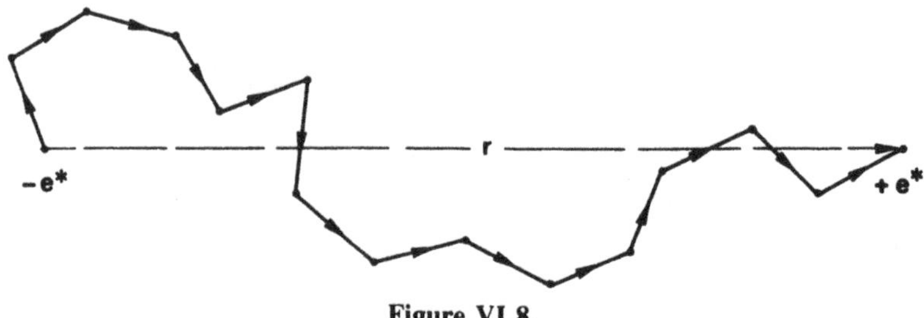

Figure VI.8.

dipole component transverse to the backbone, but these components rotate very freely and relax in a microscopic time, $\sim 10^{-10}$ sec. Thus, they are not seen in dielectric absorption measurements at radiofrequencies. There is an overall dipole $\mathbf{P} = e^\star \mathbf{r}$, where $\mathbf{r}$ is the end-to-end elongation, and $e^\star$ has the dimension of a charge. The relaxation of the dipole $\mathbf{P}$ is controlled mainly by the fundamental relaxation time τ. Measurements of τ using this method have been performed on caprolactone polyester[26]

$$\left|\!-CH_2-CH_2-CH_2-CH_2-CH_2-\underset{\underset{O}{\|}}{C}-O-\right|_N$$

They do show a good relationship between τ and $R_F{}^3$—the latter being inferred from viscosity measurements also in dilute solutions (discussed in the next paragraph).

VI.2.4. Static viscosity of dilute solutions

THE HARD SPHERE PICTURE

From the Kirkwood approximation we know that a dilute solution of polymers (with monomer concentration c or chain concentration c/N) behaves hydrodynamically like a collection of solid spheres of radius R. A calculation by Einstein[27] suggests that the viscosity η of the solution has the form

$$\eta = \eta_s [1 + 2.5 \psi] \qquad \text{(VI.50)}$$

where

$$\psi \sim \frac{c}{N} R^3 \qquad \text{(VI.51)}$$

is the volume fraction occupied by the spheres. This result is interesting because it implies that

(i) The viscosity η may be increased significantly even at low concentrations ($\psi = 1$ for $c \sim c^\star$).

(ii) The relative change $(\eta - \eta_s)/\eta_s$, at a given c and N, measures R^3 and thus gives an accurate value of the hydrodynamic radius. These viscosity measurements are cheap, require only small amounts of material, and measure a microscopic size directly!

Results are usually given in terms of

$$[\eta] \equiv \frac{\eta - \eta_s}{\eta_s c} \cong \frac{R^3}{N} \qquad \text{(VI.52)}$$

Thus the prediction is $[\eta] \sim N^{1/2}$ in Θ solvents and $[\eta] \sim N^{4/5}$ in good solvents. The experimental data agree rather well with the prediction in Θ solvents. In good solvents the situation is not as clear,[28] and the apparent exponent is often slightly smaller than 4/5. It is again tempting to interpret the deviations in terms of the des Cloizeaux-Weill effect.[21]

THE DUMBBELL PICTURE

Another qualitative approach to the viscosity is based on a direct calculation of stresses in a dilute system of springs. Counting the forces on the two sides of a given surface in the fluid (Fig. VI.9) one finds that the contribution in stress due to the springs is*

$$\delta\sigma_{\alpha\beta} = \frac{c}{N}\langle F_\alpha r_\beta \rangle = \frac{1}{2}\frac{c}{N}\langle F_\alpha r_\beta + F_\beta r_\alpha \rangle \qquad \text{(VI.53)}$$

*See for instance R. Bird, R. Armstrong, O. Hassager, *Dynamics of Polymeric Liquids*, Wiley, New York, 1976.

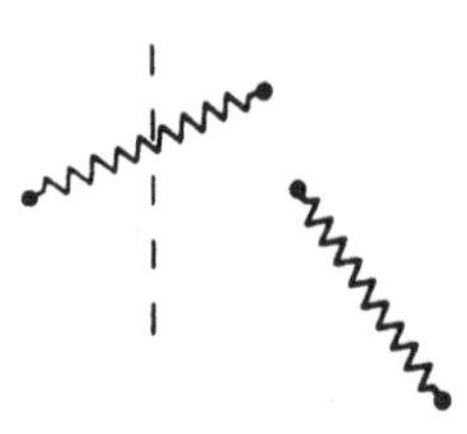

Figure VI.9.

A fictitious dividing plane (dotted line) separates the solution into two halves. The solute "springs," which contribute to the stress on this plane, are those which intersect the plane, as shown on the figure.

where $\sigma_{\alpha\beta}$ is the force (along direction α) per unit area (normal to direction β), and $F_\alpha = Kr_\alpha$ is the elastic force per spring (c/N being the number of springs per cm^3). Let us apply this basic equation to the case of a flow of the solvent with velocities

$$v_\alpha = \sum_\gamma s_{\alpha\gamma} \boldsymbol{r}_\gamma + \boldsymbol{\nu}_\alpha (0) \tag{VI.54}$$

[$\boldsymbol{\nu}$ (0) being the velocity at the origin of the dumbbell]. In eq. (VI.54) $s_{\alpha\beta}$ is the shear rate tensor. As noticed first by Kramers, it is convenient to consider flows that have no rotational component ($s_{\alpha\beta} = s_{\beta\alpha}$). For such flows the molecules reach a constant elongation. In the flow field the equation of motion [eq. (VI.45)] for the dumbbell is altered to

$$\zeta_{tot}\left(\frac{dr_\alpha}{dt} - v_\alpha\right) = - Kr_\alpha = - F_\alpha \tag{VI.55}$$

We now multiply eq. (VI.55) by r_β, symmetrize, and average. The term

$$\langle r_\alpha \frac{d}{dt} r_\beta + r_\beta \frac{d}{dt} r_\alpha\rangle = \frac{d}{dt}\langle r_\alpha r_\beta\rangle$$

vanishes since $\langle r_\alpha\ r_\beta\rangle$ is independent of time (Kramers remark). Then we are led to

$$\delta\sigma_{\alpha\beta} = \frac{c}{N}\zeta_{tot}\ s_{\alpha\gamma}\langle r_\gamma r_\beta\rangle \tag{VI.56}$$

At this stage, for small s, we may take the average $\langle r\ r\rangle$ as equal to its value in the absence of flow $\langle r_\gamma r_\beta\rangle = 1/3\ \delta_{\gamma\beta} R_F{}^2$, and we reach $\delta\sigma_{\alpha\beta} = \delta\eta s_{\alpha\beta}$, with an increment in viscosity

$$\delta\eta \cong \frac{c}{N}\zeta_{tot}\ R_F{}^2 \tag{VI.57}$$

We may also replace R_F^2 by T/K and express our result in terms of the relaxation time τ

$$\delta\eta \cong \frac{cT}{N}\tau \qquad \text{(VI.58)}$$

The two forms [eqs. (VI.57, VI.58)] are interesting.

Eq. (VI.58) expresses the viscosity increment as the product of an elastic modulus (cT/N) by a relaxation time τ. This type of relationship will often be convenient for more complex situations. In the Kirkwood approximation eq. (VI.57) does coincide with eqs. (VI.50, VI.51). If we try to go beyond the Kirkwood approximation and insert a dynamical exponent z we see from the last eq. on page 176 and eq. (VI.57) that

$$\delta\eta \sim \frac{c}{N}R^z \qquad \text{(VI.59)}$$

Values of z derived by various methods (for polystyrene in relatively good solvents) are listed below (after Delsanti[29]).

Diffusion D via light scattering[17]	$z = 2.91$
Mobility (via sedimentation coefficients)[17]	$z = 2.88$
$\Delta\omega_q$ at large q [18]	$z = 2.85$
Viscosity $\delta\eta$ [28]	$z = 2.82$
Other mechanical data (on τ) [30]	$z = 2.78$

VI.2.5. Frequency dependence of viscosities

At finite frequencies ω the viscosity increase due to solute coils $\delta\eta = \eta - \eta_s$ becomes dependent on ω and complex. Mechanical data on this function are difficult to obtain in the dilute regime, but some results have been achieved.[1] Their classical analysis was in terms of Zimm modes,[4] allowing for backflow effects but ignoring any excluded volume effects. We present here a more general scaling picture.

At low frequencies $\omega < 1/\tau$, $\delta\eta(\omega)$ is only weakly different from the static value discussed earlier [eq. (VI.59)]

$$\delta\eta \sim \frac{c}{N}R_F^z \sim cN^{\nu z-1} \qquad \text{(VI.60)}$$

At higher frequencies the complex $\delta\eta$ will have the structure

$$\delta\eta(\omega) = \delta\eta(0)\varphi_\eta(\omega\tau) \qquad \text{(VI.61)}$$

where $\varphi = \varphi' + i\varphi''$ is a complex function of $x = vt$ with the following limiting properties

$$\varphi\ (x \to 0) = 1 + (\text{constant})\ ix + \ldots \tag{VI.62}$$

$$\varphi\ (x \to \infty) = \text{constant}\ (ix)^{-m_\eta} \tag{VI.63}$$

[The fact that ix (rather than x) enters into eqs. (VI.62, VI.63) means that all dynamical equations involve $\partial/\partial t = i\omega$]. The power law structure of eq. (VI.63) is restricted by the following requirement. When $\omega \gg 1/\tau$, the only portions of the chains which can respond to the mechanical perturbation are subunits much smaller than the whole chain (the number of beads per subunit g is such that the time τ_g for *one subunit* is of order ω^{-1}). Then $\delta\eta(\omega)$ becomes independent of the total chain length (of N). Thus there must exist a simple law: $\delta\eta \cong \delta\eta(0)\ (i\omega\tau)^{m_\eta}$, where the exponent m_η ensures that the powers of N entering in $\delta\eta(0)$ and τ^{m_η} cancel. This means that $\nu z - 1 - m_\eta\ \nu z = 0$

$$m_\eta = 1 - \frac{1}{\nu z} \tag{VI.64}$$

For simplicity, let us now restrict ourselves to the Kirkwood approximation ($z = 3$) and start with a Θ solvent ($\nu = 1/2$). Then we expect $m_\eta = 1/3$, in rather good agreement with some data on polystyrene in Θ solvents.[1,13] If we now turn to a good solvent ($\nu = 3/5$), the prediction is $m_\eta = 4/9$—i.e., a slightly larger exponent. The data are not quite conclusive.[1,13]

VI.3. Special Flow Problems

VI.3.1. Deformation in strong extensional flows

Dilute solutions of polymer coils have special rheological properties: they tend to "thread." Microscopically, the chains become highly elongated in certain shear flows. The most spectacular effects occur in "extensional" flow fields such as

$$\begin{aligned} v_x &= sx \\ v_y &= -\ s/2\ y \\ v_z &= -\ s/2\ z \end{aligned} \tag{VI.65}$$

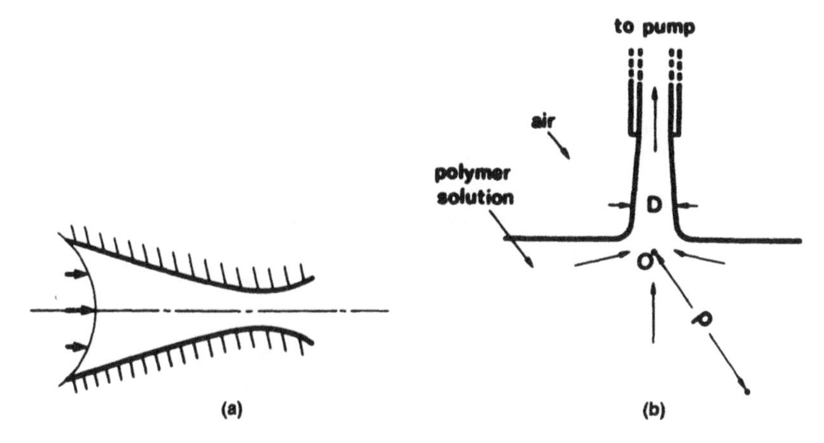

Figure VI.10.
Example of (mainly) longitudinal shear flows: (a) convergent duct, (b) the "tubeless siphon." With dilute (10^{-4}) polymer solutions one can often arrive at a column ~ 10 cm high.

This pattern corresponds to what is often called longitudinal shear. Note that div $\mathbf{v} = 0$, the flow occurs at constant density. Longitudinal shear flows occur in the axial region of a convergent duct (Fig. VI.10a) or at the entry of a capillary.

A nearly ideal situation of this type is found with the "tubeless siphon" of Fig. VI.10b. (This ascending column is easily obtained with polymer solutions because of their ability to thread.) The interest of this geometry, from the present point of view, is to provide us with a simple convergent flow (not perturbed by walls) in the bulk of the fluid *below* the siphon. (Chain behavior inside the siphon has been studied in experiments at Naples,[31] but the analysis ignored possible elongations before entry.)

In most practical realizations of longitudinal shear flows, such as (a) and (b) in Fig. VI.10, the coils have a finite transit time t; they do not remain permanently under shear. However, there are some exceptional cases where they remain trapped for a long time—e.g., near the center and in the midplane of the "four-roller experiment" shown in Fig. VI.11. This geometry has been studied at Bristol. Elongation of the molecules is important near the axis of exit yOy', and is detected* by optical birefringence measurements.[34]

We begin by discussing coils located near the center (O) of Fig. VI.11 where the gradient (s) experienced by one chain may be considered as time independent. This is exceptional but is simply a starting point. Later we

*Unfortunately, the present experiments had to be performed at concentrations $c \sim c^*$, where the picture in terms of separate coils begins to break down.

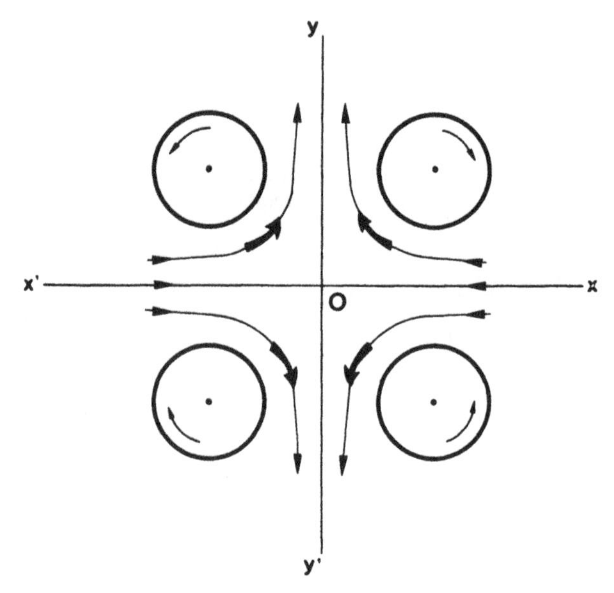

Figure VI.11.

shall proceed to cases where the transit times are comparatively short and where transient responses are essential.

Our discussion of strong flows is restricted to longitudinal gradients for the following reason. In the more familiar "transverse" situation:

$$v_x = sy$$
$$v_y = v_z = 0 \qquad \text{(VI.66)}$$

each fluid element *rotates* (at an average speed $s/2$); the coils also rotate and are extended only for a portion of each rotation cycle.[25] Thus extensional effects are less spectacular.

EFFECT OF PERMANENT GRADIENTS: COIL STRETCH TRANSITION

Let us consider one chain (with its center of gravity at the origin†) in the flow field [eq. (VI.65)] and assume that

(i) The chain remains in the same shearing flow (s = constant) for a long time t ($t \gg \tau$) where τ is the relaxation time [eq. (VI.48)] for the chain at rest.

†This eliminates an uninteresting overall translation of the chain.

(ii) The shear is strong; the dimensionless parameter $s\tau$ is much larger than unity.

We expect the chain to achieve a large end-to-end elongation **r** and to take the shape of a long cylinder. We discuss qualitatively the balance of forces in this elongated coil using the results of Chapter I for the elastic forces in the strong deformation regime, plus a simple hydrodynamic model for the friction on the cylinder.[33]

The elastic energy for a chain in a good solvent is, from eq. (I.48),

$$F_{el} \cong T\left(\frac{r}{R_F}\right)^{5/2} = T\,\lambda^{5/2} \qquad (\lambda > 1) \tag{VI.67}$$

where we have introduced a reduced deformation λ. Eq. (VI.67) holds if the chain is not fully extended—i.e., $r \ll Na$ or $\lambda < N^{2/5}$.

For the rotational flow discussed here, the friction forces can be derived from a potential, as noted first by Kramers.[35] An element of length dx of the cylinder is subjected to a force

$$dx\;\;[k\eta_s\; v_x(x)] = dx\;\;[k\eta_s\; sx] \tag{VI.68}$$

where k is a logarithmic factor, given in terms of the length r and diameter ξ of the cylinder, by the formula:[27]

$$k = \frac{2\pi}{\ln\,(r/\xi)} \tag{VI.69}$$

We are concerned only with scaling properties and treat k as a constant of order unity. The Kramers potential energy associated with the friction [eq. (VI.68)] is

$$\begin{aligned} -\int_{-r/2}^{r/2} dx\,\frac{1}{2}\,k\eta_s\,x^2 &= -\,\frac{k}{24}\,\eta_s\,sr^3 \\ &\cong -\,T\,s\tau\,\lambda^3 \end{aligned} \tag{VI.70}$$

Note the minus sign in eq. (VI.70). Large elongations lower the Kramers potential. Adding eqs. (VI.70, VI.67), we get the following scaling form for the energy

$$\frac{F_{tot}}{T} = \lambda^{5/2} - s\tau\,\lambda^3 \qquad (1 < \lambda < N^{2/5}) \tag{VI.71}$$

The resulting plot is shown in Fig. VI.12. At the high λ limit ($\lambda R_0 \sim Na$, corresponding to full elongation) the energy rises sharply above the value in eq. (VI.71). We have incorporated this extra feature in the figure. At the other end ($\lambda < 1$) the elastic energy is of order λ^2, while the Kramers term is $- s\tau\ \lambda$; we have also included these features.

Consider first the regime of large s. At strong $s\tau$ and intermediate elongations ($\lambda > 1$), the friction term dominates. There is no equilibrium elongation in this range, and the physical situation corresponds to nearly complete stretching ($r \cong Na$). Only in this region do the elastic forces become suddenly strong, and equilibrium can be reached.

Fig. VI.12 also shows the behavior expected at smaller s, which results from more extended calculations.[35] The plot is reminiscent of a phase transition; at values of s slightly larger than a critical value s_c, there are two energy minima: one at small λ (coil) and one at large λ (stretch). These minima are separated by a huge potential barrier, and the chain usually sticks to one and the same minimum. At large s ($s \gg s_c$) the "coil" minimum ceases to exist, and the chain must go to the stretched state (if the gradient s is applied over a long enough time). Related mechanical models

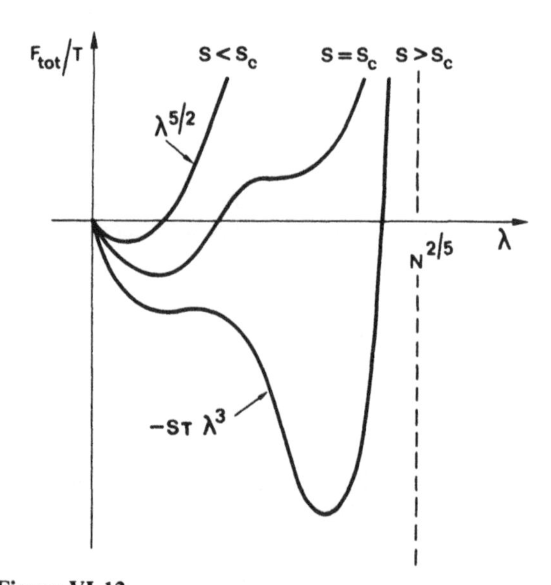

Figure VI.12.

Free energy versus relative elongation ($\lambda = r/R$) for a single polymer coil in a longitudinal shear flow.

have been analyzed by Hinch;[36] a discussion of the minima and of the barrier (for the more restrictive case of ideal chains) can be found in Ref. 35. Generalizations to other types of shear have been constructed mainly by the Bristol group.[32]

PHYSICAL BEHAVIOR UNDER TRANSIENT CONDITIONS

As noted, the permanent state discussed above is exceptional. Consider, for example, the convergent flow shown in Fig. VI.9b and more specifically a molecule approaching the tube along the axis. At a distance ρ from the center O (ρ much larger than the tube diameter D) the flow field is essentially radial,* with a velocity

$$v = \frac{J_s}{4\pi\rho^2} \qquad \text{(VI.72)}$$

where J_s is the total solvent flux entering the capillary. Eq. (VI.72) ensures the conservation of flux, $4\pi\rho^2 v = J_s$. The longitudinal gradient corresponding to eq. (VI.72) is

$$|s| = \left|\frac{dv}{d\rho}\right| = \frac{J_s}{2\pi\rho^3} \qquad \text{(VI.73)}$$

There is a certain critical radius $\rho^\star$ at which $s\tau = 1$

$$(\rho^\star)^3 \cong J_s\,\tau \qquad \text{(VI.74)}$$

The molecule arrives first in the region $\rho > \rho^\star$ (far to the left in Fig. VI.9b). It is seriously distorted only when it enters the sphere of radius $\rho^\star$. It then experiences strong deformations during its transit time t from $\rho = \rho^\star$ to $\rho \cong D$

$$t \sim \int_{D/2}^{\rho^\star} \frac{d\rho}{v(\rho)} \sim \frac{\rho^{\star 3}}{J_s} \sim \tau \qquad \text{(VI.75)}$$

Thus, *transit times are comparable with molecular relaxation times.* In most cases the assumption of a stationary state discussed earlier in this section is not adequate.

*This is the main simplification introduced by the tubeless siphon as opposed to the entry of a capillary. However, our discussion remains qualitatively valid for the latter case (provided that the solution is dilute).

A very different picture has been extracted from simple calculations on transients.[37,38] As soon as the molecule enters the sphere of radius $\rho^\star$, it stretches exactly like the surrounding fluid. This implies a stretching law (for the long axis of the chain) of the form

$$r(\rho) = R_F \left(\frac{\rho^\star}{\rho}\right)^2 \tag{VI.76}$$

(since a fluid element has its transverse dimensions scaling like ρ and has a constant volume, its longitudinal dimension must scale like ρ^{-2}).

Eq. (VI.76) will be valid from $\rho = \rho^\star$ to $\rho \sim D$ if the final length $r|_{\rho \cong D}$ is still smaller than the fully stretched value (Na). We assume this in the following, and we then write

$$r_{final} \equiv r|_{\rho \cong D} \cong R_F \left(\frac{\rho^\star}{D}\right)^2$$
$$\cong R_F \, (s_0 \, \tau)^{2/3} \tag{VI.77}$$

where $s_0 \sim J_s/D^3$ is the shear rate at entry.

The exponent 2/3 in eq. (VI.77) does not reflect excluded volume effects but expresses simply a geometric property of three-dimensional convergent flows. *Birefringence* measurements on such convergent flows do *not* give profound insight into the chain behavior; they measure principally $\rho^\star$ or τ.

Mechanical measurements may be more interesting. In particular, the radial tensions σ_d near the entry ($\rho \sim D$) are expected to be larger. From a general formula [eq. (VI.53)]

$$\sigma_d = \frac{c}{N} fr|_{\rho \sim D} \tag{VI.78}$$

where c/N is the number of chains per cm^3, and f is the elastic force inside one chain, derived from the Fisher-Pincus analysis [eq. (I.49)]. This gives

$$\sigma_d = \frac{cT}{N} (s_0 \, \tau)^{5/3} \tag{VI.79}$$

Thus σ_d increases rather rapidly with s_0. This is the origin of the "threading" observed in rheological studies. Note that the exponent in eq. (VI.79) does depend on excluded volume effects.

Transport experiments would be of interest when D is extremely small ($D < R_F$). (Capillaries with such minute sizes can be prepared using particle track etching.) It should then be possible to see if the coils are sucked in the tube, or if they are too large. Daoudi and Brochard have proposed a simple answer to this problem.[39] They assume that the coils enter if and only if their transverse dimension ($r_\perp$) after affine deformation ($\rho \sim D$) becomes smaller than the tube diameter D. The affine law for $r_\perp$ (corresponding to eq. (VI.76)) is:

$$r_\perp (D) = R_F \frac{D}{\rho^\star} \qquad \text{(VI.80)}$$

Following this idea, we expect transport when $R_F < \rho^\star$. Returning to eq. (VI.74) for $\rho^\star$ and eq. (VI.48) for τ we find a critical current for the passage of coils

$$J_c \cong \frac{T}{\eta_s} \qquad \text{(VI.81)}$$

This critical current is independent of N (at large N). Heavy molecules are stretched over longer distances $\rho^\star$, and this compensates for their size. Note also that J_c is predicted to be independent of the pore diameter D.

What about the assumption of dilute chains in the above discussion, especially in regimes where $J < J_c$? Since the chains are stuck, we might expect an increase in chain concentration near the entry of the duct; this could alter the laws of entry. In fact, a calculation of the steady-state profile suggests that the increase in c is confined to a very small region (of radius comparable with the chain size) near the entry and is not essential.

VI.3.2. Dynamics of a chain inside a cylindrical pore

Consider the motions of a chain which is squeezed inside a very thin capillary of diameter D (as in Fig. I.5). We assume that the capillary has a constant (or very slowly varying) diameter D. Thus, the effects of longitudinal gradients discussed before are negligible. The static conformation properties of the confined chain were discussed in Section I.3b. Here, we want to discuss the mobility of the chain and its coupling to solvent flow. In the language of irreversible processes[40] there are two "fluxes" of interest: the solvent flux J_s and the polymer flux J_p. The conjugate "forces" are chemical potential gradients

$$-\frac{\partial \mu_s}{\partial x} = -\mu'_s \quad \text{and} \quad -\frac{\partial \mu_p}{\partial x} = -\mu'_p$$

(where x is measured along the tube axis). For weak forces we expect linear relationships between fluxes and forces

$$\begin{aligned} J_s &= L_{ss}\,(-\mu_s') + L_{sp}\,(-\mu_p') \\ J_p &= L_{ps}\,(-\mu_s') + L_{pp}\,(-\mu_p') \end{aligned} \tag{VI.82}$$

introducing three parameters L_{ss}, L_{pp}, and $L_{ps} \sim L_{sp}$ to describe all friction processes.[40] In the dilute limit, which we discuss here, the first coefficient L_{ss} is trivial; it describes Poiseuille flow of the solvent in a tube and is given by[27]

$$L_{ss} \cong \frac{D^4}{\eta_s w_s} \tag{VI.83}$$

where w_s is the molecular volume of the solvent.

To calculate the two other Onsager coefficients, L_{sp} and L_{pp}, we consider situations where a force $(-\mu'_p)$ is applied to each monomer while no force is applied on the solvent (μ_s = constant). The basic tool is the Oseen tensor $\mathfrak{T}_{\alpha\beta}(\mathbf{r}_1\, \mathbf{r}_2)$, giving the solvent velocity at point $\mathbf{r}_1$ due to a localized force $\mathbf{f}$ applied at point $\mathbf{r}_2$

$$v_\alpha(\mathbf{r}_1) = \sum_\beta \mathfrak{T}_{\alpha\beta}(\mathbf{r}_1\, \mathbf{r}_2)\, f_\beta \tag{VI.84}$$

where $\alpha,\beta = x, y, z$ represent components. In a bulk solution we saw that $\mathfrak{T}(\mathbf{r}_1\, \mathbf{r}_2)$ decreases like $1/r_{12}\,\eta_s$. In the present tube problem (with a solvent flow velocity v which must vanish at the tube walls) $\mathfrak{T}(\mathbf{r}_1\, \mathbf{r}_2)$ decreases again like $1/r_{12}$ at small distances ($r_{12} \ll D$). At larger distances ($r_{12} \gg D$) $\mathfrak{T}(\mathbf{r}_1\, \mathbf{r}_2)$ decreases much faster (ca. exponentially). The reason for this cutoff is qualitatively explained in Fig. VI.13, where we discuss a related, but simpler, problem: the potential (at point M) due to a localized charge (at a point O) between two conducting plates.

This electrostatic problem is (roughly) similar to a hydrodynamic problem (the calculation of $\mathfrak{T}$ (O, M)) because in the absence of walls the response function decays like r^{-1} in both cases.

With the walls, the potential at M is a superposition of the direct term from 0 and of image terms, from the points A,A′, B,B′, etc. These image sources ensure that the correct boundary conditions (no tangential component of electric field) are obeyed on the two conducting walls.

Clearly, if the distance (x) between the observation point (M) and the line of images is much larger than the interimage distance (D), the effects of different charges cancel and there is essentially no field left. The Oseen tensor requires a more complicated discussion (because it describes a *vector* response to a *vector* perturbation), but the physics is the same; there is a cutoff at $x \sim D$, and this is all that we need for the scaling

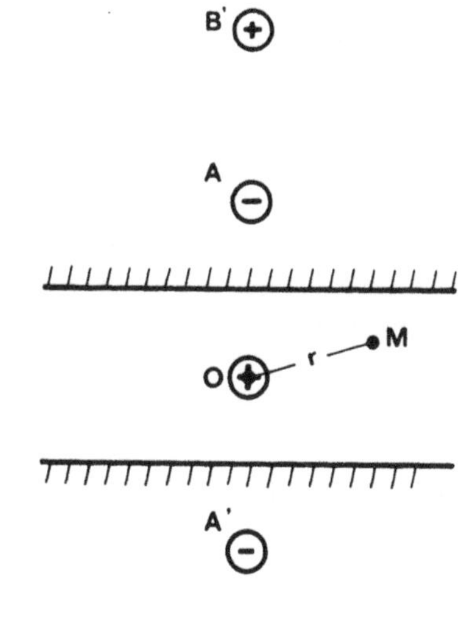

Figure VI.13.

discussion. For a component such as $\mathfrak{T}_{xx}$ $(\mathbf{r}_1\ \mathbf{r}_2)$ (which is the one of interest) we can write

$$\mathfrak{T}_x(\mathbf{r}_1, \mathbf{r}_2) = \begin{cases} \dfrac{1}{\eta_s r_{12}} & (r_{12} \ll D) \\ 0 & (r_{12} \gg D) \end{cases} \tag{VI.85}$$

Knowing this, we can compute simply the coefficient L_{sp}; to each monomer $(\mathbf{r}_n)$ is applied a force $-\mu'_p$ along x. The resulting solvent velocity at point $\mathbf{r}_1$ is

$$v_{sx}(\mathbf{r}_1) = \langle \sum_n \mathfrak{T}_{xx}(\mathbf{r}_1,\mathbf{r}_n)\rangle (-\mu'_p) \tag{VI.86}$$

where the sum is over all monomers in the solution. The average $\langle\ \rangle$ is to be taken over all locations and conformations of the chains. It gives

$$v_s(\mathbf{r}) = \int \mathfrak{T}(\mathbf{r}_1, \mathbf{r}_2)\, c(\mathbf{r}_2)\, d\mathbf{r}_2\, (-\mu'_p) \tag{VI.87}$$

where $c(\mathbf{r}_2)$ is the average concentration at $\mathbf{r}_2$. (The profile of $c(\mathbf{r}_2)$ in a cross-section is not flat, and $c(\mathbf{r}_2)$ is equal to c only after averaging over the cross-section.) Performing the integration, we get

$$v_s = \frac{1}{\eta_s D}\, cD^3\, (-\mu'_p) \tag{VI.88}$$

(since the integrand is nonvanishing only in a volume of order D^3).

Multiplying v_s by the cross-sectional area ($\sim D^2$), we obtain J_s and L_{sp}:

$$L_{sp} = J_s/(-\mu'_p) \cong \frac{cD^4}{\eta_s} \qquad \text{(VI.89)}$$

Let us now apply a similar argument to the calculation of L_{ss}. Here, with the same applied forces $(-\mu'_p)$, on each monomer ($\mathbf{r}_n$) we look at the velocity induced at the level of another monomer ($\mathbf{r}_m$)

$$v_{p_x}(\mathbf{r}_m) = \langle \sum_n \mathfrak{T}_{xx}(\mathbf{r}_{m'}\, \mathbf{r}_n)\rangle\, (-\mu'_x) \qquad \text{(VI.90)}$$

After averaging, this becomes

$$v_{p_x}(\mathbf{r}_1) = \int g(\mathbf{r}_{1'}\mathbf{r}_2)\, \mathfrak{T}_{xx}(\mathbf{r}_{2'}\mathbf{r}_2)\, d\mathbf{r}_2\, (-\mu'_p) \qquad \text{(VI.91)}$$

where $g(\mathbf{r}_1\, \mathbf{r}_2)$ is the pair correlation function for monomers inside the tube. For the intervals $\mathbf{r}_{12} \sim D$ of interest in eq. (VI.91), we have

$$g(\mathbf{r}_{1'}\mathbf{r}_2) \sim \frac{g_D}{D^3} \qquad \text{(VI.92)}$$

where g_D is the number of monomers per blob of size D, introduced in Fig. I.12. Similarly $\mathfrak{T}_{xx} \cong 1/\eta_s D$, and finally

$$v_{p_x}(\mathbf{r}_1) \sim \frac{g_D}{\eta_s D}(-\mu'_p) \qquad \text{(VI.93)}$$

Multiplying by the cross-section (D^2), this gives

$$L_{pp} \cong v_{p_x} D^2/(-\mu'_p) \cong \frac{g_D D}{\eta_s} \qquad \text{(VI.94)}$$

This completes our scaling program for the Onsager coefficients. A physical discussion will help at this point.

Eq. (VI.93) gives the polymer velocity (v_p) for a given force $(-\mu'_p)$ applied on each monomer. The total force of one coil $N(-\mu'_p) = f_{tot}$ and the coil mobility are[41]

$$\mu_{tot} = v_p/f_{tot} \cong \frac{g_D}{N\eta_s D}$$

$$\cong \frac{1}{\eta_s R_\parallel} \qquad \text{(VI.95)}$$

where $R_\parallel$ is the total length of the confined coil, discussed in eq. (I.53).

This mobility is larger than expected from naive arguments. If, following Debye,[42] we had represented the region occupied by the coil as a "sponge" of uniform density of monomers ($N/R_{\parallel}D^2$), each with a friction coefficient $\sim \eta_s a$ (where a is a monomer size), we would have arrived at a mobility

$$\mu_{Debye} \cong \frac{1}{N\eta_s a} \cong \frac{D}{a g_D} \mu_{tot} \cong \left(\frac{a}{D}\right)^{2/3} \mu_{tot} \qquad \text{(VI.96)}$$

smaller than the correct value μ_{tot} by a factor $(a/D)^{2/3}$. The assumption of uniform density overestimates the friction. In reality the monomer concentration has large fluctuations inside the tube, as shown in Fig. VI.14, and the solvent manages to pass through the "channels" in the structure.

Another important property described by the L coefficients is *coupled transport*. Assume again that $\mu'_s = 0$ but that a force $(-\mu'_p)$ is applied to the chains. They drift, and we expect them to drag some of the solvent with them. The resulting current is, from eq. (VI.82),

$$J_s = L_{sp}(-\mu'_p) = \frac{L_{sp}}{L_{pp}} J_p \qquad \text{(VI.97)}$$

$$\cong J_p \frac{cD^3}{g_D} \qquad \text{(VI.98)}$$

where we have used the scaling eqs. (VI.89, VI.94) for the Onsager coefficients. Returning to Fig. I.12 for the confined coil, we notice that

$$\frac{g_D}{D^3} = c^\star \qquad \text{(VI.99)}$$

is the concentration inside one blob and is also the concentration inside one coil (since the blobs lie side by side). Thus, we may write

$$\frac{J_s}{J_p} \cong \frac{c}{c^\star} \cong \psi \qquad \text{(VI.100)}$$

where ψ is the volume fraction occupied by coils inside the tube. Physically this means that *each coil drags its own volume of solvent* when it moves.

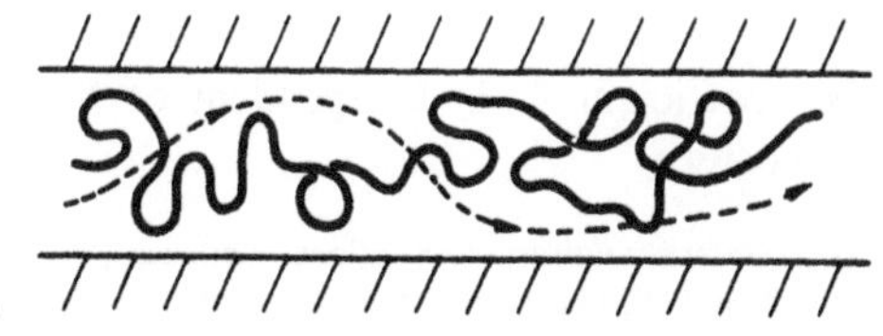

Figure VI.14.

VI.4. Problems of Internal Friction

VI.4.1. Three forms of friction

Section VI.1.2 introduced the effects of internal barriers on the friction coefficient of one "dumbbell." For a given elongation $\mathbf{r}$ (assumed small) and elongation rate $d\mathbf{r}/dt = \mathbf{V}_e$, the friction forces ($\mathbf{f}$) contain three separate contributions, to be presented below.

SOLVENT FRICTION

In the Rouse picture [eq. (VI.12)] this was additive and thus linear in N

$$\mathbf{f}_s \cong \zeta_s N \frac{d\mathbf{r}}{dt} \tag{VI.101}$$

With the more realistic Kirkwood picture, we would have [in analogy with eq. (VI.33)]*

$$\mathbf{f}_s \cong 6\pi\eta_s R_F \frac{d\mathbf{r}}{dt} \tag{VI.102}$$

BARRIER FRICTION

Energy barriers between different chain conformations (such as *trans* and *gauche* for simple carbon backbones) are responsible for a form of friction discussed by Kuhn. The result is given in eq. (VI.15).

$$\mathbf{f}_i \cong N^{-1} \zeta_i \frac{d\mathbf{r}}{dt} \tag{VI.103}$$

CERF FRICTION

The two processes above do not account entirely for the data. Very early, a third friction term was proposed by Cerf[43] with the structure

$$\mathbf{f}_c = \zeta_e \frac{d\mathbf{r}}{dt} \tag{VI.104}$$

where the coefficient ζ_e is independent of N and also independent of solvent viscosity.

*Eq. (VI.33) refers to uniform translations, while eq. (VI.101) describes an internal deformation of the dumbbell, but the two friction coefficients have the same scaling properties.

The profound difference between eq. (VI.104) and the Kuhn term [eq. (VI.103)] has not been entirely recognized by the scientific community. We now summarize the experimental evidence for the Cerf term and some possible interpretations of this form of friction.

VI.4.2. Evidence for the Cerf term

It is not easy to measure internal friction processes. For example, the contributions of eqs. (VI.103, VI.104) do *not* show up in the static viscosity for weak flows. This can be understood as follows. To discuss the viscosity increment $\delta\eta$ due to our dilute coils, we may choose any type of (weak) shear flow. It is then convenient (as noted first by Kramers[34]) to choose a longitudinal shear flow, such as the one shown in eq. (VI.65). In this situation the molecule does not rotate but simply stretches to a certain equilibrium length r (for a given shear rate s). Internal friction is involved only if the chain varies its length (or, equivalently, its conformations). In the present case the length is constant, and there is no dissipation associated with internal friction.

To find an effect of f_i and f_c on observable quantities, one must investigate finer points, such as the *relaxation times* of the dumbbell. In the absence of internal friction, we have only one relaxation rate [eq. (VI.42)]

$$1/\tau \cong T/\eta_s R_F^3$$

In the presence of internal friction, Kuhn and Kuhn noted that two distinct relaxation rates are present:[10]

(i) Rotational relaxation is unaltered ($\tau_{tot} = \tau$) since rotations do not modify the internal conformations

(ii) Extensional relaxation is slowed and contains a contribution of processes (ii) and (iii) to the friction. Using eq. (I.45) for the restoring force and equating it to the total friction force, one finds a relaxation time:

$$\tau_{ext} = \tau + \frac{R_F^2}{T}(N^{-1}\zeta_i + \zeta_c) \quad \text{(constant)} \qquad \text{(VI.105)}$$

What methods can measure τ_{ext}? In practice, the most convenient approach has been to study situations of relatively strong shear ($s\,\tau \sim 1$), choosing *transverse* shears to ensure that the molecule rotates and experiences a periodic extension/contraction cycle.[44,45] The main parameters measured were nonlinear features in the viscosity and in the flow birefringence—the latter being slightly simpler to interpret. All these nonlinear corrections finally determine the product $s\tau_{ext}$. The measurements are repeated for different solvent viscosities η_s. In eq. (VI.105) the first term τ is linear in η_s [see eq. (VI.42)], while the internal terms are, by definition, independent of η_s. Thus, by suitable extrapolations at very low η_s one may, in principle, determine the internal friction terms and follow their dependence on molecular weight (on N). This showed a contribution to τ_{ext}, behaving roughly like N to the first power, which could not be accounted for by the Kuhn term. (In our picture we would expect this new term to behave like $\zeta_c R_F{}^2 \sim N^{6/5}$, not very far from a linear law.)

This procedure is rather difficult because: 1) the nonlinear effects of a shear gradient are complex, and 2) the extrapolation toward $\eta_s \rightarrow 0$ is delicate since the residual internal friction terms are comparatively small at large N. Also, when the solvent is changed, its quality is also changed (and R_F varies slightly from one experiment to the next).

However, the patient experiments of Cerf and Leray did demonstrate the existence of internal friction effects at low frequencies in long, flexible coils and the fact that this friction is much larger than that predicted by the Kuhn process.

VI.4.3. Origin of the Cerf friction

Three microscopic explanations have been reported for the friction term f_c [eq. (VI.104)].

(a) The first is due to Freed and Adelman.[45] It is based on nonlinear coupling between modes in a flexible chain. The nonlinearity originates from the nongaussian character of the smallest subunits. According to Ref. 45, this leads to friction terms which (when translated into the "dumbbell" language) have very nearly the structure of eq. (VI.104)—i.e., they are independent of N. However, there is a fundamental objection: what is discussed here is a friction of monomers against the solvent, and the resulting friction coefficient is proportional to the solvent viscosity η_s; thus, it cannot account for the Leray observations, which were arranged to remove this type of contribution.*

(b) A recent idea from Cerf[46] is based on the mode structure computed by McInnes[47] for a Rouse chain with internal friction. The basic equation is a generalization of eq. (VI.5)

$$\dot{\mathbf{r}}_n = W_s \frac{\partial^2 \mathbf{r}}{\partial n^2} + W_s \theta \frac{\partial^2 \dot{\mathbf{r}}}{\partial n^2} \qquad \text{(VI.106)}$$

*On the other hand, at high frequencies ($\omega\tau \gg 1$) the mechanical data do point towards an added friction proportional to the solvent viscosity.

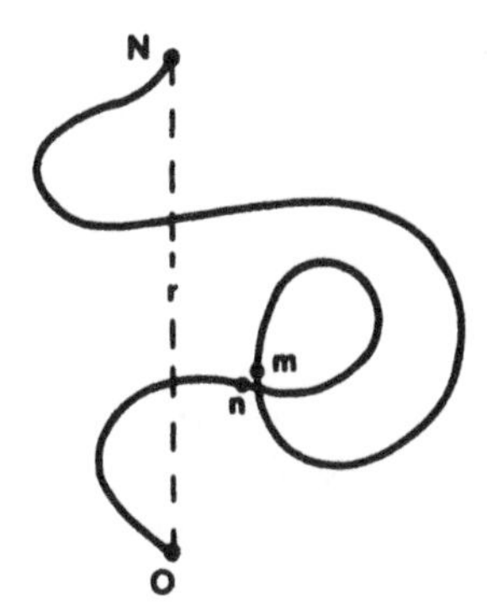

Figure VI.15.

where $W_s = 3T\,\mu/a^2$, and θ is an internal relaxation time measuring the importance of internal friction. The resulting modes have relaxation rates

$$\frac{1}{\tau_p} = W_s \left(\frac{\pi p}{N}\right)^2 \frac{1}{1 + W_s\theta\left(\frac{\pi p}{N}\right)^2} \qquad p = 1, 2, 3 \ldots \qquad \text{(VI.107)}$$

When $W_s\theta(p/N)^2$ is much larger than unity $1/\tau_p$ reaches a plateau $(1/\theta)$. At intermediate values of p/N there is an inflection point in the plot of $1/\tau_p$ versus p/N. If by accident we are operating in this region for $p = 1$, the apparent dependence of $1/\tau_p$ on $1/N$ may be linear.

(c) Another, tentative, explanation comes from the present author,[48] and is based on the notion of large loops inside a chain (Fig. VI.15).

In Fig. VI.15 we see two monomers (n) and (m) which are accidentally in close contact although the chemical distance $|n - m|$ is large. If we pull on the chain extremities, the two monomers (n) and (m) will slide onto each other, and friction will occur. This monomer–monomer friction does not involve the solvent, but will be somewhat related to the viscosity of a fluid of monomers.

We obtain a rough estimate of the friction by the following procedure. If the total elongation is r and its time derivative $\mathring{\mathbf{r}} = d\mathbf{r}/dt$, we assume that the chain deforms according to a linear law*

$$\mathring{\mathbf{r}}_n -= \mathring{\mathbf{r}}_m = \frac{(n-m)}{N}\,\mathring{\mathbf{r}} \qquad \text{(VI.108)}$$

If n and m are in direct contact (probability p_{nm}), we expect a dissipation

*In a Rouse model it can be shown that eq. (VI.108) is indeed the correct form for slow, uniform stretchings.

by monomer–monomer friction of the form $\zeta\ (\mathring{\mathbf{r}}_n - \mathring{\mathbf{r}}_m)^2$ where ζ is a molecular coefficient independent of the solvent viscosity.

Thus, the total entropy source is

$$T\mathring{S} = \frac{1}{2}(\mathring{\mathbf{r}})^2\zeta\ N^{-2}\sum_{nm} p_{nm}(n - m)^2 \qquad \text{(VI.109)}$$

There are two interesting contributions to this sum:

(i) The case where n and m are consecutive monomers ($m = n + 1$). Here, friction is associated with changes in the backbone conformation, p_{nm} is of order unity and, after summing over n, we find

$$T\dot{S}|_{backbone} \cong (\dot{\mathbf{r}})^2\zeta\ N^{-1} \qquad \text{(VI.110)}$$

Thus the resulting friction coefficient is of order N^{-1}: we recover the Kuhn theorem.

(ii) The case where n and m are partners in a large loop ($|n - m| \gg 1$) as shown in Fig. VI.15. Here, p_{nm} is small because we are in a good solvent and the monomers tend to avoid each other. The structure of p_{nm} has been investigated recently by J. des Cloizeaux (*J. Physique* **41,** 223 [1980]). The end-to-end probablility of contact (p_{1N}) was discussed in Chapter I and is given by eqs. (I.29, I.30).

$$p_{1N} \cong N^{-(\nu d+\gamma-1)} \qquad \text{(VI.111)}$$

where $d = 3$, $\nu = 3/5$, and $\gamma \sim 1.17$. Thus, to a very good approximation, we may write:

$$p_{1N} \cong N^{-1.97} \cong N^{-2} \qquad \text{(VI.112)}$$

When n and m are "well inside the chain" ($1 \ll |n - m| \ll N$) des Cloizeaux has shown that a *different* exponent appears

$$p_{nm} \cong |n - m|^{-\nu(d+\theta_2)}$$

and in three dimensions he estimates

$$\nu(d + \theta_2) = 2.18$$

In ref. 48 an approximate form was proposed for all values of n, m ($n - m \ll N$).

$$p_{nm} \cong |n - m|^{-2}$$

If we insert this form into eq. (VI.106), we see that *all pairs* (*nm*) *contribute equally*. Large loops are essential. The result is then independent of N

$$T\dot{S}|_{loops} \sim (\dot{\mathbf{r}})^2\zeta \qquad \text{(VI.113)}$$

and the friction coefficient has exactly the Cerf form.

VI.4.4. Summary

(*i*) For uniform translations, a polymer coil behaves like a Stokes sphere of the same size. (*ii*) The principal relaxation mode of the coil can be described by a dumbbell picture, where the elastic spring constant is derived from scaling and the friction constant is of the Stokes type. (*iii*) The inner modes of the coil, as probed by scattering methods at a scattering vector q, can be understood qualitatively as the principal relaxation of a subcoil having a size q^{-1}.

REFERENCES

1. J. Ferry, *Accts. Chem. Res.* **6,** 60 (1973).
2. J. Kirkwood, *Rec. Trav. Chim.* **68,** 649 (1949).
3. P. E. Rouse, *J. Chem. Phys.* **21,** 1273 (1953).
4. B. Zimm, *J. Chem. Phys.* **24,** 269 (1956).
5. P. G. de Gennes, *Proc. 30th Anniv. CRM,* Strasbourg 1977.
6. W. Stockmayer, *Fluides Moléculaires,* R. Balian and G. Weill, Eds., p. 107, Gordon & Breach, New York, 1976.
7. R. Orwoll, W. Stockmayer, *Adv. Chem. Phys.* **15,** 305 (1969).
8. P. Verdier, W. Stockmayer, *J. Chem. Phys.* **36,** 227 (1962). P. Verdier, *J. Chem. Phys.* **45,** 2122 (1966). L. Monnerie, F. Geny, *J. Polym. Sci.* **C30,** 93 (1970). L. Monnerie, F. Geny, *J. Chim. Phys.* **66,** 1708 (1969).
9. P. G. de Gennes, *Physics* **3,** 37 (1967). K. Iwata, *J. Chem. Phys.* **54,** 12 (1971).
10. W. Kuhn, H. Kuhn, *Helv. Chim. Acta* **28,** 1533 (1945). W. Kuhn, H. Kuhn, *Helv. Chim. Acta.* **29,** 71 (1946).
11. P. G. de Gennes, *Macromolecules* **9,** 587, 594 (1976).
12. P. G. de Gennes, F. Brochard, *Macromolecules* **10,** 1157 (1977).
13. J. Ferry, *Fluides Moléculaires,* R. Balian and G. Weill, Eds., p. 223, Gordon & Breach, New York, 1976.
14. K. Kawasaki, *Ann. Phys. (N.Y.)* **61,** 1 (1970). P. Hohenberg, B. Halperin, *Rev. Mod. Phys.* **49,** 435 (1977).
15. R. Ferrell, *Phys. Rev. Lett.* **24,** 1169 (1970).
16. J. Kirkwood, J. Riseman, *J. Chem. Phys.* **16,** 565 (1948).
17. G. Meyehoff, *Z. Physik. Chem.* **4,** 335 (1955). R. Mukherjea, P. Rempp, *J. Chim. Phys.* **56,** 94 (1959).

18. M. Adam, M. Delsanti, *J. Phys. (Paris)* **37,** 1045 (1976). M. Adam, M. Delsanti, *Macromolecules* **10,** 1229 (1977).
19. M. Daoud, G. Jannink, *J. Phys. (Paris)* **39,** 331 (1978).
20. J. des Cloizeaux, *J. Phys. (Paris) Lett.* **39L,** 151 (1978).
21. J. des Cloizeaux, G. Weill, *J. Phys. (Paris)* **40,** 99 (1979).
22. L. Van Hove, *Phys. Rev.* **95,** 1374 (1954).
23. E. Dubois-Violette, P. G. de Gennes, *Physics* **3,** 181 (1967).
24. J. J. Hermans, *Physica* **10,** 777 (1943). W. Kuhn, F. Grun, *Kolloid-Z.* **101,** 248 (1942). A. Peterlin, *J. Polym. Sci.* **8,** 621 (1952). A. Peterlin, *Pure Appl. Chem.* **12,** 563 (1966). A. Peterlin, *J. Elastoplastics* **4,** 112 (1972).
25. C. Tanford, *Physical Chemistry of Macromolecules,* John Wiley and Sons, New York, 1961.
26. A. Jones, G. Brehm, W. Stockmayer, *J. Polym. Sci., Polym. Symp.* **46,** 149 (1974).
27. See L. Landau, I. Lifshitz, *Fluid Mechanics,* Pergamon Press, London, 1959.
28. For polystyrene, see Ref. 17 and also W. Krigbaum, P. Flory, *J. Polym. Sci.* **11,** 37 (1953). P. Munk, M. Halbrook, *Macromolecules* **9,** 568 (1976). C. Strazielle, J. Herz, *Eur. Polym. J.* **13,** 223 (1977).
29. M. Delsanti, Ph.D. Thesis Orsay, 1978. Available from SPSRM, CEN Saclay, B. P. no. 2, Gif s/ Yvette, France.
30. D. Massa, J. Schrag, J. Ferry, *Macromolecules* **4,** 210 (1971).
31. G. Astarita, L. Nicodemo, *Chem. Eng. J.* **1,** 57 (1970). D. Acierno, G. Titomanlio, L. Nicodemo, *Rheol. Acta* **13,** 532 (1974). S. Peng, R. Landel, *J. Appl. Phys.* **47,** 4255 (1976).
32. F. C. Frank, M. Mackley, *J. Polym. Sci., Polym. Phys.* **14,** 1121 (1976).
33. P. Pincus, *Macromolecules* **10,** 210 (1977).
34. H. A. Kramers, *J. Chem. Phys.* **14,** 415 (1946).
35. P. G. de Gennes, *J. Chem. Phys.* **60,** 5030 (1974).
36. E. J. Hinch, *Proc. Symp. Polym. Lubtification,* Brest, 1974. E. J. Hinch, *J. Fluid Mech.* **74,** 317 (1976). E. J. Hinch, *J. Fluid Mech.* **75,** 765 (1976).
37. S. Daoudi, *J. Phys. (Paris)* **36,** 1285 (1975).
38. S. Daoudi, *J. Phys. (Paris) Lett.* **37L,** 41 (1976).
39. S. Daoudi, F. Brochard, *Macromolecules* **11,** 751 (1978). S. Daoudi, Ph.D. Thesis, University of Paris VI, Paris, 1978.
40. S. De Groot, P. Mazur, *Thermodynamics of Irreversible Processes,* North-Holland, Amsterdam, 1952.
41. F. Brochard, P. G. de Gennes, *J. Chem. Phys.* **67,** 52 (1977).
42. P. Debye, A. M. Bueche, *J. Chem. Phys.* **16,** 573 (1948).
43. R. Cerf, *J. Phys. Radium* **19,** 122 (1958). R. Cerf, *J. Chim. Phys.* **66,** 479 (1969).
44. R. Cerf, *C.R. Acad. Sci. Paris* **230,** 81 (1950).
45. J. Leray, Ph.D. Thesis, Strasbourg, 1959. M. Adelman, K. Freed, *J. Chem. Phys.* **67,** 1380 (1977).
46. R. Cerf, *C.R. Acad. Sci. Paris,* **286B,** 265 (1978).
47. D. McInnes, *J. Polym. Sci.* **15,** 657 (1977). D. McInnes, A. North, *Polymer* **18,** 505 (1977).
48. P. G. de Gennes, *J. Chem. Phys.* **66,** 5825 (1977).

VII

Many-Chain Systems: The Respiration Modes

VII.1. Semi-Dilute Solutions

VII.1.1. Longitudinal modes

We are concerned here with polymer–solvent or polymer–polymer systems, where the *concentration* c of one constituent fluctuates, and where we probe the dynamics of these fluctuations. It is convenient to examine a weak, sinusoidal modulation of c, as shown in Fig. VII.1 for a polymer in a solvent

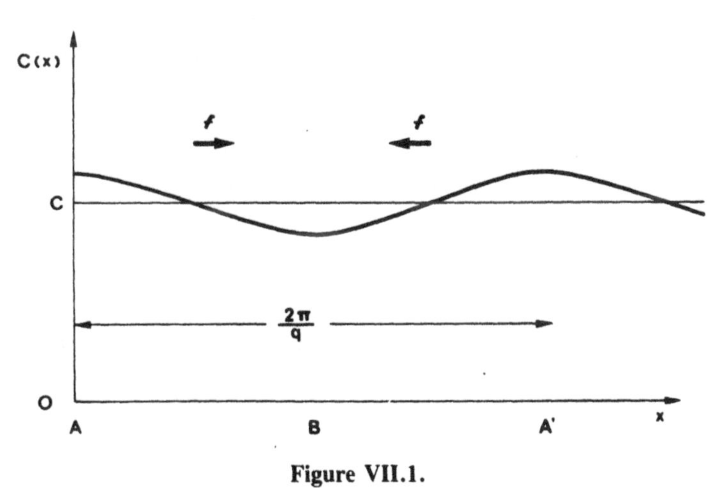

Figure VII.1.

$$c(xt) = c + \delta c_q(t) \cos qx \qquad (\delta c/c \ll 1) \tag{VII.1}$$

This sinusoidal form in space is convenient because it is self-preserving; if the profile $c(x)$ is sinusoidal at time $t = 0$, it will remain so at later times, with the same wave vector q (and a decreasing amplitude $\delta c_q(t)$. In Fig. (VII.1) we see high density regions (A, A' . . .) and low density regions (B, B' . . .). There is an elastic force f from A to B, etc., tending to equalize the concentrations. The osmotic pressure Π is largest at A, A', . . . and the force f (per cm³) is related to the gradient of Π

$$f = -\frac{\partial \Pi}{\partial x} \tag{VII.2}$$

The main problem is to find the velocities $v_p(x)$ induced by the force f in the polymer system. Both f and v_p are parallel to the direction of modulation x (more generally, we would say "parallel to the wave vector **q**"); this corresponds to what we call a *longitudinal* mode.

There is also a certain solvent flow $v_s(x)$. For longitudinal modes it is simply related to $v_p(x)$. For the slow flows of interest here, the solution may be considered incompressible. Then the average velocity $\bar{v} = \Phi_s v_s + \Phi_p v_p$ must be independent of x and must vanish if our sample is limited by impermeable walls.
Thus, we must have

$$\Phi_p v_p + \Phi_s v_s = 0 \tag{VII.3}$$

where $\Phi_p = ca^3$ is the polymer volume fraction, and $\Phi_s = 1 - \Phi_p$.

At all points the solvent velocity v_s is opposite v_p. In this section, we deal with semi-dilute solutions where $\Phi_p \ll 1$ and $\Phi_s \sim 1$; in this case v_s is much smaller than v_p and can be omitted completely.

Let us first recall the results for the *dilute limit*, obtained in Chapter VI, for the simplest case where q is small (long wavelengths). The basic variable is the current

$$J_p \sim cv_p \tag{VII.4}$$

It can be related directly to the force per cm³, f, through the coil mobility μ_{tot} (introduced in eq. (VI.32))

$$J_p = \mu_{tot} f \tag{VII.5}$$

Also, in the dilute limit, the osmotic pressure Π corresponds to a perfect gas of coils

$$\Pi = \frac{c}{N} T \tag{VII.6}$$

and eqs. (VII.2, VII.5, VII.6) give finally

$$J_p = \mu_{tot} T \left(- \frac{\partial c}{\partial x} \right) = - D \frac{\partial c}{\partial x} \tag{VII.7}$$

where D is the single coil diffusion coefficient discussed in Section VII.2.2. Writing a balance equation for the number of monomers in an interval (x_1, x_2) we have

$$\int_{x_1}^{x_2} \frac{\partial c}{\partial t} dx = - J_p (x_2) + J_p (x_1) \tag{VII.8}$$

or

$$\frac{\partial c}{\partial t} = - \frac{\partial J_p}{\partial x} \tag{VII.9}$$

Inserting this into eq. (VII.7) we recover a classical diffusion equation

$$\frac{\partial c}{\partial t} = D \frac{\partial^2 c}{\partial x^2} \quad (\text{with } D = T\mu_{tot}) \tag{VII.10}$$

If we then look for sinusoidal solutions of the form of eq. (VII.1), we find that the amplitude decays according to

$$\delta c(t) = \delta c(0) \exp (- Dq^2 t) \qquad (qR_F < 1) \tag{VII.11}$$

Eq. (VII.11) is the basis of the photon beat studies at long wavelengths, summarized in Section III.2.2. Our aim in this section is to generalize eq. (VII.11) to semi-dilute solutions. We show that there still exists a coefficient D_{coop} (c), (dependent on concentration) which describes the relaxation of concentration fluctuations. The reason for the subscript "*coop*" (cooperative) is explained below.

VII.1.2. Two diffusion coefficients

When we discuss diffusion coefficients, we must distinguish two types of experiments:

(*i*) *Dilute tracers*. If a small fraction of the coils is radioactive, we can see them spread out with a certain self-diffusion coefficient D_t (where the t means tracer).

(*ii*) *Cooperative motions*. Here all coils participate and interact. This is the case for the longitudinal modes discussed above and also for macroscopic studies where two compartments (with concentrations $c + \delta c$ and $c - \delta c$) are brought into contact, and the concentration profile is monitored as a function of time.

The diffusion coefficient measured here D_{coop} is in general completely different from D_t (they coincide only in dilute solution). We shall see that D_{coop} increases with concentration, while D_t decreases drastically.

For coil systems, the difference comes from the following fact. One labeled coil has to worm its way through a maze created by the others, while in cooperative motion all coils move together like a sponge; they do not have to disentangle. We shall discuss D_{coop} in this section, and we shall come to D_t only in the next chapter, where we attack entanglement effects.

VII.1.3. The sedimentation coefficient

Let us start with a semi-dilute solution where each monomer is subjected to a certain force $\mathbf{f}_{mono}$ (this may be a gravitational force, a centrifugal force, or an electric force if the chain is charged). We want to compute the drift velocity $\mathbf{v}_p$ of the polymer in this situation, and we write

$$\mathbf{v}_p = s\mathbf{f}_{mono} \qquad \text{(VII.12)}$$

where s is a sedimentation coefficient (written here in natural units).

To achieve this, we consider one monomer, M, that feels the flow fields from all neighboring monomers, M′. This approach is due to Kirkwood and Risemann [*J. Chem. Phys.* **16,** 565 (1948)] and is illustrated in Fig. VII.2. The velocity at M is a sum of backflow contributions from all sources of force:

$$v_\alpha(\mathrm{M}) = \sum_{\mathrm{M}'} \mathfrak{T}_{\alpha\beta}(\mathbf{r}_{\mathrm{MM}'}) f_{mono\beta} - \int \mathfrak{T}_{\alpha\beta}(\mathbf{r})\, d\mathbf{r}\; c f_{mono\beta} \qquad \text{(VII.13)}$$

The Oseen tensor $\mathfrak{T}_{\alpha\beta}$ was defined in eq. (VI.27).

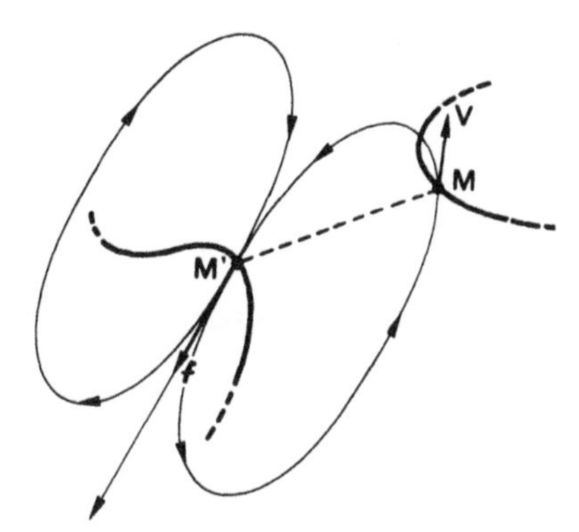

Figure VII.2.
One monomer, M', is subjected to a force $\vec{\mathbf{f}}$. It then moves in the solvent and creates a backflow indicated by the arrows. A second monomer, M, then drifts in this flow field.

Subtraction of the second term arises from the following. To maintain the solvent in the steady state, we must have a pressure gradient ∇p which exactly balances the average force ($f = cf_{mono}$ per cm^3)

$$- \nabla p + cf_{mono} = 0 \qquad \text{(VII.14)}$$

The second term in eq. (VII.4) corresponds to the effect of this average pressure gradient. If we compute the average velocity, we must weight the probability of finding monomers at M and M′ by the correlation function $\langle c\,(0)\; c\,(\mathbf{r})\rangle / c$. Thus, we arrive at

$$v_{p\alpha} \equiv \langle v_\alpha\,(\mathrm{M})\rangle = \int d\mathbf{r}\; \mathfrak{T}_{\alpha\beta}\,(\mathbf{r}) f_{mono\;\beta} \left[\frac{1}{c}\langle c\,(0)\; c\,(\mathbf{r})\rangle - c\right] \qquad \text{(VII.15)}$$

What we finally obtain is the static correlation function

$$g(\mathbf{r}) \equiv \frac{1}{c}[\langle c\,(0) c(\mathbf{r})\rangle - \langle c\rangle^2] \qquad \text{(VII.16)}$$

which was discussed in Section III.2.6. Introducing the explicit form of $\mathfrak{T}_{\alpha\beta}$ [eq. (VI.27)], we arrive at the fundamental formula for the sedimentation coefficient

$$s = \int d\mathbf{r} \frac{1}{6\pi\eta_s r}\, g(\mathbf{r}) \qquad \text{(VII.17)}$$

A similar equation was derived (for the single chain problem) in Chapter VI through the Kubo formula. The present derivation is more direct.

Let us now write the scaling form for $g(\mathbf{r})$ which is adequate for semidilute solutions—namely,

$$g(\mathbf{r}) = c\, f_g\left(\frac{r}{\xi}\right) \tag{VII.18}$$

Note that the normalization is correct

$$\int g\,(\mathbf{r})\, d\mathbf{r} \cong c\xi^3 \cong g \tag{VII.19}$$

where g is the number of monomers per blob [eq. (III.44)].

Inserting eq. (VII.18) into eq. (VII.17), we arrive at*

$$s \equiv \frac{v_p}{f_{mono}} \cong \frac{g}{6\pi\eta_s\, \xi} \cong \frac{c\xi^2}{6\pi\eta_s} \tag{VII.20}$$

Thus, the sedimentation coefficient defined as in eq. (VII.20) should scale like $g/\xi \sim \Phi^{-1/2}$. Early data on $s(\Phi)$ do give a decreasing function, but the power law was not yet entirely known. More recent data on high N polystyrene in benzene give an exponent of $-\ 0.59 \pm 0.10$.[1]

VII.1.4. Cooperative diffusion

Knowing the sedimentation coefficient, we can now return to the respiration modes, balancing the osmotic pressure gradient [eq. (VII.2)] against the viscous force described by eq. (VII.12)

$$-\frac{\partial \Pi}{\partial x} = cf_{mono} = \frac{c}{s} v_p = \frac{1}{s} J_p \tag{VII.21}$$

Using the conservation equation, eq. (VII.9), we arrive at

$$\frac{\partial c}{\partial t} = s\frac{\partial^2 \Pi}{\partial x^2} = s\ \frac{\partial \Pi}{\partial c}\ \frac{\partial^2 \Pi}{\partial x^2} \tag{VII.22}$$

[neglecting a nonlinear term $\partial^2\Pi/\partial c^2\,(\partial c/\partial x)^2$]. Ultimately we find a diffusion coefficient

$$D_{coop} = s\ \frac{\partial \Pi}{\partial c} \tag{VII.23}$$

*The coefficient 6π in eq. (VII.20) has no precise meaning but is a convenient reminder of the similarity with the Stokes law for viscous motion of a sphere.

Using the scaling form of eq. (VII.20) for s and of eq. (III.27, III.31) for Π and g, we get

$$D_{coop} \cong \frac{g}{6\pi\eta_s\,\xi}\,T\Phi^{5/4} \cong \frac{T}{6\pi\eta_s\,\xi} \qquad \text{(VII.24)}$$

Thus, D_{coop} has the same structure as a simple Stokes-Einstein diffusion coefficient for a single blob. The essential feature is that D_{coop} *increases* with concentration; the restoring forces (due to the osmotic pressure gradient) are stronger at high c.

More precisely the scaling prediction for D is $D \sim \Phi^{3/4}$. The data on polystyrene in benzene[2] give $D_c \sim \Phi^{0.67\,\pm\,0.02}$—a slightly smaller exponent.*

In discussing the limits of validity of eq. (VII.24), the entire analysis is macroscopic and is meaningful only when the wavelength $(2\pi/q)$ is much larger than the mesh size (ξ). The opposite limit $(q\xi > 1)$ is difficult to reach by optical means but is conceptually interesting. In this case we expect to return to a single chain behavior [eq. (VI.42)]—namely, to find a linewidth in inelastic scattering, at given q, of the form $\Delta\omega \cong Tq^3/\eta_s$. Note that this crosses over correctly to eqs. (VII.11, VII.24) when $q\xi \sim 1$.

Our discussion was restricted to very good (athermal) solvents. What happens in theta solvents? The question is delicate because of the stronger effects of entanglements mentioned in Section VI.1. However, on the whole, we are led to think that entanglements are not very efficiently coupled to the mode observed in light scattering.[3] Then at low q, eq. (VII.20, VII.24) for the sedimentation coefficient s (and the diffusion coefficient D_{coop}) should remain meaningful provided we reinterpret ξ and g as the blob parameters for $T = \Theta$. As discussed in Chapter VI, this corresponds to $\xi \cong a\Phi^{-1}$ and $g \sim \Phi^{-2}$. The relationship $g = c\,\xi^3$ is maintained, and $\eta_s\, s \sim c\,\xi^2$ is expected to increase when we decrease the solvent quality. This is qualitatively confirmed by the data.[1]

VII.1.5. Summary

In a semi-dilute polymer solution, there is a certain mutual friction between the coils and the solvent. The dissipated energy (per cm^3) is of the form

$$T\dot{S} \cong \eta_s \frac{v^2}{\xi^2}$$

*With a deviation from Kirkwood picture, described by an exponent z, we may improve the agreement. We would write $D_c \sim \xi^{-(z-2)} \sim \Phi^{3(z-2)}/4$ and using the dilute data on z (Chapter VI) we would arrive at a rather good theoretical value for $D_c(\Phi)$; but the whole concept of a z is doubtful.

where v is the relative velocity. This formula can be understood from the case where the polymer is at rest, and the solvent flows with velocity v: then the typical shear rates for flows in pores of size ξ are of order v/ξ.

This relative friction controls the relaxation of fluctuations of concentration in the solution: the corresponding "cooperative diffusion coefficient" D_{coop} *increases* with concentration.

VII.2. Dynamics Near a Critical Point

We consider now a polymer dissolved in a poor solvent, for which demixing may occur. The general features of the demixing critical point were discussed in Chapter IV. Here our aim is to analyze the cooperative diffusion D_{coop}. We follow the classical description of Kawasaki and Ferrell for simple binary mixtures, incorporating the (few) special features required for polymer systems.

The starting point is a general formula relating D_{coop} to the static correlations in the solution, which we have already written in eqs. (VII.22) and (VII.17). We have

$$D_{coop} = s\frac{\partial \Pi}{\partial c} \tag{VII.25}$$

$$= \frac{\partial \Pi}{\partial c} \int d\mathbf{r} g(\mathbf{r}) \frac{1}{6\pi \eta_s r} \tag{VII.26}$$

Eq. (VII.26) is not entirely rigorous because it neglects certain anomalies in the viscosity of the solution which renormalize η_s in the denominator; however, these corrections are minor. It is also possible to express the osmotic compressibility $\partial c/\partial \Pi$ in terms of the pair correlation $g(\mathbf{r})$ through the sum rule

$$\int g(\mathbf{r}) d\mathbf{r} = T\frac{\partial c}{\partial \Pi} \tag{VII.27}$$

Thus, we arrive at the striking form

$$D_{coop} = \frac{\int d\mathbf{r} g(\mathbf{r}) \dfrac{T}{6\pi \eta_s r}}{\int d\mathbf{r} g(\mathbf{r})} \tag{VII.28}$$

When we get close to the critical point, $g(\mathbf{r})$ describes correlations extending to a very large distance ξ_s; ξ_s is then much larger than the size of one chain (the latter being $\sim R_0 \sim N^{1/2}a$ near the Θ point). To an excellent approximation, it is sufficient to describe $g(\mathbf{r})$ by a classical Ornstein-Zernike form

$$g(\mathbf{r}) \cong \frac{1}{a^2\, r} \exp(-\, r/\xi_s) \tag{VII.29}$$

Note that for $r \leqslant R_0$, $g(r)$ crosses over smoothly to the single chain correlation [eq. (I.17)] for an ideal chain. If we choose a concentration exactly equal to the critical concentration and vary the temperature above T_c, we expect the following scaling form for the correlation length ξ_s

$$\xi_s \cong R_0 \left[\frac{\Theta - T_c}{T - T_c}\right]^{\nu_s} \tag{VII.30}$$

where ν_s is an exponent near 2/3. For $T = \Theta$, eq. (VII.30) gives the correct form $\xi_s \cong R_0$. Within the Ornstein-Zernike approximation, it is also possible to relate ξ_s to the osmotic compressibility through eq. (VII.29). The result is

$$\xi_s \cong a\left(T \frac{\partial c}{\partial \Pi}\right)^{1/2} \quad \text{(Ornstein-Zernike)} \tag{VII.31}$$

Both forms [eqs. (VII.30, VII.31) are nearly equivalent, and either can be used, depending on the available data. We can now return to eq. (VII.28) for D_{coop}, and we find

$$D_{coop} = \frac{1}{6\pi\eta_s} \frac{\int_0^\infty dr\, r^2 \frac{1}{r^2} \exp(-\, r/\xi_s)}{\int_0^\infty dr\, r^2 \frac{1}{r} \exp(-\, r/\xi_s)}$$

$$= \frac{T}{6\pi\eta_s\xi_s} \tag{VII.32}$$

Eq. (VII.32) is the Kawasaki-Ferrell result.[4] The only special feature of polymer solutions is the magnitude of ξ_s; the prefactor R_0 in eq. (VII.30) is a coil size, and the temperature factors multiply it by a large number. Thus, we expect D_{coop} to be small, and to vanish as $\Delta T^{2/3}$ when we get near the critical point. This has been found in laser scattering experiments at Stony Brook.[5]

The macroscopic analysis leading to a relaxation rate $1/\tau_q = D_{coop}\, q^2$

[eq. (VII.11)] holds only at very small q ($q\xi_s < 1$). At large q values the linewidth $\Delta\omega_q$ is expected to become independent of ξ_d and to have the form[4]

$$\Delta\omega_q \sim \frac{T}{6\pi\eta_s} q^3 \qquad (q\xi_s > 1) \tag{VII.33}$$

This crosses over correctly at $q\xi_s = 1$ and also merges into the single chain spectrum [eq. (VI.41)] at large q values ($qR_0 > 1$)

Entanglements are probably not essential at the *polymer–solvent* critical point for the following reason. At $c = c_{critical}$ the coils are essentially adjacent, as explained in Chapter IV. They do not overlap very much, and they are relatively free to move side by side. On the other hand, for *polymer-polymer* systems (without solvent), entanglements are essential. Their dynamics near a critical point are considered in the next chapter.

VII.3. Dynamics of Gels

VII.3.1. Longitudinal modes of swollen gels

Section V.3 showed that a swollen gel is very similar (regarding scaling properties) to a *solution* (without crosslinks) at the overlap concentration $c = c^\star$. This analogy is also present in certain dynamical properties. Our general formulas [eqs. (VII.17, VII.23)] express the sedimentation coefficient and the diffusion constant D_{coop} in terms of static correlations. All that we need is to insert for $g(\mathbf{r})$ the correct form for gels [eq. (V.30)]. More concisely, since we are dealing with a system where $c \cong c^\star$, we must replace—in all the dynamical formulas of Section VII.1—the mesh size ξ by R_F and the number of monomers per blob g, by N (N being the degree of polymerization of the constituent chains).

This leads to a sedimentation coefficient

$$s \cong \frac{N}{6\pi\eta_s R_F} \sim N^{2/5} \tag{VII.34}$$

and to a diffusion coefficient

$$D_{coop} \cong \frac{T}{6\pi\eta_s R_F} \sim N^{-3/5} \tag{VII.35}$$

Measurements of D_{coop} by inelastic scattering of laser light were initiated at M.I.T.[6] Systematic studies on gels with variable N values were performed later in Strasbourg.[7] When comparing different gels (with the same solvent and solute but variable chain lengths) it is convenient to plot the results not as a function of N but in terms of the (more directly measurable) concentration c (or Φ) in the gel. The relationship between $\Phi \cong \Phi^{\star}$ and N is given in eq. (V.23). For very good solvents ($v \sim a^3$)

$$\Phi^{\star} \sim N^{-4/5} \qquad \text{(VII.36)}$$

One then expects

$$\begin{aligned} s &\sim \Phi^{-1/2} \\ D_{coop} &\sim \Phi^{3/4} \end{aligned} \qquad \text{(VII.37)}$$

The photon beat results on D_{coop} agree roughly with this prediction. However, the experimental exponents are systematically slightly smaller than 0.75, except for special chains such as siloxanes.[8]

The numerical coefficients in eqs. (VII.34, VII.35) depend on the functionality of the crosslinks and on the details of gel preparation. Some discussion on these points can be found in Ref. 7.

VII.3.2. Slow motions near the spinodal threshold

In Section V.3.4 we discussed experiments where a gel is rapidly cooled. This corresponds (usually) to a decrease in the quality of the solvent. The gel would like to contract, but it cannot during the experimental time available. This type of metastable state can exist only above a certain spinodal temperature T_s. When T is decreased to T_s, an instability appears. At T_s, the Young modulus E of the gel vanishes.

In this section, we discuss longitudinal modes in the gel at temperatures T slightly above T_s. We see that the cooperative diffusion coefficient becomes vanishing small when we get down to the spinodal. This has been shown by the M.I.T. group.[9] Their original interest was connected with studies on eye cataracts, where the occurrence of opacity upon cooling was announced by a drop in the cooperative diffusion coefficient.[10] Since the eye lens is a highly complex system, these workers turned to model systems and studied polyacrylamide gels, where they found a similar phenomenon. They proposed an interpretation in terms of spinodal decompositions of the gel.[9]

The essential points are:

(*i*) For $T \geqslant T_c$, the system is metastable, but it is still in local equilibrium, and all our basic equations remain valid. In particular the Kawasaki-Ferrell result for D_{coop} in terms of pair correlations [eq. (VII.28)] and the compressibility sum rule for the correlations [eq. (VII.27)]

(ii) Near the spinodal, the pair correlation $g(r)$ becomes of long range and is well approximated by an Ornstein-Zernike form:

$$g(\mathbf{r}) \cong \frac{N}{R_F^2\, r} \exp(-r/\xi_s) \qquad \text{(VII.38)}$$

where ξ_s is a correlation range which diverges for $T \to T_s$. In the M.I.T. analysis ξ_s was extracted from a mean field theory and was assumed to behave like $(T - T_s)^{-1/2}$. More generally one may relate ξ_s to Young's modulus using the compressibility sum rule. It is also interesting that eq. (VII.38) agrees with the requirement $g(\mathbf{r}) \sim c = N/R_F^3$ for $r \sim R_F$. If we compare eq. (VII.38) with our earlier result for the gel in complete equilibrium [eq. (V.30)], we see that they agree at $r \sim R_F$, but that a long range tail at large r has appeared when $T \to T_s$.

(iii) Finally, from eq. (VII.28), we are led to expect a diffusion coefficient of the form

$$D_{coop} \sim \frac{T}{6\pi\eta_s\, \xi_s} \qquad \text{(VII.39)}$$

and this will vanish for $T \to T_s$ since $\xi_s \to \infty$. Qualitatively all these behaviors have been verified by the M.I.T. group on polyacrylamide gels, using both intensity measurements (giving E^{-1}) and elastic measurements (giving D_{coop}). Perhaps there are some significant deviations from mean field behavior, but the complications brought in by a gel may well obscure that point.

VII.3.3. Dynamics at the sol-gel transition

Light scattering studies on longitudinal modes near the sol-gel transition would not be very interesting for the following reason. In the reaction mixture, where monomers begin to build up large connected clusters, the fluctuations of concentration are not large near the transition point. A single cluster would give a large intensity, but the overlapping set of clusters gives a much smaller intensity because of destructive interference between neighboring clusters.

Static pair correlation functions are not very sensitive to the establishment of long range connections. This implies that dynamic parameters, such as the diffusion coefficient D_{coop}, are not very sensitive to the establishment of a weak gel phase.

A more interesting mechanical parameter is the *viscosity* of the reaction bath, when the fraction p of connected bonds increases and becomes very close to the critical value p_c. Clearly the viscosity will diverge when $\Delta p = p_c - p \rightarrow 0$—i.e., when we arrive at very large clusters.

In Chapter V we related one mechanical parameter of the *gel* phase (the elastic modulus E) to certain electrical problems on a percolation network. A related analogy can be established for the viscosity of the *sol* phase. Here, we assume that the bonds (with probability p) correspond to superconducting links. The voltages X at both ends of such a link must be equal $X_i = X_j$. On the other hand, when the pair (ij) is not linked (probability $1 - p$), we assume that there is a capacitance C_e between (i) and (j). This corresponds to a current

$$J_{ij} = C_e (\dot{X}_i - \dot{X}_j) \tag{VII.40}$$

As explained in Chapter V, the mechanical viewpoint amounts to taking X_i as the displacement of monomer (i) and J_{ij} as the force due to (i) and acting on (j). Eq. (VII.40) gives a force proportional to the relative velocity and correctly describes viscous effects between clusters of all sizes (for very small, point-like clusters, eq. (VII.40) does lead to the Navier-Stokes function in a fluid of monomers).

Returning to the electrical problem, when we increase p toward the threshold value p_c, we expect that the macroscopic dielectric constant, or the polarizability (α) of the system will tend to diverge with a certain exponent s

$$\alpha = \text{constant } C_e (p_c - p)^{-s} \tag{VII.41}$$

This is corroborated by numerical studies of Straley,[11] which suggest $s = 0.7 \pm 0.1$ for three-dimensional systems. The macroscopic polarization P is related to the field $= - \partial X/\partial x$ by

$$P = \alpha\left(-\frac{\partial X}{\partial x}\right) \tag{VII.42}$$

If we differentiate this relationship with respect to time, we get the polarization current $\dot{P} = J$

$$J = -\alpha \frac{\partial}{\partial x} (\dot{X}) \tag{VII.43}$$

As explained in Chapter V, the macroscopic current is the analog of the stress; thus, eq. (VII.42) implies a linear relationship between shear rate and stress, and the viscosity thus coincides with α.

We do not have many reliable data on the exponent s in eq. (VII.41) for the sol viscosity, but new experiments on this question are being attempted.

On the theoretical side, the exponent s can be calculated rather simply at high dimensionalities[12,13] (between $d = 4$ and $d = 6$). In polymer language, this amounts to computing the friction inside each cluster in a Rouse approximation (ignoring all backflow effects). The viscosity is then proportional to the (weight average) square gyration radius of the clusters [eq. (V.9)], and $s = 2\nu - \beta$. In dimensions lower than 4, the backflow terms become essential, and how to include them remains a problem.

REFERENCES

1. M. Jacob, R. Varoqui, S. Klenine, M. Daune, *J. Chim. Phys.* (1962), p. 865. P. Mijnlieff, W. Jaspers, *Trans. Faraday Soc.* **67,** 1837 (1971). C. Destor, F. Rondelez, *Polym. Lett.* (in press).
2. M. Adam, M. Delsanti, *Macromolecules* **10,** 1229 (1977).
3. F. Brochard, P. G. de Gennes, *Macromolecules* **10,** 1157 (1977).
4. K. Kawasaki, *Ann. Phys. (N.Y.)* **61,** 1 (1970). R. Perl, R. Ferrell, *Phys. Rev.* **A6,** 2538 (1972).
5. Q. Lao, B. Chu, N. Kuwahara, *J. Chem. Phys.* **62,** 2039 (1975).
6. T. Tanaka, L. Hocker, G. Benedek, *J. Chem. Phys.* **59,** 5151 (1973).
7. J. P. Munch *et al., J. Polym. Sci.* **14,** 1097 (1976). J. P. Munch *et al., J. Phys. (Paris)* **38,** 1499 (1977).
8. Most of the Strasbourg data concern siloxanes in solutions (not in gels), but they do show a trend toward more ideal scaling behavior.
9. T. Tanaka, *Phys. Rev.* **A17,** 763 (1978). T. Tanaka, *Phys. Rev. Lett.* **40,** 820 (1978).
10. T. Tanaka, G. Benedek, *Invest. Ophthalmol.* **14,** 449 (1975).
11. J. Straley, *Phys. Rev.* **B17,** 4444 (1978). The Straley analogies refer to the *conductance* of a superconducting normal mixture, but a simple change from normal conductances to capacitances can be performed on the equations without altering their structure.
12. P. G. de Gennes, *C.R. Acad. Sci. Paris* **286B,** 131 (1978).
13. M. Stephen, *Phys. Rev.* **B17,** 4444 (1978).

VIII

Entanglement Effects

VIII.1. Dynamics of Melts and Concentrated Solutions

Concentrated polymer systems have fascinating motions. They show a unique combination of viscous and elastic behavior, which has been studied in careful experiments and is analyzed in a classical book by J. Ferry.[1] Theoretically, the situation is less brilliant; the dynamics of entangled chains (which can slip onto each other but cannot cross each other) is still poorly understood. The main ideas are described in a review by W. Graessley.[2] In this chapter, we first summarize the concepts extracted from the mechanical data on melts. Then we proceed to the simpler problem of one chain which is moving inside a crosslinked network. Here a relatively plausible picture of the motions can be constructed and is known as the "reptation model." Finally we return to the melts and discuss some generalizations of reptation for these systems. This third area, however, is largely conjectural.

VIII.1.1. Rubber-like and liquid-like behaviors

The fundamental experiment on weak mechanical perturbations can be presented as follows.[1] Starting with a fluid at rest, we apply at all times ($t > 0$) a small perturbative stress σ (the detailed specification of this stress—extension, or transverse shear—is not essential for our discussion). We now look at the strain $e(t)$ induced by this step increase in stress. For small σ, the strain is a linear function of the stress

$$e(t) = \sigma J(t) \tag{VIII.1}$$

The function $J(t)$ is called the *creep compliance* of the polymer. For a melt (or concentrated solution) of *long* chains it has the structure shown in Fig. VIII.1.

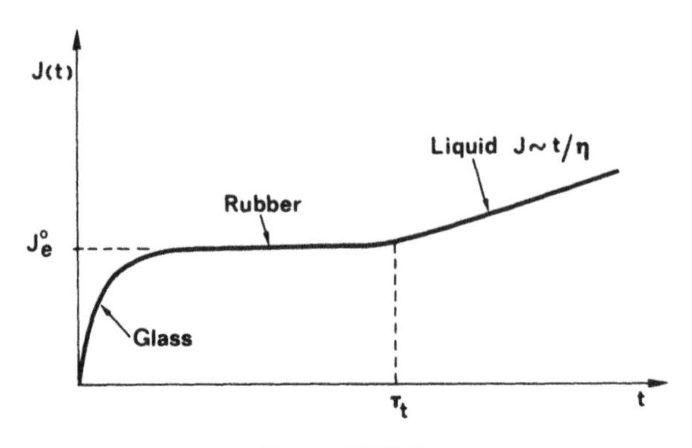

Figure VIII.1.

At short times, the conformations (*trans, gauche,* etc.) in all the chains do not adjust, and the mechanical response is similar to what one measures in a polymeric glass (where the conformations are permanently frozen). At longer times the conformations do change, and the strain becomes more important. However, there is a large span of time ($t < \tau_t$) where the chains remain entangled and behave like a rubber network, as shown in Fig. VIII.2.

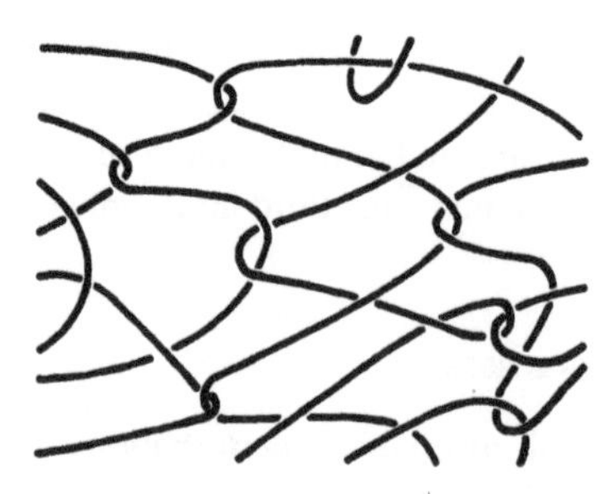

Figure VIII.2.

In this range of t values, the strain has a plateau

$$J(t) \to J_e^\circ \qquad (t < \tau_t) \tag{VIII.2}$$

J_e° is called the *steady-state compliance*. The inverse

$$E = (J_e^\circ)^{-1}$$

can be called the elastic modulus of the transient network.

Finally, at very high times the creep compliance starts to increase linearly with time. This is the normal steady-state behavior for a liquid, where the strain *rate* (de/dt) is proportional to the stress σ. The ratio between the two is the viscosity, η

$$\eta \frac{de}{dt} = \sigma \qquad (t > \tau_t) \tag{VIII.3}$$

The time τ_t when we cross from eq. (VIII.2) to eq. (VIII.3) is called the *terminal relaxation time*. It is the longest relaxation time observed in mechanical measurements. There seems to be a scaling form for the creep compliance that covers both the rubber and liquid regions

$$J(t) = J_e^\circ \varphi_J (t/\tau_t) \tag{VIII.4}$$

where $\varphi_J (x)$ is a universal function (for monodisperse chains) with the limiting behaviors

$$\begin{aligned} \varphi_J(x \ll 1) &= 1 \\ \varphi_J(x \geq 1) &\cong x \end{aligned} \tag{VIII.5}$$

If we compare the latter form with the viscous flow equation [eq. (VIII.3)], we are led to an important scaling relationship for the viscosity

$$\eta \cong E\tau_t \tag{VIII.6}$$

which is obeyed well in practice. We now summarize the data on the three constants E, η, and τ_t.

VIII.1.2. Elastic modulus of the transient network

For long chains, the plateau modulus E is independent of chain length. When the chains are fully flexible, E essentially measures the *number of entanglement points* per unit volume in the transient network. Often the modulus in a melt is written as

$$E = cT/N_e \tag{VIII.7}$$

where c ($\sim 1/a^3$) is the concentration in the melt, and N_e represents the average interval between entanglement points along one chain. Typical values of N_e are of order 100.[2] Our discussion here is restricted to long chains $N \gg N_e$, so that entanglements are indeed essential. The value of N_e must be sensitive to the local rigidity of the chain and to the monomer-monomer correlations. Unfortunately, little is known about the temperature dependence of N_e, and no realistic calculation of N_e is available.

Another question of interest is the concentration dependence of E if we go from melts ($\Phi = 1$) to less concentrated solutions. The region $\Phi \sim 0.1$ to 1 is probably complicated by many nonuniversal features, related to local monomer–monomer correlations. In the semi-dilute limit ($\Phi^* \ll \Phi \ll 1$) we could expect a scaling form

$$E \cong \frac{1}{N_e}\frac{cT}{g} \cong \frac{T}{N_e \xi^3} \qquad \text{(VIII.8)}$$

where g is the number of monomers per blob.

The idea underlying eq. (VIII.8) is that if we consider blobs as the fundamental units, we are led back to a completely dense system, and eq. (VIII.7) can still be applied: c/g is the number of blobs per unit volume. However, this idea may not be valid. Although blobs and monomers are essentially equivalent in terms of thermodynamic properties, they may differ in their modes of entanglements (blobs may entangle more easily than monomers). This would show up in a direct dependence of N_e on Φ in eq. (VIII.8). We shall meet this problem repeatedly in this chapter.

VIII.1.3. Viscosity and terminal time

The viscosity η is very sensitive to chain length. Both η and τ_t increase as a power of N

$$\eta \sim \tau_t \sim N^{m_\eta} \qquad (N \gg N_c) \qquad \text{(VIII.9)}$$

where m_η is of order 3.3 to 3.4.[2]

The exponent m_η represents one of the major unsolved problems of polymer physics. An early attempt to derive m_η is due to Bueche and is based on the notion of one chain dragging other chains.[3] His analysis led to $\eta \sim N^{3.5}$, which is good, but to a value of the terminal time $\tau_t \sim N^{4.5}$, which is clearly ruled out by the experiments. More refined discussions along similar lines were analyzed and criticized by Graessley.[2]

Another work, by Edwards and Grant,[4] attempted to give a self-consistent description of a chain trapped in a tube due to neighboring chains. This calculation gave $\eta \sim N^3$. It suffered, however, from two defects:

(i) The discussion of stresses included only the stresses inside one chain and did not incorporate interchain constraints. Thus, the elastic modulus E was underestimated. The authors write $E \sim cT/N$ rather than cT/N_e.

(ii) The self-consistent relaxation time τ_t was found to be very large $\tau_t \sim N^4$. Thus, in this calculation also, τ_t and η do not scale in the same way, and this is incompatible with the mechanical data.

Another group of models is based on the notion of a transient network, introducing a finite lifetime for crosslinks into the classical rubber elasticity.[5] This contains some excellent physical points. However, the dependence of τ_t on the molecular mass is not attacked and remains unexplained in this picture. Thus, we do not describe the transient network models.

A different approach, based on the reptation concept, was initiated by the present author[6] and was recently augmented by Edwards and Doi.[7] As explained in the next section, it leads to $\eta \sim \tau_t \sim N^3$, and there is no obvious correction which might increase the exponent from 3 to the experimental value of 3.3. Thus, the present situation is still unsatisfactory.

VIII.2. Reptation of a Single Chain

VIII.2.1. Coils trapped in a network

Here we look at a system which is simpler than a polymer melt but which still shows some nontrivial entanglement effects. This corresponds to a single, ideal, polymeric chain P (with N monomers) trapped in a three-dimensional network.

Incorporation of the chain into the network is difficult. It is not enough to put a swollen gel into contact with a solution of chains. Even if thermodynamic equilibrium allows for a finite concentration of chains in the gel, the kinetics of chain diffusion are usually too slow. However, it is possible to prepare a mixture of chains P with other chains C, and to crosslink the C chains in a second stage. This was done by the Wisconsin group using two different pathways: 1) with P = polyisobutylene and C = butyl rubber;[8] and 2) with P = saturated ethylene propylene copolymer, C = ethylene propylene terpolymer [+ crosslinking agent (sulfur)].[9]

Another possibility is obtained with block copolymers AB to which one adds some chains (P) which are chemically identical to the A group. Starting with a disordered solution or melt, by suitable changes in temperature or solvent concentration, one can reach a final state where the B group segregates into micronodules. The A chains are then crosslinked by these nodules, and the P chains are trapped in a network.[10]

The best idea is probably to replace the network by a *melt* of chains that are chemically identical to P but much longer. Then the chains do not disentangle in the times of interest and behave effectively as a network. Also, the system is simple to prepare!

In all these processes, it is essential that the P chains never become attached to the network. As shown later, the behavior of a long chain which is attached at one end is very different from that of an unattached chain. It is also important (for similar reasons) that the P chain be unbranched.

Assume now that we do have one linear P chain moving in a given network, as shown in Fig. VIII.3 below (for a two-dimensional example). The

Figure VIII.3.

network is described by fixed obstacles O_1, O_2, etc. The chain P is not allowed to cross any of them, but it can move in between in a wormlike fashion. We call this "reptation." The basic reptation process is shown in Fig. VIII.4. Reptation is similar to unraveling a knot. We begin by accumulating a stored length in one portion of the knot, and then we circulate it to different loops, up to the moment we have relaxed an essential constraint.

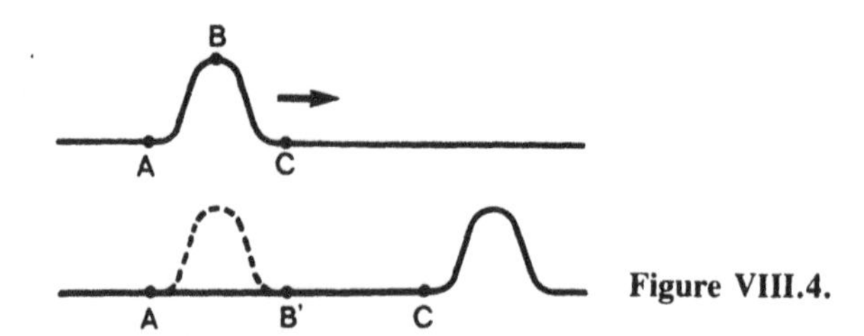

Figure VIII.4.

VIII.2.2. The terminal time, τ_t

To understand the effect of the obstacles at one moment, it is convenient to think of the chain as being trapped in a certain *tube* (Fig. VIII.5). The

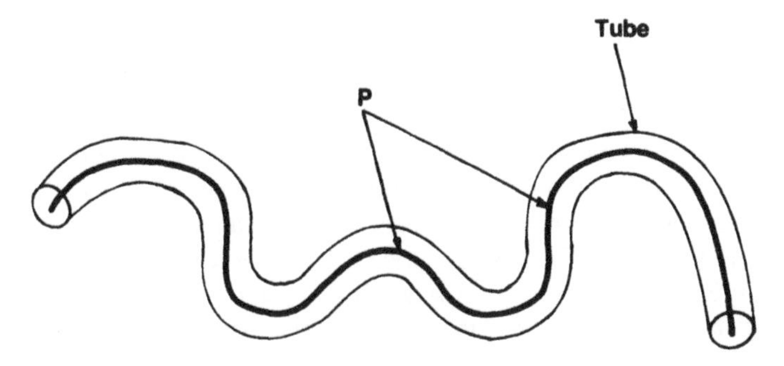

Figure VIII.5.

notion of a tube is due to S. F. Edwards.[11] Of course, the chain progresses by reptation. It leaves some parts of the tube, and it "creates" some new parts as shown in Fig. VIII.6.

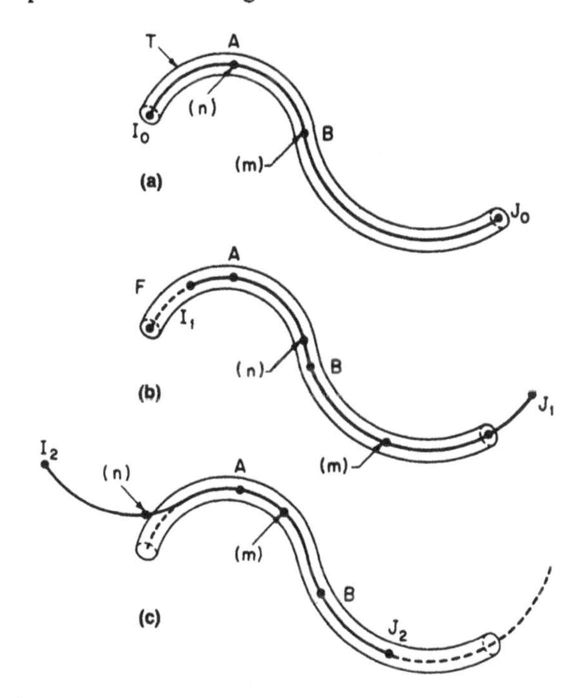

Figure VIII.6.

Free successive situations for a reptating chain: (a) the chain is trapped in its original tube; (b) the chain moves to the right and a certain portion (I,F) of the original tube disappears; (c) the chain moves to the left, and a portion (J_0J_2) of the original tube disappears.

A detailed statistical description of the fluctuations in the stored length and of their dynamic effects is available.[6,7] Here we present a simplified discussion, which allows us to predict the main scaling laws.

The terminal time τ_t is essentially the *time required for complete renewal of the tube*. We can derive the N dependence of τ_t by a simple argument. Let us assume for a moment that our chain is trapped inside one infinitely long tube. For motions along the tube we can introduce a "tube mobility" for the chain μ_{tube}. This is defined by applying a steady force f to the chain (along the tube direction) and measuring the resulting chain velocity along the tube $v = \mu_{tube} f$. We assume that long range backflow effects are negligible (which is correct for our dense systems). Then the friction force v/μ_{tube} must be essentially proportional to the length of the chain—i.e., to N. Thus we are led to

$$\mu_{tube} = \frac{\mu_1}{N} \qquad \text{(VIII.10)}$$

where μ_1 is independent of N. A similar property holds for the "tube diffusion coefficient," D_{tube}, which is related to μ_{tube} by an Einstein relationship

$$D_{tube} = \frac{\mu_1 T}{N} = \frac{D_1}{N} \qquad \text{(VIII.11)}$$

As is clear on Fig. VIII.6, to eliminate its original tube, the chain must progress by tube diffusion over a length comparable with its overall length L. The corresponding time is:

$$\tau_t \cong L^2/D_{tube} \cong \frac{NL^2}{D_1} \qquad \text{(VIII.12)}$$

and since L is linear in N, we expect the scaling form

$$\tau_t = \tau_1 N^3 \qquad \text{(VIII.13)}$$

How does this compare with mechanical measurements of chains trapped inside a network? Of course, the measurements are difficult since one must take the *difference* between the response of the doped and pure networks. However, a certain number of data have been taken, on the second and third systems listed at the beginning of this section.[8,9,10] The first results suggested $\tau \sim N^3$, but recent experiments give a higher exponent $\tau \sim N^{3.3}$—similar to what is measured in solutions. We hope that future measurements on the last class of systems may clarify the situation. Again an essential feature is to avoid all branching of the mobile chains, especially at high N.

Finally, notice how large the reptation time [eq. (VII.13)] can be. Assuming that our melt is far above any glass transition temperature T_g, τ_1 may be of order 10^{-11} sec. If we have a long chain ($N = 10^4$), this leads to $\tau_t \sim 10$ sec—i.e., to very macroscopic times.* Thus the reptation concept does give us a plausible feeling for the viscoelastic behavior of polymers.

VIII.2.3. Translational diffusion

In a time τ_t a reptating chain has moved along its tube by a length $L \sim Na$, but in space this motion corresponds to a much smaller displacement because the tubes are contorted; assuming an ideal chain, the displacement is $R_0 \cong N^{1/2} a$. After one such time τ_t all memory of the original conformation is lost, and successive time intervals of length τ_t are statistically independent. Thus, we can immediately estimate the translational diffusion coefficient of the reptating chain

$$D_{rep} \cong \frac{R_0^2}{\tau_t} = D_1 N^{-2} \qquad \text{(VIII.14)}$$

This diffusion coefficient is expected to be very small. The prefactor D_1 should be comparable with the diffusion coefficient in a liquid of monomers and of order 10^{-5} to 10^{-6} cm²/sec. If N is 100, this would lead us to $D \sim 10^{-9}$ cm²/sec, corresponding to a diffusion length $\sqrt{D_{rep}\, t}$ of only 0.3 mm for a diffustion time t of 10 days. If we go to higher N values, the process becomes even slower.

Experimentally, we do not have any data on the diffusion of trapped chains in a *network*. What is available is the diffusion coefficient of a labeled chain in polyethylene *melts,* measured by two techniques: 1) from nuclear spin resonance data,[12] and 2) using deuterated chains as labels, the local concentrations of deuterated/protonated species being probed by infrared measurements.[13]

The first experiment gives the dependence, $D_{chain} \sim N^{-5/3}$. The second experiment gives $D_{chain} \sim N^{-2\pm 0.1}$. In both cases N is small ($< 10^3$).

VIII.2.4. Reptation in swollen systems

Our discussion of reptation has been limited to completely dense systems—e.g., a dry network with closely spaced entanglement points or a polymer melt of very long chains, incorporating one extra "test chain." To

*However, some important prefactors may modify the value of τ_t: see the discussion after eq. (VIII.29).

avoid relaxation times which might become prohibitively long at high N, it may be convenient to work with a good solvent that is added the system. Then the structure will be more open, and all motions will accelerate.

Do we expect simple scaling laws for the dependencies on *concentration* of the terminal time, the self-diffusion constant, etc.? The answer is dubious. For static properties, we know from Chapter III that a semi-dilute solution can be pictured as a "melt of blobs." But it is not certain that this picture can also be applied to entanglement properties. The crucial point is the following: in a melt, we know from the mechanical data that a rather large number ($N_e \sim 200$) of consecutive monomers along one chain is required to make one entanglement (or one knot). But if we go to a solution at concentration c, the number $N_e(c)$ of *blobs* required to make an entanglement may decrease at low c. Even a single blob has a finite chance to become entangled, while this has been ruled out for a monomer.

Thus the equivalence between blobs and monomers is imperfect; and scaling laws for entanglements may be absent, or may be confined to unphysically high values of N.

In the following lines, we shall, however, describe the scaling laws which would occur in a *melt of blobs,* if the entanglement abilities of the blobs were *identical* to those of monomers. We know that this is too crude, but it may be a useful reference for comparison.

Also, for simplification, we shall not include explicitly the number N_e in our discussion, but assume that N_e is constant and ignore al powers of N_e. This is criminal, since $N_e \sim 200$, but it simplifies the presentation! (An improved discussion for melts, incorporating N_e, is summarized at the end of Section VIII.3.1.)

SOLUTIONS

We discuss first the case where the "network" is replaced by a semi-dilute solution of concentration c, composed of chains with a large degree of polymerization N_l. In this solution we add one "test chain" (N) and ask for the reptation time of the test chain $\tau_t(c)$. For simplicity we choose an athermal solvent ($a^3 = v$, or $\chi = 0$).

As usual, we can relate the semi-dilute region to the concentrated regime using blobs as our fundamental units. If, as in Chapter III, we call $g = (ca^3)^{-5/4}$ the number of monomers per blob, the number of blobs in the test chain is N/g, and the reptation time is, by a natural extension of eq. (VIII.13),

$$\tau_t(c) = \left(\frac{N}{g}\right)^3 \tau_{blob} \qquad \text{(VIII.15)}$$

where τ_{blob} is the relaxation time associated with one blob, of size ξ, and is given, in analogy with single coil properties, by a transposition of eq. (VI.48):

$$\tau_{blob} \cong \frac{\eta_s \xi^3}{T} \tag{VIII.16}$$

This leads to:

$$\tau_t(c) \cong \left(\frac{N\xi}{g}\right)^3 \frac{\eta_s}{T} = \frac{L_t^3 \eta_s}{T} \tag{VIII.17}$$

where $L_t = (N/g)\xi$ is the "primitive length of the tube" (made of N/g portions each of length ξ). Using the results of Chapter III for $\xi(c)$ and $g(c)$ the reptation time can be cast in the form

$$\tau_t(c) \cong \frac{\eta_s a^3}{T} N^3 (ca^3)^{3/2} \tag{VIII.18}$$

Eq. (VIII.18) can also be obtained by a direct scaling argument, imposing the restriction that τ_t be proportional to N^3 and to a certain power of c and demanding that $\tau_t(c^\star)$ be proportional to the single coil relaxation time [eq. (VI.48)]. However, the derivation used above has more physical significance. Reptation studies on solutions at variable c have been started recently.*

REPTATION IN A GEL

A similar law is expected to hold for swollen gels. Here the analog of g is the number of monomers per chain N' in the network, and the analog of ξ is the Flory radius $R_F\,(N')$ given in terms of the excluded volume parameter v (measuring the quality of the solvent) by eq. (IV.49). We take $N \gg N'$ to ensure that the chain is indeed trapped into a tube. Then we expect a reptation time of the form:

$$\tau_t \cong \frac{\eta_s R_F^3(N')}{T} \left(\frac{N}{N'}\right)^3 \tag{VIII.19}$$

$$\cong \frac{\eta_s a^3}{T} \left(\frac{v}{a^3}\right)^{3/5} \frac{N^3}{(N')^{6/5}} \tag{VIII.20}$$

*L. Léger, H. Hervet, F. Rondelez, *Macromolecules* **14,** 1732 (1981).

Since $N/R_F^3(N')$ scales like the concentration c in the gel, this may also be expressed in terms of c. The result is simple:

$$\tau_t = \frac{\eta_s a^3}{T} (cv)^{3/2} N^3 \qquad \text{(VIII.21)}$$

Eq. (VIII.21) has the same structure as eq. (VIII.18) but is slightly more general because here we allow for solvents of variable quality (variable v)—a useful feature for swollen gels. [In all applications of eq. (VIII.21) where the quality of the solvent is changed but N is fixed, we must recall that c is a decreasing function of v, as explained in Chapter V.]

VIII.2.5. Reptation of a branched chain

We now show that when the reptating chain is branched or has some long side groups, its motions are strongly quenched. A typical situation is shown in Fig. VIII.7 where only one side group CE_1 (carrying N_s monomers) is present. Let us assume that with an external force, we have pulled the main chain to the right, displacing the branch point from C_0 to C.

The distance C_0C, measured in units of the mesh size, is δ. The displacement $C_0 \rightarrow C$ reduces the entropy of the side group by an amount $\delta \ln z$, where z is the number of "gates" surrounding one unit cell in the network. As soon as p exceeds a few units, this entropy defect is large, and there is a strong elastic force that tends to bring C back to C_0.

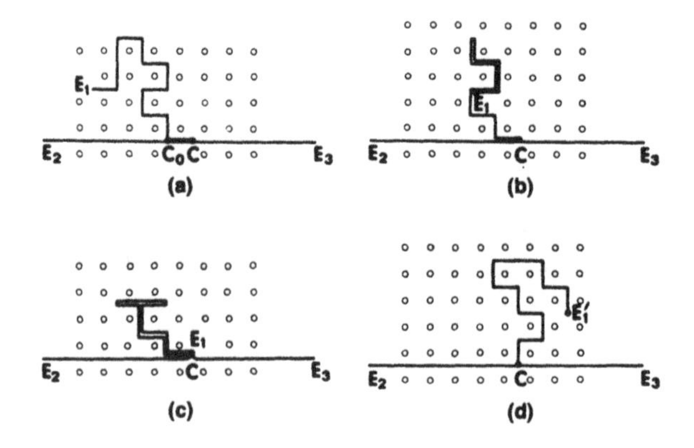

Figure VIII.7.

A reptating chain (E_2F_3) with a long side group (CE_1). When the chain is suddenly moved to the right (C_0 going to C), it takes a long time for the side group to relax (through steps b, c, d).

Let us fix p at some value for which the entropy defect is finite (say $\delta = 1$) and maintain the branch point at C by some external means. How long will it take for the lateral chain CE_1 to return to equilibrium? This requires that the original conformation CE_{10} be transformed into a completely different one, such as CE_1'. If the lateral chain were independent and free to reptate, it would achieve this during one reptation time $\tau_t(N_s) = \tau_1 N_s^3$, but in our problem C is fixed; the single pathway available from conformation CE_{10} to CE_1' amounts to having E_1 return to C, following exactly the original tube that surrounded (CE_1). After E_1 has retraced its steps to C, it will be able to start again and to generate a new tube CE_1'. It is only after all these steps are completed that the entropy defect of the lateral group will be eliminated.

This leads us to compute the probability that E_1 retraces its steps exactly. The total number of paths with N_s steps starting from C is z^{N_s}, where z is the number of nearest neighbors to one site on the lattice in Fig. VIII.7. We want to count the fraction P of these paths for which: 1) the end point is at the origin, and 2) the path is topologically equivalent to zero—i.e., it has the "tree" structure shown in Fig. VIII.8.

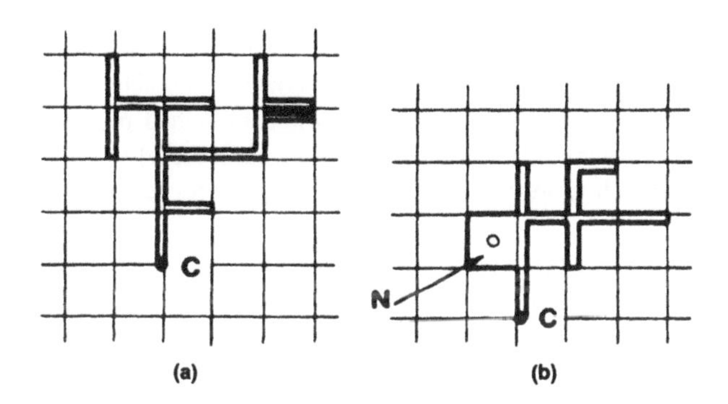

Figure VIII.8.

The quantity P can be calculated[14] and has the structure

$$P = A(N_s)\exp(-\alpha N_s) \qquad \text{(VIII.22)}$$

where $A(N_s)$ is a prefactor with only a weak dependence on N_s (a power law), and the leading factor is the exponential. The quantity α is a numerical constant that depends on the lattice structure and is of order unity.*

*The detailed calculation of α described in Ref. 14 by the present author is wrong.

We can assume that the tube mobility μ_{tube} for the backbone is essentially reduced by a factor P, or $\exp(-\alpha N_s)$, (omitting all weak prefactors). Thus, as soon as N_s is large, the reptation time for a chain with one (or more) side groups becomes exponentially long

$$\tau_{rep} \rightarrow \tau(N,N_s) \exp(\alpha N_s) \qquad \text{(VIII.23)}$$

where the prefactor $\tau(N,N_s)$ is not known precisely but is not essential for our discussion. The difference between eq. (VIII.13) for linear chains and eq. (VIII.23) for branched chains is spectacular. As soon as the extended length of the lateral chain exceeds a few mesh units of the network, it quenches all reptation.

This theoretical conclusion has not been confirmed by direct reptation experiments, but it has some implications. Mechanical measurements on strongly entangled, high molecular weight chains may be completely dominated by the presence of a *few* branch points. If exponential laws such as eq. (VIII.23) are involved, we need only a small fraction of branch points, and such fractions cannot be detected by standard physicochemical methods. We conclude that mechanical measurements in long chain systems can be extremely sensitive to certain chemical defects. Unfortunately, we do not have reptation data on controlled branched polymers. We do have data on mechanical properties of branched melts,[15] but the melt problem is much more complex than the reptation problem, as shown in next section.

A similar problem occurs when we have *long dangling ends* in the gel such as shown in Fig. VIII.9. The above discussion on renewal of conformations applies also to this type of chain which is attached at one end. Again we expect chain relaxation to be severely quenched whenever the dangling end is much longer that the mesh size.

When the relaxation time τ_{rep} becomes exponentially large (either for branched chains or for long dangling ends), the behavior at frequencies larger than $1/\tau_{rep}$ becomes important. Consider, for example, a network with one dangling end and assume that a constant *strain* e is applied from time $t = 0$ on. The network is strained; at early times the dangling chain is strained exactly in the same way and wishes to relax toward a more isotropic distribution of orientations. After a time t a number l of monomers, near the free extremity of the chain, have relaxed by processes similar to those in Fig. VIII.8. The relationship between l and t is the analog of eq. (VIII.23)

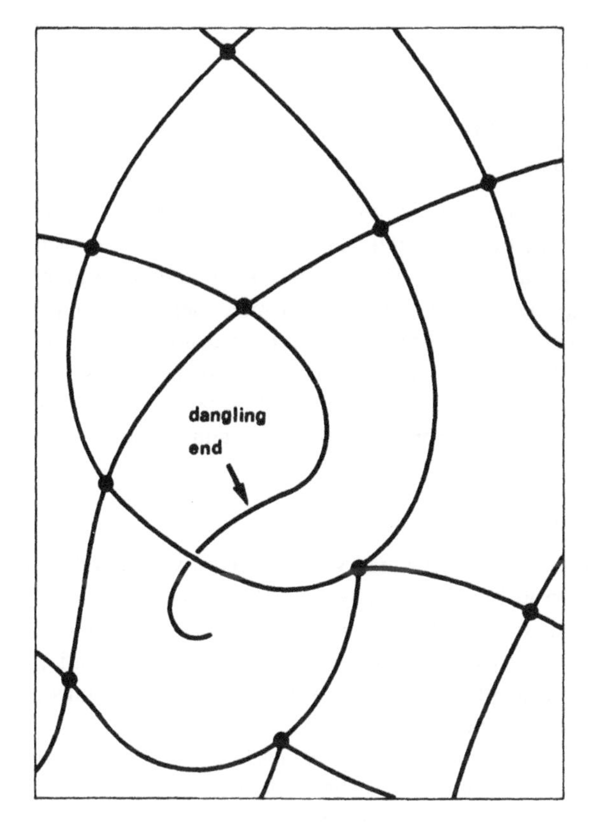

Figure VIII.9.

$$t \cong \tau_1 \exp(\alpha l)$$
$$l = \alpha^{-1} \ln\left(\frac{t}{\tau_1}\right) \qquad \text{(VIII.24)}$$

It is plausible to assume that the mechanical stress $\sigma(t)$ at time t will have a component proportional to $l(t)$

$$\sigma(t) = e\,[E_0 - kl(t)] \qquad \text{(VIII.25)}$$

[where k is proportional to the concentration of dangling ends].

Thus, we expect that with long dangling ends, the stress response to a stepwise strain relaxes logarithmically. This type of stress relaxation study may provide a direct check on the effects of topological constraints on branched structures.

VIII.3. Conjectures on Polymer Melts

VIII.3.1. One long chain in a melt of shorter chains

To understand the dynamics of one chain in a melt, it is convenient to start from a slightly different problem. We consider one "test chain" of N_1 monomers, embedded in a monodisperse melt of the same chemical species, with a number N of monomers per chain. We consider three types of motion for the test chain: reptation, "tube renewal," and Stokes-Einstein friction. We first describe tube renewal and show that this is probably negligible for most practical purposes. Then we discuss competition between reptation and Stokes-Einstein friction.

TUBE RENEWAL

The basic process associated with this word was introduced in Ref. 14 and is shown in Fig. VIII.10, where we see one entanglement constraint being altered: when one of the ambient chains (Γ) has an extremity in the immediate vicinity of the test chain (Γ_1), the relative positions of (Γ) and (Γ_1) may change qualitatively in a very short time. This may be viewed as a modification of the tube.

We then try to picture the tube itself as a Rouse chain, following early ideas of Edwards.[11] The basic parameter in this description is the microscopic jump frequency of one unit of tube—the analog of the con-

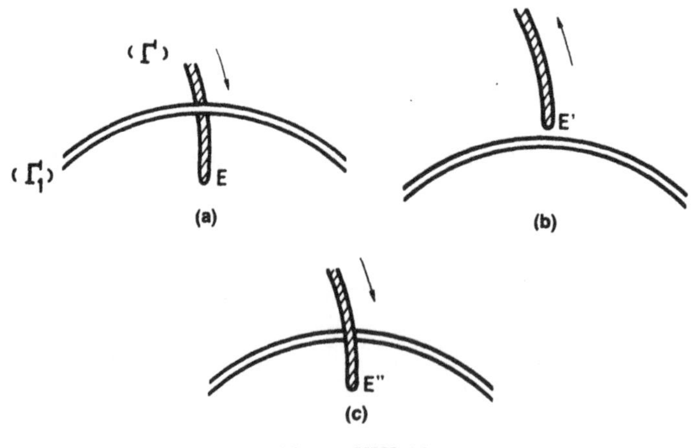

Figure VIII.10.

stant W in eq. (VI.8). Since the process shown in Fig. VIII.10 is a rare event, we expect this frequency to be much smaller than for a free chain: $W \rightarrow \tilde{W} \ll W$.

In an early attempt,[14] the present author assumed that W was simply reduced by a factor Ψ equal to the fraction of tube sites which are near the extremity of a Γ chain

$$W \rightarrow \tilde{W} = W\Psi$$
$$\Psi = \frac{2}{N} \qquad \text{(VIII.26)}$$

(since each Γ chain has N beads, out of which only two are at the ends). However, closer examination shows that the reduction in W is much stronger than suggested by eq. (VIII.26). Each costraint, due to one (Γ) chain, is relaxed only after one reptation time $\tau_{rep}(N)$. Thus, it is probably more correct to write, following an idea of J. Klein,*

$$\tilde{W} = \frac{1}{\tau_{rep}(N)} \qquad \text{(VIII.27)}$$

A more detailed justification of eq. (VIII.27) has been given recently by Daoud and the present author (to be published in *J. Polym. Sci.*). The frequency $\tilde{W}$ is very low. For example, it leads to a diffusion coefficient of the long chain

$$D_{ren}(N_1) = \tilde{W}a^2\frac{1}{N_1} = D_1\frac{1}{N_1N^3} \qquad \text{(VIII.28)}$$

This could dominate over the reptation value $D_{rep}(N_1) = D_1N_1^{-2}$ only if $N_1 > N^3$. For most practical situations the opposite inequality holds—i.e., *tube renewal is not observable.*

STOKES-EINSTEIN DIFFUSION

It must also be realized that for very large N_1, reptation itself is not the dominant mobility process for the long chain. For $N_1 \rightarrow \infty$ we can think of the N chains as forming a solvent of comparatively small molecules, with a certain viscosity η_N. This viscosity is discussed in the next section, and, in the reptation model, it scales like $\eta_N = \eta_1N^3$. Then the diffusion constant is given by the Stokes-Einstein equation

***Macromolecules* **11,** 852 (1978).

$$D_{Stokes}(N_1) = \frac{T}{6n\eta_N R(N_1)} \tag{VIII.29}$$

Using the ideal value $R(N_1) = 1/aN_1^2$, we can compare $D_{Stokes}(N_1)$ with $D_{rep}(N_1)$. We find that reptation dominates only when $N_1 < N^2$. Thus, we are led to expect two regimes for the test chain

(*i*) $N_1 < N^2$. The test chain is statistically ideal and moves by reptation.

(*ii*) $N_1 > N^2$. The test chain is swollen, and moves as a hydrodynamic sphere, dragging the solvent inside it.

PREFACTORS

Our scaling discussion ignores all detailed prefactors in formulas such as eq. (VIII.13) for reptation or eq. (VIII.23) for tube renewal. There is, however, a numerical constant which is important in melts and must be displayed explicitly. This is the average number of monomers between consecutive entanglement points N_e introduced in Section VIII.1.2. This number depends on the detailed geometry and flexibility of the monomers, but in most cases it is large ($N_e \sim 200$), and thus powers of N_e should be stated explicitly.

Elastic modulus	$E \cong T/(a^3\, N_e)$	(eq. VIII.8)
Reptation times	$T_{rep}\,(N) \cong \tau_{mono}\, N^3/N_e$	
Melt viscosity	$\eta \sim \eta_{mono} N^3/N_e^2$	

Here τ_{mono} and η_{mono} are microscopic parameters defined at the monomer scale. These equations can be derived either from a microscopic study at the scale of one chain portion between entanglements or more directly, by requiring that for N decreasing to N_e, there must be a crossover toward Rouse behavior.[16] A more detailed discussion is contained in recent papers by W. Graessley and co-workers.[15]

VIII.3.2. Newtonian viscosities in a homodisperse melt

As explained in Section VIII.1 the viscosity (η) appears as the product of an elastic modulus E and a relaxation time τ_t. In the reptation picture E is independent of N, while $\tau_t \sim N^3$, and η should thus scale like N^3 for a polymer melt of *linear* chains.

What happens if we go to branched chains? This has been discussed recently by Graessley et al.[15] Two effects come into play:

(i) For a given molecular weight, a branched chain is more compact than a linear chain; this tends to reduce the viscosity.

(ii) Entanglement effects are enhanced by chemical branching; this tends to increase the viscosity.

After a close scrutiny of the experimental data, Graessley et al.[15] were led to assume that relaxation is controlled by reptation, and that the reptation times for the branched species increase exponentially, as in eq. (VIII.23). Again the complete solution to these problems will probably involve experiments on dilute mixtures of branched chains in a linear matrix and vice-versa.

VIII.3.3. Behavior in strong transverse shear flows

Since the relaxation times τ_t are very long, it is relatively easy to realize shear rates (s) larger than $1/\tau_t$ and to measure the resulting properties. It is convenient to work in situations of permanent (s), where the molecules reach a steady state with finite distortions. This can be achieved only in transverse shears. For dilute solutions (Chapter VII) we discarded the transverse shear situations because they led to small (and thus complex) effects.[17,18] For concentrated solutions, the distortions become strong even in transverse shear flows, and we can restrict our attention to this more common case.

For example, in simple shear ($v_x = sy \cdot v_y = v_z = 0$) performed at finite s we may still define an effective viscosity $\eta(s)$. The literature in this area is vast and diffuse. However, there seems to be a convergence toward the following results.[2,18] The viscosity $\eta(s)$ appears to follow a universal scaling law

$$\eta(s) = \eta(0) f_\eta(s\tau_t(0)) \qquad \text{(VIII.30)}$$

where $\tau_t(0)$ is the terminal time in the absence of flow. The dimensionless function $f_\eta(x)$ has the following limits

$$\begin{aligned} f_\eta(0) &= 1 \\ f_\eta(x)\big|_{x \gg 1} &\cong x^{-p_\eta} \end{aligned} \qquad \text{(VIII.31)}$$

where p_η is of order 0.8.

The relaxation time $\tau_t(s)$ is much more difficult to measure, but it is plausible to assume that it follows a similar scaling relationship:

$$\tau_t(s) = \tau_t(0) f_\tau(s\tau_t(0)) \qquad \text{(VIII.32)}$$

where $f_\tau(x)$ is another dimensionless function.

If correct, eq. (VIII.30) represents an essential simplification for hydrodynamic studies. It implies that $s\tau_t(s)$ is a function only of $s\tau_t(0)$, and thus that a single dimensionless parameter $s\tau_t(0)$ is required to characterize non-newtonian effects in the flow.* At large x we find $f_T(x) \sim x^{-p_T}$ with p_T values which are of order 0.8.

No fundamental justification is available for these scaling laws. An interesting discussion of transient entanglements is given by Graessley.[19] Another useful phenomenological picture, based on the kinetics of entanglement points, was constructed by Marucci *et al.*[20]

VIII.3.4. Critical dynamics in entangled binary mixtures

We consider here a solution (or melt) with two types of polymer chains, A and B, and a slight incompatibility between A and B. We want to investigate the dynamics of the fluctuations of concentration $\delta\Phi$ near the critical point in the one-phase region. Experimentally it will help if we add a fraction of solvent to the AB mixture, thereby decreasing the viscosities and relaxation times. However, for simplicity here we consider only the case without solvent and a symmetrical situation ($N_A = N_B = N$).

Dynamically, there is an essential difference between this case and the more common problem of one polymer plus one solvent. In the latter problem, near the critical point, backflow is essential, while entanglements are minor; but with a molten mixture of chains, the situation is reversed: backflow is negligible, and entanglements are essential.

In practice what is measured is a cooperative diffusion coefficient D_c which controls the fluctuations of Φ according to:

$$\frac{\partial \Phi}{\partial t} = D_c \nabla^2 \Phi \qquad \text{(VIII.33)}$$

Our discussion of D_c proceeds in the two steps described below.

When the Flory interaction parameter χ vanishes, the concentration is equalized by reptation and by tube renewal, and we expect to have, according to eq. (VIII.14)

$$D_c\ (\chi = 0) = D_1\ N^{-2} \qquad \text{(VIII.34)}$$

*The above presentation is restricted to flows where each molecule sees a constant shear rate. However, there are reasons to believe that only one "Deborah number" $U\tau/L$ (where U and L are characteristic velocities and lengths in the flow) is required for more general steady flows.

where D_1 is a microscopic constant, independent of N but dependent on temperature.

When χ is finite and close to the critical value $\chi_c = 2/N$ [eq. (IV.11)], the restoring force which tends to equalize the concentrations becomes very small. This can be understood from the structure of the free energy per site [eqs. (IV.6, IV.12)]

$$\left.\frac{F}{T}\right|_{site} = (\chi_c - \chi)\, \delta\Phi^2 + O\,(\delta\Phi^4) \qquad \text{(VIII.35)}$$

where $\delta\Phi = \Phi - 1/2$ measures the fluctuations from the critical concentration. The restoring force (conjugate to Φ) is

$$-\frac{\partial F}{\partial \Phi} = -\,T(\chi_c - \chi)\, 2\delta\Phi \qquad \text{(VIII.36)}$$

and is very small near $\chi = \chi_c$. To see how this is reflected in the diffusion equation we write the diffusion current of A monomers in the form

$$J = -\,L\nabla\mu \qquad \text{(VIII.37)}$$

where $\mu = \partial F/\partial\Phi$ is the exchange chemical potential, and L is a transport coefficient. Our central assumption, originally proposed by Van Hove[22] is to assume that L has no singularities near the critical point—i.e., all critical properties are included in μ [eq. (VIII.36)]. Inserting eq. (VIII.36) into eq. (VIII.37), we find

$$J = -\,LT\,(\chi_c - \chi)\, 2\nabla\Phi \qquad \text{(VIII.38)}$$

and we can identify the coefficient with D_c

$$D_c = 2LT\,(\chi_c - \chi)\, a^{-3} \qquad \text{(VIII.39)}$$

We can eliminate the constant L by returning to a situation of zero χ described by eq. (VIII.34). The result is

$$D_c\,(\chi, T) = D_1\,(T)\, N^{-2}\, \frac{\chi_c - \chi}{\chi_c} \qquad \text{(VIII.40)}$$

Eq. (VIII.40) includes three major factors that act on diffusion: 1) temperature through $D_1\,(T)$, 2) entanglements through N^{-2}, and 3) weakness of restoring forces near $\chi = \chi_c$ through the last factor $(\chi_c - \chi)/\chi_c$. This third feature is typical of critical phenomena and has been called "thermodynamic slowing down."[23,24]

We do not have any data to compare with eq. (VIII.40). However some remarks may be useful at this stage. The diffusion described by eq. (VIII.40) is extremely slow. Factors 2 and 3 tend to make D_c small. Further, the Van Hove assumption is not rigorous (it neglects some weak singularities that are known theoretically for simpler cases,[24] but is should be adequate for the first studies on these difficult systems. Finally, why are backflow corrections negligible? In analogy with eq. (VII.33) we could think of backflow contributions to D_c of order $T/6\pi\eta\xi_s$ where ξ_s is the large correlation length observed in critical phenomena [eq. (IV.24)], but here the viscosity η is the viscosity of an entangled system and is very large. Using $\eta \sim \eta_s N^3$ and comparing with eq. (VIII.40) we can check that backflow may be omitted at large N:

$$\frac{D_{backflow}}{D_c} \sim N^{-3/2}\left(\frac{\chi_c}{\chi_c - \chi}\right)^{1/2} \ll 1 \qquad \text{(VIII.41)}$$

VIII.3.5. Summary

In an entangled melt of chains the fundamental relaxation time τ_t scales like a strong power of the degree of polymerization N ($\tau_t \sim N^{3.3}$). The reptation model attempts to describe τ_t by a calculation of the wiggling motions of one chain inside of a "tube" formed by its neighbors. It leads to a somewhat weaker exponent $\tau_t \sim N^3$. The discrepancy is unexplained.

REFERENCES

1. J. D. Ferry, *Viscoelastic Properties of Polymers,* 2nd ed., John Wiley and Sons, New York, 1970. J. D. Ferry, *Molecular Fluids,* R. Balian and G. Weill, Eds., Gordon & Breach, New York, 1976.
2. W. Graessley, *Adv. Polym. Sci.* **16** (1974).
3. F. Bueche, *J. Chem. Phys.* **25,** 599 (1956).
4. S. F. Edwards, J. W. Grant, *J. Phys.* **A6,** 1169, 1186 (1973).
5. M. Green, A. Tobolsky, *J. Chem. Phys.* **14,** 80 (1946). A. Lodge, *Trans. Faraday Soc.* **52,** 120 (1956). M. Yamamoto, *J. Phys. Soc. Japan* **11,** 413 (1956). G. Ronca, *Rheol. Acta* **15,** 149, 156 (1976).
6. P. G. de Gennes, *J. Chem. Phys.* **55,** 572 (1971).
7. M. Doi, S. F. Edwards, *J.C.S., Faraday Trans.,* II, **74,** 1789, 1802, 1818 (1978).
8. O. Kramer, R. Greco, R. Neira, J. D. Ferry, *J. Polym. Sci., Polym. Phys.* **12,** 2361 (1974).

9. O. Kramer, R. Greco, J. D. Ferry, *J. Polym. Sci., Polym. Phys.* **13,** 1675 (1975).
10. L. Toy, M. Miinomi, M. Shen, *J. Macromol. Sci.* **B11,** 281 (1975). G. Kraus, K. Rollman, *J. Polym. Sci., Polym. Phys* **15,** 385 (1977).
11. For a review see S. F. Edwards, *Molecular Fluids,* R. Balian and G. Weill, Eds., Gordon & Breach, New York, 1976.
12. D. McCall, D. Douglass, E. Anderson, *J. Chem. Phys.* **30,** 771 (1959).
13. J. Klein, *Nature* **271,** 143 (1978).
14. P. G. de Gennes, *J. Phys. (Paris)* **36,** 1199 (1975).
15. G. Berry, T. Fox, *Adv. Polym. Sci.* **5,** 261 (1968). T. Masuda, Y. Ohta, S. Onogi, *Macromolecules* **4,** 763 (1971). L. Utracki, J. Roovers, *Macromolecules* **6,** 366, 373 (1973). M. Suzuki, Ph.D. Thesis, Strasbourg, 1972. W. Graessley, T. Masuda, J. Roovers, N. Hajichristidis, *Macromolecules* **9,** 127 (1976).
16. M. Doi, *Chem. Phys. Lett.* **26,** 269 (1974).
17. A. Peterlin, *Adv. Macromol. Chem.* **1,** 225 (1968).
18. *Proc. Vth Int. Conf. Rheol.,* Tokyo University Press, Tokyo.
19. W. Graessley, *J. Chem. Phys* **43,** 2696 (1965).
20. G. Marruci, G. Titomanlio, G. Sarti, *Rheol. Acta* **12,** 269 (1973).
21. P. E. Rouse, *J. Chem. Phys.* **21,** 1273 (1953).
22. L. Van Hove, *Phys. Rev.* **95,** 1374 (1954).
23. The word was introduced in the present author's Ph.D. work (Paris, 1957).
24. For a general discussion see P. Hohenberg, B. Halperin, *Rev. Mod. Phys.* **49,** 435 (1977).

Part C

CALCULATION METHODS

IX

Self-Consistent Fields and Random Phase Approximation

IX.1. General Program

Our aim is to find a relatively simple description for one (or more) interacting chains. The chains can also be restricted by certain boundary conditions—e.g., they may be confined in a pore or attracted to an adsorbing surface or attached at one end at a given point, and so forth.

The natural approach—initiated by the classic workers (Kuhn, Hermans, Flory, etc.) and formalized later by Edwards[1,2]—is based on the idea of a self-consistent field. We describe it for a typical case where: 1) all monomers are chemically identical, and 2) the interactions are repulsive and local (no long range forces). We write the interaction between monomers (i) and (j) in the form $vT\delta(\mathbf{r}_{ij})$, where v is the excluded volume parameter defined in eq. (III.10). This form is adequate for uncharged molecules in semi-dilute (or dilute) solutions with good solvents.

The procedure is then as follows. One assumes a certain concentration profile $c(\mathbf{r})$. To this profile one associates an average repulsive potential, proportional to the local concentration:

$$U(\mathbf{r}) = Tvc\,(\mathbf{r}) \tag{IX.1}$$

One then describes each chain as an ideal chain subjected to an external potential $U(\mathbf{r})$. This type of calculation is feasible, and is discussed in Section IX.1.2. At the end of this stage, one can compute all chain properties and in particular derive a new value $c'(\mathbf{r})$ for the local concentration.

One then iterates the procedure, defining a new potential $U' = Tvc'$ and solving again. We hope that the sequence of approximations $c(\mathbf{r}) \rightarrow c'(\mathbf{r}) \rightarrow c''(\mathbf{r}) \rightarrow \ldots$ converges to a stable solution, $c_{final}(\mathbf{r})$. The potential $U_{final} = vTc_{final}$ is then self-consistent.

The entire calculation has some analogies with the Hartree method for electrons in atoms or molecules,[3] and the level of complexity is roughly the same. In this chapter, we do not discuss any numerical calculations, but we do give some exercises where the self-consistent method can be applied directly.

In the theory of many-body systems, the Hartree approximations can generally be augmented by an application of self-consistent methods, not to one body concentration $c(\mathbf{r})$ but to *two-body properties*, such as the pair correlations $g(\mathbf{r}_1\ \mathbf{r}_2)$ which we have discussed often. This corresponds to the "random phase approximation" (RPA) introduced by Bohm, Pines, and Nozières.[4] In electron systems, RPA is useful mainly for *nearly free electrons*. Similarly, for our chain systems, RPA will work for *nearly ideal chains*—i.e., in melts. The corresponding experiments are essentially based on neutron scattering with labeled molecules and are summarized briefly in Chapter II.

IX.2. Self-Consistent Fields

IX.2.1. An ideal chain under external potentials

For definiteness, we inscribe our chain on a Flory-Huggins lattice of parameter a. It is then described by a walk of N steps linking the lattice points $\mathbf{r}_1 \ldots \mathbf{r}_N$. If a potential $U(\mathbf{r})$ acts on each monomer, the statistical weight associated with this particular realization is

$$\exp\left(-\frac{1}{T}[U(\mathbf{r}_1) + U(\mathbf{r}_2) + \ldots + U(\mathbf{r}_N)]\right) \qquad \text{(IX.2)}$$

Let us consider the sum of all such weights on paths with fixed ends ($r_1 = \mathbf{r}'$ and $r_N = \mathbf{r}$, fixed). We call this sum

$$z^N\, G_N\, (\mathbf{r}',\mathbf{r}) \qquad \text{(IX.3)}$$

where z is the number of neighbors of one site on the lattice. (It is convenient to extract the factor z^{-N}, which is simply the total number of paths of N steps.) The function G_N $(\mathbf{r}',\mathbf{r})$ is real, positive, and symmetric $G_N(\mathbf{r}',\mathbf{r}) \equiv G_N(\mathbf{r},\mathbf{r}')$. Of course, if N is zero, G_N reduces to a delta function*

$$G_o\ (\mathbf{r}',\mathbf{r}) = \delta_{\mathbf{r}'\,\mathbf{r}} \tag{IX.4}$$

It is easy to find an equation for G_N by adding one unit to the chain:

$$G_{N+1}\ (\mathbf{r}',\mathbf{r}) = \frac{1}{z}\sum_{\mathbf{r}''}{}' \ G_N\ (\mathbf{r}',\mathbf{r}'') \exp(-U(\mathbf{r})/T) \tag{IX.5}$$

where the sum Σ' is over all sites which are neighbors of ($\mathbf{r}$). Eq. (IX.5) says that any path, of $N + 1$ steps, reaching $\mathbf{r}$, must have reached one of these neighbors at the next-to-the-last step. The factor $1/z$ results from the normalization chosen in eq. (IX.3).

In most applications we are primarily interested in spatial scales which are much larger than the lattice parameter a. Thus, we may assume that G is a slowly varying function (of N and of $\mathbf{r}$). We also assume that U/T is small. In most cases of interest this may be verified *a posteriori* (what is of order unity is NU/T rather than U/T). With these simplifications we write eq. (IX.5) in the form

$$G_{N+1}\ (\mathbf{r}',\mathbf{r}) = z^{-1}\left(1 - \frac{U(\mathbf{r})}{T}\right) \sum_{\mathbf{r}''}{}' \left\{ G_N\ (\mathbf{r}',\mathbf{r}) + (\mathbf{r} - \mathbf{r}'')\ \frac{\partial G_N}{\partial \mathbf{r}} + \frac{1}{2}(\mathbf{r} - \mathbf{r}'')_\alpha\ (\mathbf{r} - \mathbf{r}'')_\beta \frac{\partial^2 G}{\partial \mathbf{r}_\alpha \partial \mathbf{r}_\beta} + \ldots \right\} \tag{IX.6}$$

On the right side, the terms linear in $\mathbf{r} - \mathbf{r}''$ vanish for a centrosymmetric lattice, and we get:

$$\begin{aligned} G_{N+1}\ (\mathbf{r}',\mathbf{r}) - G_N\ (\mathbf{r}',\mathbf{r}) &\cong \frac{\partial G_N}{\partial N}\ (\mathbf{r}',\mathbf{r}) \\ &= -\frac{U(r)}{T}\ G_N\ (\mathbf{r}',\mathbf{r}) + \frac{a^2}{6}\ \nabla^2 G_N\ (\mathbf{r}',\mathbf{r}) + \ldots \end{aligned} \tag{IX.7}$$

where we have used the geometrical sum property (for a cubic lattice in three dimensions)

*When we deal with a discrete lattice we use a discrete delta function which is 1 if $\mathbf{r}' = \mathbf{r}$ and zero for different sites. When we go to a continuous limit, this becomes $\delta(\mathbf{r} - \mathbf{r}')\ a^3$.

$$\frac{1}{z}\sum_{\mathbf{r}}{}' \; (\mathbf{r} - \mathbf{r}'')_\alpha (\mathbf{r} - \mathbf{r}'')_\beta = \delta_{\alpha\beta}\frac{a^2}{3} \tag{IX.8}$$

It is usual to rewrite eq. (IX.7) as:

$$-\frac{\partial G}{\partial N} = \frac{-a^2}{6}\nabla^2 G + \frac{U(\mathbf{r})}{T} G \tag{IX.9}$$

because eq. (IX.9) has a remarkable similarity to another celebrated equation of theoretical physics—namely, the Schrödinger equation for a non-relativistic particle of wavefunction $\psi(\mathbf{r}, t)$. The latter is

$$-i\hbar\frac{\partial\psi}{\partial t} = -\frac{\hbar^2}{2m}\nabla^2\psi + V(\mathbf{r})\psi \tag{IX.10}$$

where t is the time, $\hbar = h/2\pi$ is the reduced Planck constant, m is the mass of the particle, and $V(\mathbf{r})$ is the potential energy. Note that $V(\mathbf{r})$ is the analog of $U(\mathbf{r})/T$. Potential energies play similar roles in both problems. Note also that N is (apart from a factor i) the analog of time. One particular chain conformation corresponds to one particular path for the particle, and the wavefunction appears as a coherent superposition of amplitudes for different paths.[5]

This analogy—between ideal chain statistics under external potentials and a quantum mechanical problem—is often useful because 50 years of manipulations of the Schrödinger equation have given us a wide spectrum of solution methods.[6] However, in this book, we will not assume any detailed knowledge of quantum mechanics by the reader.

Eq. (IX.9), supplemented by the boundary condition of eq. (IX.4), defines the statistical weight $G_N(\mathbf{r}',\mathbf{r})$ completely. In some cases one should solve eq. (IX.9) directly. In many other cases, however, it is more interesting to use an *expansion in eigenfunctions* which we now describe.

We introduce a linear operator $\mathfrak{H}$ corresponding to the right side of eq. (IX.9)

$$\mathfrak{H} = -\frac{a^2}{6}\nabla^2 + \frac{U(\mathbf{r})}{T} \tag{IX.11}$$

Then we introduce the set of eigenfunctions $u_1(\mathbf{r}), u_2(\mathbf{r}), \ldots u_k(\mathbf{r}) \ldots$* They are such that $\mathfrak{H}u_k$ is proportional to u_k

*It is possible to restrict the functions u_k to being real, but it is sometimes more convenient to allow them to be complex—just as $\exp(ikx)$ is more convenient than $\cos(kx)$ for many wave problems.

$$\mathfrak{H} u_k(\mathbf{r}) = \epsilon_k u_k(\mathbf{r}) \qquad \text{(IX.12)}$$

The numbers ϵ_k for a sequence, starting from a minimal value ϵ_0 which we call the ground state. It may be shown that the u_k values satisfy two important relationships:[6]

$$\int u_k^\star(\mathbf{r})\, u_l(\mathbf{r})\, d\mathbf{r} = \delta_{kl} \quad \text{(orthogonality)} \qquad \text{(IX.13)}$$

$$\sum_k u_k^\star(\mathbf{r}')\, u_k(\mathbf{r}) = \delta(\mathbf{r} - \mathbf{r}') \quad \text{(closure)} \qquad \text{(IX.14)}$$

The explicit form of $G_N(r',r)$ is an expansion in the eigenfunctions u_k:

$$G_N(\mathbf{r}',\mathbf{r}) = a^3 \sum_k u_k^\star(\mathbf{r}')\, u_k(\mathbf{r}) \exp(-N\epsilon_k) \qquad \text{(IX.15)}$$

The reader may check that this G_N satisfies the differential equation [eq. (IX.9)]. It also obeys the boundary condition of eq. (IX.4) as seen from the closure property, eq. (IX.4). Thus, it is indeed the solution.

One essential property of the weights $G_N(\mathbf{r}',\mathbf{r})$ (for ideal chains under external forces) is the *composition law,* which we write

$$G_N(\mathbf{r}',\mathbf{r}) = \sum_{\mathbf{s}} G_{N'}(\mathbf{r}',\mathbf{s})\, G_{N-N'}(\mathbf{s},\mathbf{r}) \qquad \text{(IX.16)}$$

Physically this means that we can divide the chain into two consecutive sections (N') and $(N - N')$ with a certain junction point $\mathbf{s}$. We then compute the weights for the first section extending from $\mathbf{r}'$ to $\mathbf{s}$, and for the second section, extending from $\mathbf{s}$ to $\mathbf{r}$. The product of these two weights, summed over all possible positions of the junction,* gives us $G_N(\mathbf{r}',\mathbf{r})$. [We have already used eq. (IX.16) in eq. (IX.5), where we separated a chain of $N + 1$ links into (N) and (1).]

The composition law [eq. (IX.16)] is very specific for noninteracting chains. If we had interactions between N' and $N - N'$, the weights could not be factored into two terms. Technically we can verify eq. (IX.16) on the eigenfunction expansion [eq. (IX.15)] using the orthogonality of the u_k values.

In a certain sense, eq. (IX.15) solves our problem. If we look at *one* chain under the external potential $U(\mathbf{r})$, we first derive the eigenfunctions $u_k(r)$ by direct calculations. Then we can construct $G_N(\mathbf{r}',\mathbf{r})$ and derive any physical property of interest. If we wish to know the concentration $c(\mathbf{s})$

*In the continuous limit we write $\sum_{\mathbf{s}} \to \frac{1}{a^3}\int ds$.

at one point s, we look at the statistical weight for a chain, starting from an arbitrary point $\mathbf{r}'$, reaching s and then extending to an arbitrary point $\mathbf{r}$

$$\Phi(\mathbf{s}) = a^3\, c(\mathbf{s}) = \frac{\sum_{\mathbf{r}'} \sum_{\mathbf{r}} \sum_{N'=0}^{N} G_{N'}(\mathbf{r}',\mathbf{s})\, G_{N-N'}(\mathbf{s},\mathbf{r})}{\sum_{\mathbf{r}'} \sum_{\mathbf{r}} G_N(\mathbf{r}',\mathbf{r})} \qquad \text{(IX.17)}$$

The denominator of eq. (IX.17) is the sum of all weights and ensures the normalization. If we have many independent chains in the same external potential $U(\mathbf{r})$, we simply multiply eq. (IX.17) by the total number of chains.

IX.2.2. Situations of ground state dominance

The eigenfunction expansion [eq. (IX.15)] contains a factor $exp(-\epsilon_k N)$ that tends to give maximum weight to the ground state wavefunction $u_0(\mathbf{r})$ for which $\epsilon_k = \epsilon_0$ is minimum. For certain situations, it may be enough to retain only the term $k = 0$ in the expansion. We then say that the ground state is dominant, and we give a special name to the corresponding eigenfunction $u_0(\mathbf{r}) = \psi(\mathbf{r})$. If we retain only $\psi(r)$ in eq. (IX.15) and if we use eq. (IX.17), we arrive at the very formula:

$$c(\mathbf{s}) = N\,|\psi(\mathbf{s})|^2 \qquad \text{(IX.18)}$$

Note that c is proportional to the *square* of ψ; this is because *two* chain portions converge at point s, each carrying a factor ψ. The normalization in eq. (IX.18) is easily checked since $\int c(\mathbf{s})ds = N$ for a one-chain problem.

The validity of the truncation is, of course, limited. One has to check that the intervals $\epsilon_k - \epsilon_0$ are large enough to make the ground state dominant, and the answer may depend also on the particular quantity which is computed. Below we give some examples where ground state dominance can be justified.

Exercise 1: One ideal chain confined between two strongly repulsive walls (separation $D < R_0$). Compute: 1) the reduction of entropy, and 2) the concentration profile (E. Cassasa).[7]

Answer: If we put the walls at $x = 0$ and $x = D$, the eigenfunctions u_k must vanish at both walls. Inside the interval they satisfy $\epsilon_k\, u_k = (-a^2/6)\, \nabla^2 u_k$. The ground state is simply

$$\psi = u_0 = \text{constant} \sin\left(\frac{\pi x}{D}\right) \tag{IX.19}$$

and the corresponding eigenvalue is

$$\epsilon_0 = \frac{1}{6}\left(\frac{\pi a}{D}\right)^2 \tag{IX.20}$$

The main factor in the statistical weight is $\exp(-\epsilon_0 N)$, giving an entropy reduction

$$\Delta S = -\frac{\pi^2}{6}\frac{R_0^{\,2}}{D^2} \tag{IX.21}$$

in agreement with the scaling form of eq. (I.12).

The concentration profile is deduced from eq. (IX.18):

$$c = \text{constant} \sin^2\left(\frac{\pi x}{D}\right) \tag{IX.22}$$

It vanishes quadratically near both plates (see Fig. IX.1).

If we investigate the validity of ground state dominance for these two specific questions (ΔS and $c(x)$), we find that there is a whole branch of states near the ground state; the general structure of the eigenfunctions is

$$u_{n,q}(xyz) = (\text{constant}) \sin\left(\frac{\pi n x}{D}\right) \exp(i(q_y y + q_z z)) \tag{IX.23}$$

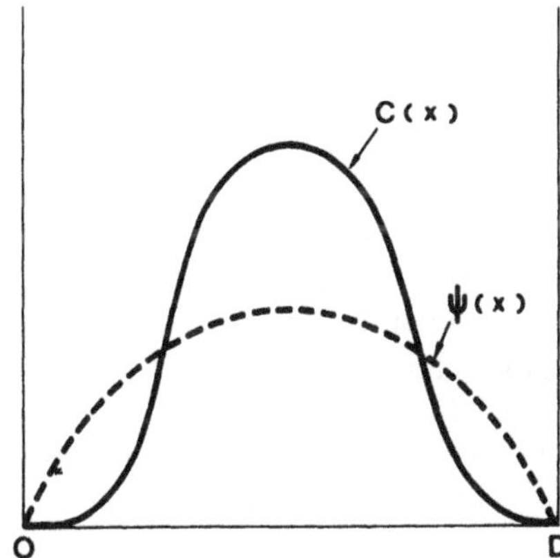

Figure IX.1.

where $n = 1, 2, 3$, etc. The eigenvalues are:

$$\epsilon_{n,q} = \frac{a^2}{6}\left[\left(\frac{\pi}{D}\right)^2 n^2 + q^2\right] \tag{IX.24}$$

Thus near $q = 0$ there are many eigenstates $(u_{1,q})$ with eigenvalues near ϵ_0; they would be important for a complete calculation of $G_N(\mathbf{r}',\mathbf{r})$, but fortunately they are not important for ΔS and for $c(x)$. What is essential here is the interval between the eigenvalues for $n = 1$ and higher n. The error (on ΔS or c) due to the assumption of ground state dominance is of the order

$$\exp\left[-N(\epsilon_{10} - \epsilon_{00})\right] \sim \exp\left[-\text{(constant)}\frac{R_0^2}{D^2}\right] \tag{IX.25}$$

and it is small if $R_0 \gg D$.

Exercise 2: An ideal chain is weakly adsorbed on a flat surface. Find the gain in free energy and the concentration profile.

Answer: The chain sees a potential of the form shown in Fig. IX.2, where the range b of interaction is assumed to be small (of order a). Outside of this range the ground state wavefunction is ruled by the simple equation

$$-\frac{a^2}{6}\frac{d^2\psi}{dx^2} = \epsilon_0 \psi \tag{IX.26}$$

The effect of the potential is essentially to impose a boundary condition[2]

$$\frac{1}{\psi}\frac{d\psi}{dx}\bigg|_{x \cong 0} = -\kappa \tag{IX.27}$$

where κ is a positive parameter when the attraction dominates. For a given potential shape, κ may be computed explicitly by solving the eigenvalue equation in the attractive region. [This part of the calculation cannot be done in the continuous approximation because the potential varies rapidly in space near the surface; return to eq. (IX.5).] We are mainly interested in the case of small κ ($\kappa a \ll 1$). This defines weak adsorption. The solution ψ outside of the attractive wall is then

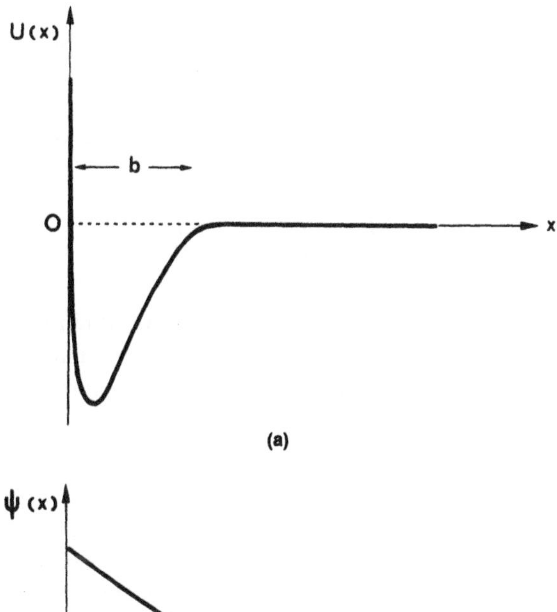

Figure IX.2.

$$\psi = (\text{constant}) \exp(-\kappa x) \qquad (x > b) \tag{IX.28}$$

with an eigenvalue

$$\epsilon_0 = -\frac{a^2}{6}\kappa^2 \tag{IX.29}$$

We conclude that: 1) the concentration decreases as $\exp(-2\kappa x)$, and 2) the change in free energy due to adsorption is (per chain)

$$\Delta F = -T\frac{Na^2}{6}\kappa^2 \tag{IX.30}$$

In this case we can check that the assumption of ground state dominance is correct when $\kappa R_0 \ll 1$—i.e., when the chain is confined within a thickness much smaller than its natural size. Note that the result for ΔF for ideal chains agrees with our scaling arguments of Chapter I, where we looked for confinement in a thickness $D = \kappa^{-1}$ and wrote the free energy as a sum of localization and attraction contributions:

$$\frac{\Delta F}{T} = \frac{R_0^2}{D^2} - N\delta \frac{a}{D}$$

($T\delta$ is an effective attractive toward the surface.) Minimizing ΔF gave $D \cong a\delta^{-1}$ or $\delta \cong \kappa a$. This gives a microscopic meaning to δ; from a precise calculation of the ground state wavefunction near the surface and of the resulting κ, we can calculate δ.

IX.2.3. Self-consistency with ground state dominance[8]

Whenever ground state dominance holds, it is possible to simplify considerably the self-consistent program sketched in the introduction to this chapter. Fundamentally, what we have is a concentration proportional to $|\psi|^2$. In this section we normalize ψ differently, replacing eq. (IX.18) by

$$c = |\psi|^2 \tag{IX.31}$$

Then the self-consistent potential is

$$U(\mathbf{r}) = Tvc(\mathbf{r}) = Tv\, |\psi(\mathbf{r})|^2 \tag{IX.32}$$

Consider the following integral

$$I = \int \left\{ \frac{a^2}{6} |\nabla\psi|^2 + \frac{1}{2} v\, |\psi|^4 \right\} d\mathbf{r} = F/T \tag{IX.33}$$

Physically it contains an entropy term $(\nabla\psi)^2$ and an interaction term, and can be interpreted as a free energy F (divided by T). Let us demand that I is a minimum for all variations of ψ which keep the total number of monomers constant

$$N_{tot} = \int |\psi|^2 d\mathbf{r} \tag{IX.34}$$

The latter condition is included through a Lagrange multiplier, writing that the variations δI and δN_{tot} be related by

$$\delta I = \epsilon \delta N_{tot} \tag{IX.35}$$

writing δI and δN_{tot} for arbitrary variations of ψ, $(\delta\psi(\mathbf{r}))$, at each point, we find

$$\epsilon\psi = -\frac{a^2}{6}\nabla^2\psi + v|\psi|^2\psi \tag{IX.36}$$

which is exactly the eigenvalue equation (IX.15) but with the self-consistency condition [eq. (IX.20)] imposed on the potential.

The program is then much simpler. We try to solve the nonlinear equation [eq. (IX.36)] and, if we succeed, we know the self-consistent potential.

Exercise 1: Find the density profile for a semi-dilute solution, in good solvent, near a repulsive wall (in the self-consistent field approximation).

Answer: If the wall is located at $x = 0$, the boundary conditions on $\psi(x)$ are

$$\begin{aligned} \psi(0) &= 0 \\ \psi(x \to \infty) &= c^{1/2} \end{aligned} \tag{IX.37}$$

where c is the bulk concentration. We set $\psi = c^{1/2} f(x)$.

Eq. (IX.36) then becomes

$$\frac{a^2}{6}\frac{d^2f}{dx^2} = -\epsilon f + vcf^3 \tag{IX.38}$$

Far from the wall we must have $f = 1$ and $d^2f/dx^2 = 0$. This means that

$$\epsilon = vc \tag{IX.39}$$

We then multiply both sides of the equation by df/dx and integrate, obtaining

$$\frac{a^2}{12}\left(\frac{df}{dx}\right)^2 = \frac{vc}{4}(f^4 - 2f^2 + 1) \tag{IX.40}$$

where the last constant is chosen to ensure that $df/dx = 0$ when $f = 1$. Both sides are exact squares, and this allows for the simple form

$$\frac{dx}{\xi} = \frac{df}{1 - f^2} \tag{IX.41}$$

where

$$\xi = \frac{a}{\sqrt{3\, vc}} \tag{IX.42}$$

is a correlation length (calculated here in a self-consistent field approximation).

Eq. (IX.41) is integrated easily and gives

$$f(x) = \tanh\left(\frac{x}{\xi}\right) \tag{IX.43}$$

We conclude that:

(i) There is a depletion layer of thickness ξ, the concentration profile being

$$c(x) = c \tanh^2\left(\frac{x}{\xi}\right) \tag{IX.44}$$

(ii) The concentration on the first layer is

$$c_1(a) = c\frac{a^2}{\xi^2} \sim c^2 \tag{IX.45}$$

(iii) The interfacial energy is

$$\gamma = T\int\left(\frac{a^2}{6}\,(\nabla\psi)^2 + \frac{1}{2}\,v\,|\psi|^4\right)dx \cong cT\frac{a^2}{\xi} \cong c^{3/2} \tag{IX.46}$$

In practice all these results are wrong. The correct powers of c entering in ξ, in $c(a)$, and in γ are discussed in Chapter III, and are different.

Exercise 2: Find the concentration profile $c(x)$ for chains in a good solvent near a weakly adsorbing wall, in the self-consistent field approximation (Jones and Richmond[9]).

Answer: The difference from the earlier exercise on adsorption is that now we have repulsion between the monomers. The self-consistent potential has the shape shown in Fig. IX.3a, where the short range, attractive part is maintained but where a repulsive part arises from the finite chain concentration at some distance from the wall. The nonlinear equation [eq. (IX.36)] becomes

$$\frac{a^2}{6}\frac{d^2\psi}{dx^2} = -\epsilon\psi + v\psi^3 \qquad \text{(IX.47)}$$

and the effects of the short range attraction ($x < b$) can again be replaced by the boundary condition [eq. (IX.27)].

If we fix the concentration in the bulk $c(x \to \infty) = c$, this means that $\epsilon = vc$. We then go to reduced variables ($\psi = c^{1/2} f$) and find the same equations [eqs. (IX.38, IX.40)] for f than in the previous exercise. There is one essential difference, however. Now we have $f > 1$, and the solution is a cotanh (instead of a tanh)—a decreasing function of x

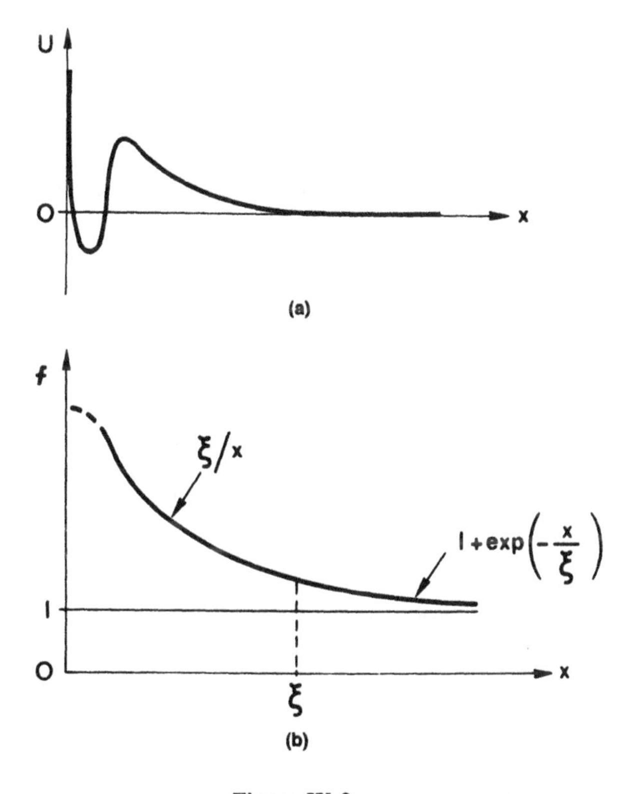

Figure IX.3.

$$f = \text{cotanh}\left(\frac{x + x_0}{\xi}\right) \tag{IX.48}$$

Here we have kept an integration constant x_0 which will fit the boundary condition at the adsorbing plane ($f'/f = -\kappa$). This gives

$$\frac{-1 + f^2(0)}{f(0)} = \kappa\xi \qquad \left(f(0) = \text{cotanh}\frac{x_0}{\xi}\right) \tag{IX.49}$$

There are two simple regimes:

(*i*) If $\kappa\xi$ is small, $f(0)$ is not much larger than unity, and $f(x) - 1$ is essentially an exponential relaxing in one length ξ. This limit corresponds to bulk solutions which have a large c, for which the attractive wall gives only a weak perturbation.

(*ii*) If $\kappa\xi$ is large (low concentration), the effect of the wall is more spectacular, and $f(0)$ is much larger than 1. At distances x smaller than ξ, the decay is relatively slow

$$f(x) = \frac{\kappa\xi}{1 + \kappa x} \tag{IX.50}$$

and in the range $\kappa^{-1} < x < \xi$, $f(x) \sim 1/x$. However, once again, this self-consistent result is incorrect; the correct scaling form is $f(x) \sim x^{-\beta/\nu}$ where β and ν are exponents (defined in the next chapter), and $\beta/\nu \sim 1/2$ in three dimensions:*

Apart from these scaling modifications there is a serious physical limitation to the above discussion—namely, the assumption that the wall attraction (measured by κ) is independent of the surface concentration. In reality the first layer begins to be filled, and the effective value of κ should decrease

$$\kappa_{eff} = \kappa[1 - (\text{constant})\ \psi^2(0)] \tag{IX.51}$$

This leads to more complicated self-consistent equations.

IX.3. The Random Phase Approximation for Dense Chains

We wish to compute all correlation functions in a dense mixture of strongly interacting polymer chains. This seems to be a formidable task,

*Note also that in the correct scaling approach the concentration profile $c(x)$ is *not* proportional to $f^2 = x^{-2\beta/\nu}$. The correct law is $c(x) \sim x^{-4/3}$ ensuring that $c(\xi) \cong c_{Bulk}$.

but on closer inspection it is not unfeasible. Some of the simplifications that we found in polymer melts (Chapter II) are still helpful here—i.e., the chains are nearly ideal.

This allows us to use a simple scheme, the "random phase approximation" (RPA). This name was introduced by Bohm, Pines, and Nozières[4] in connection with electron problems in metals. However, the main concepts are in a certain sense older and appear, for example, in the Debye-Hückel theory of electrolytes.[10] The first application to polymers is from S. F. Edwards.[11] The general idea as follows.

What is computed is a response function $S(\mathbf{r},\mathbf{r}')$. This is defined by applying a weak perturbing potential W at one point ($\mathbf{r}'$) and looking at the resulting changes in concentration $\delta\Phi$ at a different point ($\mathbf{r}$). It may be shown that, apart from normalized factors, $S(\mathbf{r},\mathbf{r}')$ is identical to the correlation function $\langle\delta\Phi(\mathbf{r}')\,\delta\Phi(\mathbf{r})\rangle$. Thus, a calculation of S will give us the correlations.

To compute the response S, we treat each chain as ideal but subjected to a potential which contains two parts. One is the external potential $W(\mathbf{r}')$, the other is an internal, self-consistent, potential, due to the surrounding chains, and includes terms linear in $\delta\Phi$. In many cases there is a strong tendency toward cancellation between the two parts; this we call this screening. Finally we arrive at a self-consistent prediction for S, which is rather accurate for concentrated chains. Thus RPA is *a self-consistent field calculation* for pair correlations.

IX.3.1. Definition of response functions

As our example we choose a dense system of chains that fills all sites of a Flory-Huggins lattice. Each chain has a sequence of monomers $(1, 2, \ldots, n, \ldots N)$, and they are all chemically identical. However, since we are interested in scattering experiments where some of the units are labeled, we treat the different monomers separately. For example, we introduce a (dimensionless) concentration $\Phi_n(\mathbf{r})$ which gives the average number of monomers of rank n on site $\mathbf{r}$. The average value of Φ_n in a filled lattice is $\Phi/N = 1/N$, but they may be local deviations $\delta\Phi_n$ from this average.

Let us now apply a set of weak perturbing potentials to the various sites ($\mathbf{r}'$); we assume that they depend not only on the site but also on the index n of the monomer which probes the potential. Then the perturbation is characterized by N functions $W_1(\mathbf{r}')\ldots W_n(\mathbf{r}')\ldots W_N(\mathbf{r}')$. We ask, what are the average changes $\delta\Phi_n$ at another $\mathbf{r}$ which will be induced by the perturbations $W_m(\mathbf{r}')$. For small W_s, this response must be a linear function of W_m. The most general form is

$$\delta\Phi_n(\mathbf{r}) = -\frac{1}{T}\sum_{\mathbf{r}'}\sum_{m} S_{nm}(\mathbf{rr}')\ W_m(\mathbf{r}') \qquad \text{(IX.52)}$$

We call $S_{nm}(\mathbf{r},\mathbf{r}')$ a *response function*. The choice of the minus sign in eq. (IX.52) is natural. A positive (repulsive) potential tends to deplete the concentrations. Also the factor $1/T$ makes S dimensionless.

In bulk systems, which are invariant by translation, the functions $S_{nm}(\mathbf{r},\mathbf{r}')$ depend only on the relative distance $\mathbf{r} - \mathbf{r}'$. It is then convenient to introduce Fourier transforms for all quantities of interest:

$$W_n(\mathbf{q}) = \sum_{\mathbf{r}'} W_n(\mathbf{r}') \exp(-\mathbf{i}q\cdot\mathbf{r}') = \frac{1}{a^3}\int d\mathbf{r}'\ W(\mathbf{r}') \exp(-\mathbf{i}q\cdot\mathbf{r}')$$

$$S_{nm}(\mathbf{q}) = \sum_{\mathbf{r}'} S_{nm}(\mathbf{r} - \mathbf{r}') \exp(\mathbf{i}q\cdot(\mathbf{r}' - \mathbf{r}) \quad \text{etc.} \qquad \text{(IX.53)}$$

The composition theorem for Fourier transforms gives a simple form to eq. (IX.52)

$$\delta\Phi_n(\mathbf{q}) = -\frac{1}{T}\sum_{m} S_{nm}(\mathbf{q})\ W_m(\mathbf{q}) \qquad \text{(IX.54)}$$

Having defined the response functions, let us now discuss the relationship between them and correlation functions. This is contained in a classic theorem of Yvon.[12] With our notation, it reads

$$\langle\delta\Phi_n(\mathbf{r})\ \delta\Phi_m(\mathbf{r}')\rangle = S_{nm}(\mathbf{r},\mathbf{r}') \qquad \text{(IX.55)}$$

where the left side is a correlation function at equilibrium, defined in the absence of any perturbing potential. The only assumption underlying eq. (IX.55) is that quantum effects are negligible. This is entirely correct for our systems. (The quantum effects mentioned here derive from the uncertainty principle; it is not possible to localize an atom, or a monomer, exactly. In our case the resulting "zero-point motion" is much smaller than thermal motion and is negligible).

IX.3.2. Response functions for noninteracting chains

We first define the response functions for a noninteracting system of chains (which would fill the lattice at random, each site then being allowed to carry more than one monomer). Later we introduce the interactions. Consider one ideal chain, and ask for the correlation

$$\langle \delta\Phi_n(0)\, \delta\Phi_m(\mathbf{r})\rangle_{ideal} = S^{\circ}_{nm}(\mathbf{r}) \tag{IX.56}$$

Its Fourier transform may be presented as

$$S^{\circ}_{nm}(q) = \langle \exp(\mathbf{i}q\cdot(\mathbf{r}_n - \mathbf{r}_m)\rangle_{ideal}$$

where the average is taken over all conformations of the ideal chain. We know that (for large $n - m$) the interval $\mathbf{r}_n - \mathbf{r}_m$ then has a gaussian distribution. For any gaussian variable x, we can use the theorem

$$\langle \exp(iqx)\rangle = \exp\left(-\frac{1}{2}q^2\langle x^2\rangle\right) \tag{IX.57}$$

In our case, the average square of one component of the vector $\mathbf{r}_n - \mathbf{r}_m$ is

$$\langle (x_n - x_m)^2\rangle = \frac{1}{3}|n - m|\, a^2 \tag{IX.58}$$

and the noninteracting response function is (for $qa < 1$):

$$S^{\circ}_{nm}(q) = \exp\left[-|n - m|\, q^2a^2/6\right] \tag{IX.59}$$

At this point we may note the relationship with the Debye scattering function $g_D(q)$ introduced in Chapter I. This function arises when we superpose equal scattering amplitudes on all monomers. With the normalization of Chapter I, it is

$$g_D(q) = N^{-1}\sum_{nm} S^{\circ}_{nm}(q) \tag{IX.60}$$

$$= \frac{2}{u}\left[1 - \frac{1}{u}(1 - \exp(-u))\right] \tag{IX.61}$$

where $u = Nq^2a^2/6$. It is easy to check that $g_D(q)$, as given by eq. (IX.60), has the limiting behaviors described in Chapter I.

We also find it useful to consider sums of the responses S_{nm} over one index, defining

$$S^{\circ}_n(q) = \sum_m S^{\circ}_{nm}(q) =$$

$$\frac{6}{q^2a^2}\left\{2 - \exp\left[-n\frac{q^2a^2}{6}\right] - \exp\left[-(N-n)\frac{q^2a^2}{6}\right]\right\} \tag{IX.62}$$

Physically S_n° describes the response of the nth monomer to perturbations which act equally on all monomers in an ideal chain.

Now we have listed all our tools. The next problem is to proceed from ideal chains to strongly interacting chains.

IX.3.3. Self-consistent calculation of responses

We now return to the problem of a dense chain system where weak perturbations $W_m(\mathbf{r}')$ are applied on all sites. Our assumption will be to write the responses $\delta\Phi_n(\mathbf{r})$ as the response of ideal chains, with the chains experiencing not only external potentials W but also a self-consistent potential U (which will be a linear function of $\delta\Phi$).

For a semi-dilute solution we would write

$$U(\mathbf{r}') = Tv\,\delta c\,(\mathbf{r}') = Tva^{-3} \sum_n \delta\Phi_n\,(\mathbf{r}') \qquad \text{(IX.63)}$$

However, the main domain of application of the RPA is not the semi-dilute regime (where inside each blob, an ideal chain picture is not acceptable) but rather the melt regime, where the total concentration is fixed

$$\sum_n \delta\Phi_n\,(\mathbf{r}') \equiv 0 \qquad \text{(IX.64)}$$

U will be defined so as to maintain the identity of eq. (IX.64).

Let us first state our assumption in detail, writing $\delta\Phi_n$ as a function of W_m

$$\delta\Phi_n\,(\mathbf{r}) = -\frac{1}{T}\sum_m \sum_{\mathbf{r}'} S_{nm}^\circ(\mathbf{r}\mathbf{r}')[W_m\,(\mathbf{r}') + U\,(\mathbf{r}')] \qquad \text{(IX.65)}$$

or in terms of Fourier transforms

$$\delta\Phi_n\,(\mathbf{q}) = -\frac{1}{T}\sum_m S^\circ(\mathbf{q})[W_m\,(\mathbf{q}) + U(\mathbf{q})] \qquad \text{(IX.66)}$$

It is essential to note that the self-consistent potential U is *the same for all n* (just as it is in the semi-dilute case of eq. (IX.63)). This expresses the chemical identity of all monomers and reduces our self-consistent problem considerably. There is only one unknown function U, and we can obtain it explicitly from the condition of constant total concentrations eq. (IX.64). Inserting eq. (IX.64) into eq. (IX.66) we get:

$$U(\mathbf{q}) \sum_{nm} S^{\circ}_{nm}(\mathbf{q}) = - \sum_{nm} S^{\circ}_{nm}(\mathbf{q})\, W_m(\mathbf{q}) \qquad \text{(IX.67)}$$

or with the notation introduced above:

$$U(\mathbf{q}) = - \frac{1}{N g_D(\mathbf{q})} \sum_m S^{\circ}_m(\mathbf{q}) W_m(\mathbf{q}) \qquad \text{(IX.68)}$$

Finally, if we use this self-consistent potential in eq. (IX.66), we arrive at the explicit form for the response functions

$$\delta\Phi_n(\mathbf{q}) = - \frac{1}{T} \sum_m S_{nm}(\mathbf{q}) W_m(\mathbf{q})$$

$$S_{nm}(\mathbf{q}) = S^{\circ}_{nm}(\mathbf{q}) - \frac{S^{\circ}_n(\mathbf{q}) S^{\circ}_m(\mathbf{q})}{N g_D(\mathbf{q})} \qquad \text{(IX.69)}$$

Eq. (IX.69) is the central RPA result for dense chain systems.[13] It allows for detailed calculations of scattering by partly labeled chains. If the scattering amplitude of the nth monomer is α_n, the intensity scattered at a wave vector q has the form

$$I(\mathbf{q}) = \text{constant} \sum_{nm} \alpha_n\, \alpha_m\, S_{nm}(\mathbf{q}) \qquad \text{(IX.70)}$$

We discussed $I(\mathbf{q})$ in Chapter II. Here we simply note that for $q \rightarrow 0$, $I(q)$ always vanishes: for $q = 0$, $S_{nm} = 1$, and $S_n = S_m = g_D = N$. Physically this means that on large scales the concentrations Φ_n cannot fluctuate because they become simply proportional to the total concentration

$$\Phi_n \rightarrow \frac{\Phi}{N} = \frac{1}{N} = \text{constant} \qquad (q \rightarrow 0)$$

However, at larger q we do get some scattering because although the total Φ is constant, we can have local fluctuations where monomers of a particular n become more numerous in one small region.

Quantitative experiments on partly labeled polystyrenes (with molecular weights $\sim 10^5$) have been performed by Cotton and co-workers.[14] They show that eqs. (IX.69, IX.70) give a rather accurate fit to the data, with no adjustable parameter (see Fig. IX.4).

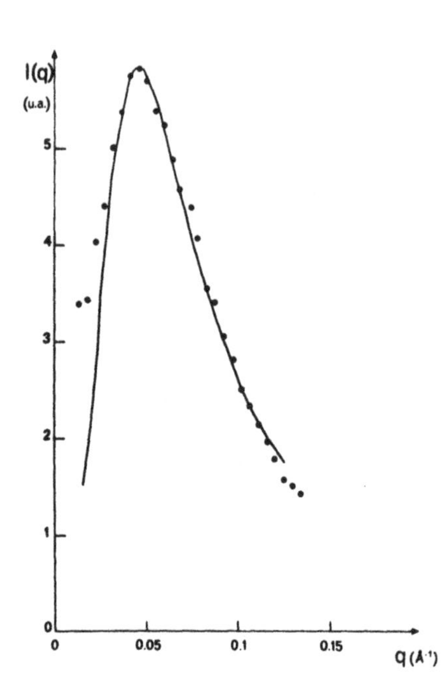

Figure IX.4.
Small-angle scattering of neutrons by a melt of *triblock copolymers:* polystyrene H—polystyrene D—polystyrene H (where D stands for deuterated). Each block has a molecular weight of 13000. $q = 4\pi/\lambda \sin \theta/2$ is the scattering vector. The points give the scattered intensity $I(q)$ in arbitrary units. The continuous curve is the prediction of eqs. (IX 69, 70), computed for this case by J. P. Cotton. The only adjustable parameter was the unperturbed size R_0 of the chain, or equivalently its radius of gyration R_g: the curve corresponds to R_g = 56 Å. The R value expected from separate neutron studies on *PS* melts containing a few labeled chains is 60 Å. After F. Boue *et al.*, *Neutron Inelastic Scattering 1977*, International Atomic Energy Agency, Vienna, 1978.

REFERENCES

1. S. F. Edwards, *Proc. Phys. Soc. (London)* **85,** 613 (1965).
2. P. G. de Gennes, *Rep. Prog. Phys.* **32,** 187 (1969).
3. F. Wiegel, *Phys. Rep.* **16,** 59 (1975). K. Freed, *J. Chem. Phys.* **55,** 3910 (1971). H. Gilys, K. Freed, *J. Phys. (Paris) Lett.* **A7L,** 116, (1974).
4. D. Pines, *Elementary Excitations in Solids,* W. A. Benjamin, New York, 1963.
5. R. P. Feynman, R. Leighton, M. Sands, *Feynman Lectures on Physics,* Vol. III, Addison-Wesley, New York, 1963.
6. P. Morse, H. Feshbach, *Methods of Theoretical Physics,* McGraw-Hill, 1953.
7. See Refs. 4, 5 of Chapter I.
8. M. Moore, *J. Phys. (Paris)* **A10,** 305 (1977).
9. I. S. Jones, P. Richmond, *J. C. S., Faraday Trans. II* **73,** 1062 (1977).
10. R. Robinson, R. Stokes, *Electrolyte Solutions,* 2nd ed., Butterworths, London, 1959.
11. S. F. Edwards, *Proc. Phys. Soc. (London)* **88,** 265 (1966).
12. J. Yvon, *Les Corrélations et l'Entropie en Mécanique Statistique Classique,* Dunod, Paris, 1965.
13. P. G. de Gennes, *J. Phys. (Paris),* **31,** 235 (1970).
14. F. Boue *et al.*, *Neutron Inelastic Scattering 1977,* Vol. I, p. 563, International Atomic Energy Agency, Vienna, 1978.

X

Relationships Between Polymer Statistics and Critical Phenomena

X.1. Basic Features of Critical Points

X.1.1. Large correlated regions

There is a strong analogy between the statistics of linear, flexible polymers and various features of critical phenomena. To make it clear, we first describe some essential aspects of ferromagnetic transition points—ferromagnets are the best example for our purpose. For a more detailed introduction, the classic reference is the book of H. E. Stanley.[1]

From a macroscopic point of view, ferromagnets are characterized by a *magnetization* **M**. This is a vector with a number n of independent components. The case $n = 3$ is frequent, but other values of n are also important. For example, if the magnetic moments are necessarily parallel to one axis (uniaxial ferromagnets), we have $n = 1$. If the moments are restricted to an "easy plane" of magnetization, we must put $n = 2$. The average magnetization M is a function of the temperature τ and of the magnetic field H. (We use τ to represent the temperature of the magnetic system. When we shall relate this to a polymer problem, the temperature T of the polymer system will *not* be equal to τ.) In zero field, we might guess naively that the magnetization must vanish by symmetry, the moments having equal chances to be "up" or "down." This is correct at high temperatures, but at low temperatures, the situation is different. The plot of free energy F as a function of M has two equivalent minima (Fig. X.1), and the system will reach one of them. Then we measure a finite

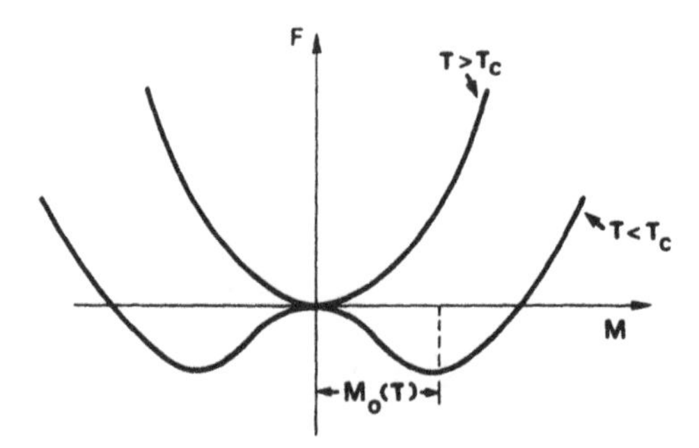

Figure X.1.

average magnetization M_0 (T). The plot of M_0 versus (τ_c) is shown in Fig. X.2.

A critical temperature τ_c (first studied by P. Curie) separates the two regimes. The following discussion is concerned with the immediate vicinity of τ_c for the following reason. Consider a temperature $\tau = \tau_c(1 + \epsilon)$ with ϵ small and positive. Since we are above τ_c, the average M is zero. However, if we look at the local distribution of $M(\mathbf{r})$, (as is possible by neutron scattering techniques[2]) we find that for small ϵ, there are regions where M does not average to zero. The characteristic size of these regions is called the correlation length ξ, and it obeys a scaling law of the form

$$\xi \cong a|\epsilon|^{-\nu} \qquad (\epsilon \rightarrow 0) \tag{X.1}$$

where a is the distance between neighboring atoms, and ν is a certain "critical exponent." The essential feature is that when ϵ is small, ξ is much larger than a. In solids we can typically achieve values of ξ of a few hundred Angströms. The correlated regions are much larger than a lattice unit, and all details of the lattice structure, of the couplings, and so forth, become irrelevant. We reach a very *universal* regime, where only two essential parameters remain; one is the dimensionality (d), the other is the number of equivalent components (n). It turns out that all critical exponents such as ν depend only on d and n.

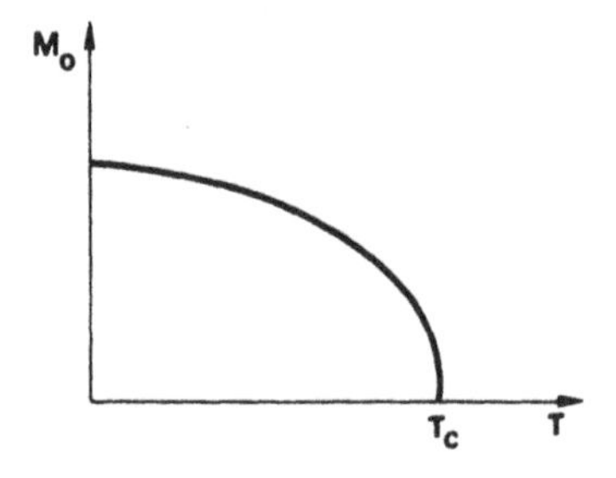

Figure X.2.

This trend toward universality is identical to what we found in Chapter I for polymer chains of large N, where certain laws become universal and independent of the details of the monomer structure. In fact if we think of the single-chain problem in a good solvent, the analogy is very close. Here we have a Flory radius, R_F, which is also in the range of a few hundred Angströms, and which has the scaling form:

$$R_F \cong aN^{\nu} \qquad (N \rightarrow \infty) \tag{X.2}$$

Comparing this with eq. (X.1), we notice a correspondence between N^{-1} and ϵ. To reach a high degree of universality, we want $\epsilon \rightarrow 0$ or $N \rightarrow \infty$. This correspondence can be cast in a precise theorem (discussed below).

X.1.2. Critical exponents for a ferromagnet

Let us first consider the high temperature side of the transition and determine the main effects that indicate the onset of ferromagnetic order. A first method applies a small magnetic field H to the system and measures the average magnetization M induced by H. For small H this must be of the form $M = \chi_M H$ where $\chi_M H$ is called the susceptibility (not to be confused with the Flory interaction parameter). χ_M depends on temperature and diverges when $\epsilon = (\tau - \tau_c)/\tau$ becomes very small:

$$\chi_M = \chi_0 \, |\epsilon|^{-\gamma} \tag{X.3}$$

A second, more local method of probing was mentioned above—i.e., neutron scattering—which is sensitive to the size ξ of the correlated regions, described by the exponent ν [eq. (X.1)].

A third approach is based on the specific heat C. In some ferromagnets one finds a singularity in C (the specific heat in zero field) which can be represented as

$$C = C_0 \, |\epsilon|^{-\alpha} \qquad (\epsilon \rightarrow 0) \tag{X.4}$$

where α is a (small) exponent.

We now go to the low temperature side ($\epsilon < 0$). We again find singularities and exponents for a correlation length, for the specific heat, and so forth, with the *same* values α, ν, and so forth (although the prefactors such as C_0 in eq. (X.4) are different below τ_c). Another exponent of interest is obtained from the magnetization law in zero field (Fig. X.2). At small ϵ we have

$$M_0(T) = M_1|\epsilon|^\beta \tag{X.5}$$

X.1.3. Relations among exponents

The above list seems to indicate a very large number of mysterious exponents. However, there are relations among them, and only two exponents are independent. These relations are:

$$\alpha + 2\beta + \gamma = 2 \quad \text{(Widom)} \tag{X.6}$$

$$\alpha = 2 - \nu d \quad \text{(Kadanoff)} \tag{X.7}$$

The Widom relation can be related to a simple scaling structure for the free energy $F(M, \tau, H)$ (per atom)

$$F(M, \tau, H) = F_0 + \Gamma(M) - MH$$

$$\Gamma(M) = \epsilon^{2-\alpha} f_F\left(\frac{M}{|\epsilon|^\beta}\right) \tag{X.8}$$

In eq. (X.8) the first term F_0 is regular at $T = T_c$ and is unimportant for our purposes. The second term $\Gamma(M)$ is represented in Fig. X.1. It gives a minimum at $M = M_0 \sim \epsilon^\beta$ when we are below T_c. This forces the function f_F to depend only on M/M_0. The factor $\epsilon^{2-\alpha}$ before f_F is required to give us the correct singularity in the specific heat above T_c. Here $M = 0$, and f_F is a constant. If we differentiate F twice with respect to temperature, we get the entropy first and then the specific heat; the latter must behave as eq. (X.4). The third term in eq. (X.8) is simply the coupling between moments and field.

If we again choose $\epsilon > 0$ and H small, we expect the equilibrium M to be small, and we may then expand $f_F(x)$ in powers of $x = M/M_0$. As is clear in Fig. X.1, $f_F(x)$ is an even function of x, and the expansion starts with a term in x^2

$$F = F(x = 0) + (\text{constant})\ \epsilon^{2-\alpha}\left(\frac{M^2}{\epsilon^{2\beta}}\right) - MH$$

Minimizing this with respect to M, we get the susceptibility

$$\frac{M}{H} = \text{constant}\ \epsilon^{2\beta+\alpha-2}$$

and comparing with the definition of exponent γ [eq. (X.3)], we are led to the Widom relation [eq. (X.6)].

The Kadanoff relation [eq. (X.7)] is more delicate (and may even require some slight corrections for $d = 3$). However, we may get some insight into it by the following procedure. Instead of thinking of individual atomic moments, it is more realistic, near τ_c, to choose consecutive regions of size ξ as basic units. Inside one such region, the correlations are strong, and the only independent variable is the total moment of the region. If we then compute a partition function and a free energy ΔF for these total moments, we expect ΔF to be extensive—i.e., proportional to the number of regions per unit volume—which is (in d dimensions)

$$\frac{1}{\xi^d} = \epsilon^{\nu d}$$

Thus we are led to a $\Delta F(M = 0)$ proportional $\epsilon^{\nu d}$. Comparing this with eq. (X.8) we obtain to the Kadanoff relation [eq. (X.7)].

Our presentation of the scaling relationships has been strictly phenomenological. A more fundamental approach, based on renormalization group ideas, can be found in various advanced texts.[3]

X.1.4. Correlation functions

We have introduced the notion of correlated regions near the Curie point. Now we make this discussion more precise. We choose a temperature τ slightly above τ_c, in the absence of any external field. A good measure of correlation properties is given by the following thermal average, involving the magnetization M measured locally at two points separated by a distance r:

$$\langle \mathbf{M}(0) \cdot \mathbf{M}(\mathbf{r}) \rangle = \text{correlation function}$$

An enormous amount of information—theoretical and experimental—has been accumulated on these correlations and is summarized below.

SPATIAL SCALING

The correlations at temperature close to the Curie point can be written as

$$\langle \mathbf{M}(0) \cdot \mathbf{M}(\mathbf{r}) \rangle = \frac{1}{r^{d-2+\eta}} f_M \left(\frac{r}{\xi}\right) \qquad \text{(X.9)}$$

where η is another critical exponent, first introduced by Fisher,[4] ξ is the correlation length, and f_M is a dimensionless function that satisfies:

$$f(0) = 1$$
$$f(x) \cong x^{\eta} \exp(-x) \qquad (x \gg 1)$$

The limit $x \gg 1$ corresponds to small correlated regions (small ξ); in this region a simple Ornstein-Zernike picture becomes valid,[1] and the spatial decay of correlations has the form $1/r \exp(-r/\xi)$ (for $d = 3$). The limit $x = 0$ corresponds to $\tau = \tau_c$ (ξ infinite); at $\tau = \tau_c$, the correlation decays slowly, like a power law $r^{-(d-2+\eta)}$. We return to the region of small x in the section, "Connection with the Specific Heat."

CONNECTION WITH THE MAGNETIC SUSCEPTIBILITY

The exponent η can be related to the other exponents through a general thermodynamic theorem connecting the space integral of the correlation function to the susceptibility[1]

$$\tau\chi_M = n^{-1} \int \langle \mathbf{M}(0) \cdot \mathbf{M}(\mathbf{r}) \rangle \, d\mathbf{r} \qquad \text{(X.10)}$$

Using eq. (X.9) and switching to the dimensionless variable r/ξ, this gives

$$\chi_M \sim \xi^{2-\eta}$$
$$\epsilon^{-\gamma} \sim \epsilon^{-\nu(2-\eta)}$$

and hence

$$\gamma = \nu(2 - \eta) \qquad \text{(X.11)}$$

In practice γ is only slightly larger than 2ν and η is small for all three-dimensional systems.

CONNECTION WITH THE SPECIFIC HEAT[5]

If we choose two neighboring points $r = a$, the correlation $\langle M(0)\, M(a) \rangle$ is expected to measure the coupling energy which is responsible for magnetic order. This energy must contain a term of order $\epsilon^{1-\alpha}$ since by differentiation it gives the specific heat ($\sim \epsilon^{-\alpha}$). Thus, we must have

$$\langle M(0)\, M(a) \rangle = \langle M(0)\, M(a) \rangle_{\tau=\tau_c} - \text{constant } \epsilon^{1-\alpha} \qquad \text{(X.12)}$$

Comparison with eq. (X.9) implies that the function $f(x)$ for small x must behave according to the law

$$1 - f(x) \cong x^{(1-\alpha)/\nu}$$

This property will be useful when transposed to polymer problems.

X.1.5. The n vector model

We now describe the system of atomic moments or "spins" S_i in more detail on a specific model, the n vector model. We assume that the magnetic atoms are located on a periodic lattice. Each magnetic atom (i) carries a spin $\mathbf{S}_i$; this is a vector, with n components $S_{i1}, S_{i2} \ldots S_{in}$. In our considerations, we ignore all quantum effects; the components $S_{i\alpha}$ are just numbers. There is one constraint—i.e., the total length S of each spin is fixed. We choose the following normalization:

$$S^2 \equiv \sum_{\alpha=1}^{n} S_{i\alpha}^2 = n \tag{X.13}$$

Neighboring spins are coupled, and their energy is minimized when they are parallel. The coupling energy, or "Hamiltonian" $\mathfrak{H}$, has the form

$$\mathfrak{H} = -\sum_{i>j} K_{ij}\, \mathbf{S}_i \cdot \mathbf{S}_j - \sum_i \mathbf{H} \cdot \mathbf{S}_i \tag{X.14}$$

The constant K_{ij} is positive, $K_{ij} = K$, for nearest-neighbor pairs (ij) and vanishes for all other choices of i and j. In eq. (X.12) we have also incorporated terms ($-\mathbf{H}\cdot\mathbf{S}_i$) describing the effect of an external field $\mathbf{H}$. The partition function of the spin system is

$$Z = \prod_i \int d\Omega_i \exp(-\mathfrak{H}/\tau) \tag{X.15}$$

where $\int d\Omega_i$ represents an integration over all allowed orientations of spin $\mathbf{S}_i$. (For example, with $n = 3$, $\delta\Omega_i = \sin\theta_i d\theta_i d\varphi_i$, where θ_i and φ_i are the polar angles of spin $\mathbf{S}_i$.)

It has been a traditional temptation to expand the partition function in powers of the coupling energy K_{ij} for each pair:

$$\exp(-K_{ij}(\mathbf{S}_i\cdot\mathbf{S}_j))/\tau = 1 - \frac{K_{ij}}{\tau}(\mathbf{S}_i\cdot\mathbf{S}_j) + \frac{1}{2}\left(\frac{K_{ij}}{\tau}\right)^2 (\mathbf{S}_i\cdot\mathbf{S}_j)^2 + \ldots \tag{X.16}$$

Usually when such expansions are inserted into eq. (X.15) for Z, they result in a complicated structure. There is one case, however, when they become simple—namely, if we go to the limit $n = 0$. This is a very formal step since our definition of n implied that it was a positive integer. However, this step can be performed and is useful. We shall see that when $n = 0$ the expansion [eq. (X.16)] leads to a problem of self-avoiding chains on a Flory-Huggins lattice.

X.2. The Single Chain Problem

X.2.1. The limit $n = 0$

We begin by a purely geometric discussion of vector orientations in an n-dimensional space and present it so that the calculations are meaningful when n is not a positive integer. Our approach follows the appendix of Ref. 6 and is due to G. Sarma.

AVERAGE OVER ORIENTATIONS

Let us first define an average over all orientations (equally weighted) of each spin. This is the analog of the integration over solid angles (for $n = 3$). We denote this average by $\langle\ \rangle_0$.

The subscript (o) emphasizes the difference between this type of average (where all states are equally weighted) and a thermal average (where they are weighted by the Boltzmann exponential $\exp(-\mathfrak{H}/\tau)$; thermal averages are written $\langle\ \rangle$ (without a subscript). The relationship between the two types of averages is, for any function $G(\mathbf{S}_1 \ldots \mathbf{S}_i \ldots)$ of the spins:

$$\langle G \rangle = \frac{\langle \exp(-\mathfrak{H}/\tau) G \rangle_0}{\langle \exp(-\mathfrak{H}/\tau) \rangle_0} \tag{X.17}$$

The partition function is from eq. (X.15)

$$Z = \Omega \langle \exp(-\mathfrak{H}/\tau) \rangle_0 \tag{X.18}$$

where $\Omega = \Pi_i \int d\Omega_i$ is the total volume of the phase space for the spins, and is an uninteresting factor.

THE MOMENT THEOREM

We now focus on one of the vectors $\mathbf{S}_i$ (which we call $\mathbf{S}$ for simplicity). If we perform an expansion of the partition function Z following eq. (X.14) and use the averaging rule [eq. (X.18)], we are led to consider averages such as:

$$\langle S_\alpha \rangle_0$$
$$\langle S_\alpha S_\beta \rangle_0$$
$$\langle S_\alpha S_\beta S_\gamma \rangle_0 \text{ etc.}$$

where α, β, etc. are component subscripts. We now show that when $n = 0$, only one type of average is nonvanishing—namely, the quadratic term

$$\langle S_\alpha S_\beta \rangle_0 = \delta_{\alpha\beta} \tag{X.19}$$

Eq. (X.19) is not surprising; the diagonal terms must all be equal, and their sum is equal to n as can be seen from the normalization condition of eq. (X.13). The nondiagonal terms vanish by symmetry.

The real surprise comes when we look at higher moments and find them all equal to zero—e.g.,

$$\langle S_\alpha^4 \rangle_0 = 0 \tag{X.20}$$

Clearly, this will bring enormous simplifications to all thermodynamic calculations.

The proof proceeds as follows. We start with an integral (positive) value of n and with our spin length normalized in agreement with eq. (X.13). We then introduce the "characteristic function" $f(\mathbf{k})$ of the variables S_α; this is a function of a vector $\mathbf{k}$, also with n components $\mathbf{k}_\alpha$, and is defined by

$$f(\mathbf{k}) = \langle \exp(i\mathbf{k}\cdot\mathbf{S}) \rangle_0 \tag{X.21}$$

From this function all the moments of the distribution of $\mathbf{S}$ for random orientations can be extracted. For example, the second moment is:

$$\langle S_\alpha S_\beta \rangle_0 = \left(-i\frac{\partial}{\partial k_\alpha}\right)\left(-i\frac{\partial}{\partial k_\beta}\right) f(\mathbf{k})\Big|_{k=0} \tag{X.22}$$

and similar formulas hold for all moments. Let us now try to construct $f(\mathbf{k})$ explicitly. Clearly, f depends only on the length of $\mathbf{k}$ since it represents an average over all (equally weighted) orientations. If we start from the definition of eq. (X.21) and differentiate twice in $\mathbf{k}$ space, we get

$$\nabla^2 f \equiv \sum_\alpha \frac{\partial^2 f}{\partial k_\alpha^2} = -\sum_\alpha \langle S_\alpha^2 \exp(i\mathbf{k}\cdot\mathbf{S})\rangle_0 \qquad \text{(X.23)}$$

and with the normalization [eq. (X.13)], this gives

$$\nabla^2 f = -nf \qquad \text{(X.24)}$$

It is convenient now to use $|\mathbf{k}| = k$ as our variable since our function f depends only on k. Writing for such a function that $kdk = \sum_\alpha k_\alpha dk_\alpha$

$$\frac{\partial f}{\partial k_\alpha} = \frac{k_\alpha}{k}\frac{\partial f}{\partial k}$$

$$\frac{\partial^2 f}{\partial k_\alpha^2} = \frac{1}{k}\frac{\partial f}{\partial k} + \frac{k_\alpha^2}{k}\frac{\partial}{\partial k}\left(\frac{1}{k}\frac{\partial f}{\partial k}\right)$$

$$\sum_\alpha \frac{\partial^2}{\partial k_\alpha^2} = \frac{n}{k}\frac{\partial f}{\partial k} + k\frac{\partial}{\partial k}\left(\frac{1}{k}\frac{\partial f}{\partial k}\right) = \left(\frac{n-1}{k}\right)\frac{\partial f}{\partial k} + \frac{\partial^2 f}{\partial k^2}$$

Inserting this into eq. (X.22), we arrive at the final equation for $f(k)$

$$\frac{\partial^2 f}{\partial k^2} + \left(\frac{n-1}{k}\right)\frac{\partial f}{\partial k} + nf = 0 \qquad \text{(X.25)}$$

This is supplemented by boundary conditions at $k = 0$, which are

$$f(k=0) = 1$$
$$\frac{\partial f}{\partial k}(k=0) = 0 \qquad \text{(X.26)}$$

The second condition says that $f(k)$ is even and regular at small k.

What is important in eqs. (X.25, X.26) is that they remain valid for all n values (not necessarily positive integers). Starting with them, we can construct a function $f(k)$ which is adequate for any n. This function will represent the analytic continuation of the standard forms known for $n = 1, 2$, etc. to more general values of n.

In what follows we specialize in cases where $n = 0$, which turns out to be the interesting choice for our purposes. Eq. (X.25) then becomes

$$\frac{d^2 f}{dk^2} - \frac{1}{k}\frac{df}{dk} = 0$$

and the solution (satisfying the boundary conditions of eq. (X.26)) is a parabolic form

$$f(k) = 1 - \frac{1}{2}k^2 \tag{X.27}$$

There are no powers higher than k^2. If we return to the equations for the moments (e.g., eq. (X.22)), this implies that all moments involving 3, 4, etc. components of S must vanish; this is how we prove equations such as eq. (X.20).

X.2.2. The magnetic partition function expanded in self-avoiding loops

Now we return to the partition function Z and more specifically to the case of zero external field. We can rewrite eqs. (X.13, X.16) in the form

$$\frac{Z}{\Omega} = \langle \prod_{i>j} \exp\left(\frac{K_{ij}}{T}\sum_{\alpha} S_{i\alpha}S_{j\alpha}\right)\rangle_0 \tag{X.28}$$

$$= \langle \prod_{i>j} \left[1 + \frac{K_{ij}}{T}\sum_{\alpha} S_{i\alpha}S_{j\alpha} + \frac{1}{2}\left(\frac{K_{ij}}{T}\right)^2 \sum_{\alpha\beta} S_{\beta j}S_{\beta j}\right]\rangle_0 \tag{X.29}$$

The essential feature is that all higher terms in the expansion of the exponential vanish because of the moment theorem. Successive terms of Z/Ω may be represented by graphs on the lattice. To each nearest neighbor link K_{ij} is associated a continuous line. To each site i must be associated two spin components $S_{i\alpha}$ $S_{i\alpha}$ (to obtain a nonzero average). These rules mean that the only allowed graphs are *closed loops,* such as shown in Fig. X.3. The loop can never intersect itself. If it did, at one site i, this would

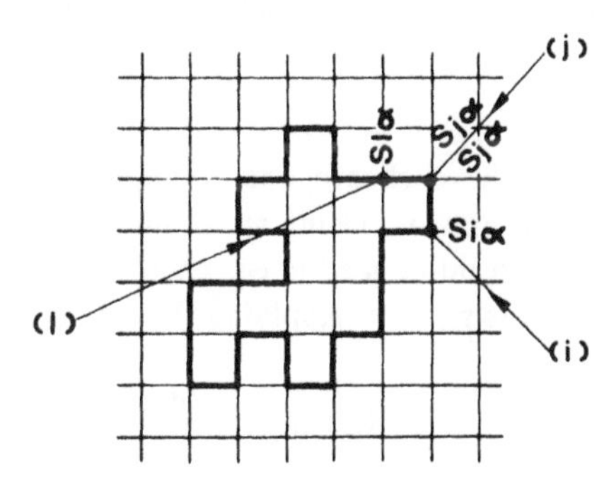

Figure X.3.

imply an average $\langle S_i^4 \rangle_0$ which vanishes by the moment theorem. At this point we begin to see the connection between magnetism and self-avoiding chains.

First let us settle some technical points concerning the partition function. The quadratic terms $(K_{ij}/\tau)^2$ in the expansion simply correspond to the smallest loop, and could be drawn as

$i \qquad j$

Note that each loop has a single value of the component index α occurring at all its sites. This comes from eq. (X.19) and means that the two factors S_i at one point involve the same component.

When we sum over the component index α for one loop, we get

$$(K/\tau)^N n$$

where N is the number of bonds in the loop, and n is the component index. Because $n = 0$ in our case, the contribution of all loops ultimately vanishes, and we can write

$$\frac{Z}{\Omega} \equiv 1 \qquad (n = 0) \tag{X.30}$$

Thus, after all this effort, we get a trivial result for the partition function. However, we have progressed and can now proceed to more interesting subjects such as correlation functions.

X.2.3. Spin correlations and the one-chain problem

Consider the spin-spin correlation function in zero field:

$$\langle S_{il} S_{jl} \rangle$$

(where we have chosen one component (l)). Apart from normalization conditions, this is identical to the magnetization correlation function $\langle \mathbf{M}(0)\, \mathbf{M}(\mathbf{r}) \rangle$ introduced in eq. (X.9); the distance between the two points (i) and (j) is $\mathbf{r}_{ij} = \mathbf{r}$. Applying the rule of averages [eq. (X.17)], we get

$$\langle S_{il}\ S_{jl} \rangle = \frac{\langle \exp(-\mathfrak{H}/\tau)\ S_{il}\ S_{jl} \rangle_0}{\langle \exp(-\mathfrak{H}/\tau) \rangle_0} \tag{X.31}$$

Note that the denominator of eq. (X.31) is equal to 1 (from eq. X.30). Then expand the exponential in the numerator. Again, for $n = 0$, the only graphs which contribute are self-avoiding paths, but here they are not closed loops because eq. (X.31) contains two extra spin factors $S_i\ S_j$. In fact what we have is a sum over all self-avoiding walks linking sites i and j (Fig. X.4). If the walk involves N steps, the resulting contribution to eq. (X.31) is simply

$$\left(\frac{K}{T}\right)^N$$

All along the walk the component index α must be equal to the chosen value ($\alpha = 1$). There is no summation to be carried on α.

Finally we are led to the fundamental theorem:

$$\langle S_{i1}S_{j1}\rangle\Big|_{n=0} = \sum_N \mathfrak{N}_N(ij)\left(\frac{K}{T}\right)^N \quad (X.32)$$

where $\mathfrak{N}_N(ij)$ is the number of self-avoiding walks of N steps linking points (i) and (j) on the lattice (discussed in Chapter I). Eq. (X.32) is the basic link between chains and magnets. We now present some of its applications.

X.2.4. Properties of self-avoiding walks (SAW)

TOTAL NUMBER OF WALKS

The total number of SAWs of N steps, starting from point (i) is:

$$\mathfrak{N}_{N\ (total)} = \sum_j \mathfrak{N}_N(ij) \quad (X.33)$$

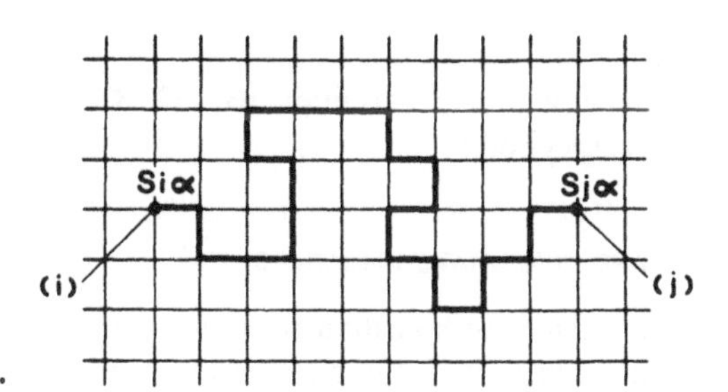

Figure X.4.

The asymptotic form of $\mathfrak{N}_N$ was given in eq. (I.21): $\mathfrak{N}_N \cong \bar{z}^N N^{\gamma-1}$. We now check that this agrees with the definition [eq. (X.3)] of χ as a susceptibility exponent. The susceptibility χ_M may be expressed in terms of magnetic correlations by eq. (X.10) or, in terms of correlations for one component (S_{i1}) of the spins ($\mathbf{S}_i$):

$$\chi_M = \frac{1}{\tau} \sum_j \langle S_{i1} S_{j1} \rangle \tag{X.34}$$

$$= \frac{1}{\tau} \sum_N \mathfrak{N}_{N\ (total)} \left(\frac{K}{\tau}\right)^N$$

$$\cong \frac{1}{\tau} \sum_N \left(\frac{K\bar{z}}{\tau}\right)^N N^{\gamma-1} \tag{X.35}$$

This series converges at large τ and diverges when τ reaches the critical value

$$\tau_c = K\bar{z} \tag{X.36}$$

If we consider temperatures τ slightly above τ_c, we can write

$$\tau = \tau_c (1 + \epsilon) \cong \tau_c \exp(\epsilon)$$

$$\chi_M \cong \frac{1}{\tau_c} \sum_N \exp(-N\epsilon)\ N^{\gamma-1} \tag{X.37}$$

and replacing the sum by an integral $\int_0^\infty dN$, we get

$$\chi_M \cong \frac{1}{\tau_c} \epsilon^{-\gamma} \tag{X.38}$$

in agreement with the definition [eq. (X.3) of the singularity for the magnetic susceptibility.

N AND ϵ ARE CONJUGATE VARIABLES

Having located the transition point, we return to the general correlation $\langle S_{i1} S_{j1} \rangle$ and rewrite it (for small ϵ) in the form

$$\langle S_{i1} S_{j1} \rangle = \sum_N \exp(-N\epsilon)\, \mathfrak{N}_N(ij) \tag{X.39}$$

Thus the relationship between number of paths $\mathfrak{N}_N(ij)$ and magnetic correlations is of the Laplace transform type. We can say that ϵ and N are conjugate variables. This is the precise form of our statement in Section X.1.1, according to which a small ϵ corresponds to a large N.

SCALING LAW FOR SELF-AVOIDING WALKS

All the properties discussed in Section X.1.4. for magnetic correlations now have their counterpart for self-avoiding walks. The first and most essential is the existence of a single correlation length $\xi \sim \epsilon^{-\nu}$ in eq. (X.9). The analog of ξ is the range of the self-avoiding walks $R_F(N) \sim N^\nu$. The spatial scaling law which corresponds to the existence of one single characteristic length is eq. (I.24)

$$\frac{\mathfrak{N}_N(ij)}{\mathfrak{N}_N(tot)} = p(r) = \frac{1}{R_F{}^d}\, \varphi_p \left(\frac{r}{R_F}\right) \qquad (r_{ij} \equiv r) \tag{X.40}$$

ASYMPTOTIC FORM AT LARGE R

The limiting behavior of the correlations at large distances is given by an Ornstein-Zernike form (introduced in Section X.1.4). Inverting the Laplace transformation [eq. (X.39)], it is then possible to find the asymptotic form of $\mathfrak{N}_N$ or of φ_p in eq. (X.40) at large r. The result is:[7]

$$\varphi_p \sim \exp - \left(\frac{r}{R_p}\right)^{1/(1-\nu)} \tag{X.41}$$

(where we omit all power laws that enter before the exponential). Eq. (X.41) was justified in Chapter I, using the simpler Pincus argument. Thus, we do not give the full Laplace transform calculation here.

SELF-AVOIDING WALKS RETURNING TO THE ORIGIN

In Chapter I we introduced the number of closed polygons of $(N + 1)$ steps and pointed out that this was equal to z (number of neighbors of one site) times $\mathfrak{N}_N a$. We announced in eq. (I.28) that

$$\mathfrak{N}_N(a) \cong N^{-\nu d} = N^{-2+\alpha} \tag{X.42}$$

The magnetic analog of $\mathfrak{N}_N(a)$ is the correlation function between nearest-neighbor spins and is related directly to the average energy per site E. From the definition [eq. (X.14)] of the couplings, we can write

$$E = -\frac{1}{2} zK \langle \mathbf{S}_i \cdot \mathbf{S}_{jk} \rangle \tag{X.43}$$

i and j being nearest neighbors; the factor $z/2$ counts each pair only once.

Let us now check that eq. (X.42) agrees with the scaling properties of the energy E. From the basic theorem of eq. (X.32) we have:

$$\begin{aligned} E &= -\frac{1}{2} zK \sum_N \exp(-N\epsilon)\, \mathfrak{N}_N a \\ &\cong -K \sum_N \exp(-N\epsilon)\, N^{-2+\alpha} \end{aligned} \tag{X.44}$$

We split eq. (X.44) into two parts

$$E(\epsilon) = E(0) + \sum_N [1 - \exp(-\epsilon N)]\, N^{-2+\alpha} \tag{X.45}$$

or replacing the last sum by an integral

$$E = E(0) + \int_0^\infty dN (1 - \exp(-\epsilon N)) N^{-2+\alpha} \tag{X.46}$$

$E(0)$ is a finite number, and the second integral has no singularities either for $N \rightarrow 0$ or for $N \rightarrow \infty$. Then we may set $\epsilon N = t$ and write

$$\int_0^\infty dN N^{-2+\alpha} (1 - \exp(-\epsilon N)) = \epsilon^{1-\alpha} \int_0^\infty dt\, t^{-2+\alpha} (1 - \exp(-t)) \cong \epsilon^{1-\alpha}$$

Thus, we arrive at an equation equivalent to eq. (X.12)

$$E(\epsilon) = E(0) - \text{constant } \epsilon^{1-\alpha} \tag{X.47}$$

This completes our proof; α is indeed the specific heat exponent defined in eq. (X.4).

SUMMARY

All properties of one self-avoiding walk on a lattice can be related to the spin correlation of a ferromagnet with an n-component magnetization when we formally set $n = 0$. This introduces a link between the exponents for self-avoiding walks and critical exponents.

Both sets of exponents become simple above $d = 4$. Then we have $\nu = 1/2$ and $\gamma = 1$. For phase transitions, this is called mean field behavior. For self-avoiding walks, this corresponds to ideal chain behavior.

The temperature T of the polymer system is not related to the temperature τ of the magnetic system, but $(\tau - \tau_c)/\tau_c = \epsilon$ is the conjugate variable of the degree of polymerization, N.

X.3. Many Chains in a Good Solvent

X.3.1. The des Cloizeaux trick

Des Cloizeaux was the first to notice that certain problems involving *many* mutually avoiding chains also had a magnetic counterpart.[8] The trick is to apply a magnetic field, H, along one direction (say $\alpha = 1$), to the associated spin system and to expand the partition function in powers of H. The zero-order term is trivial $Z(H = 0) = \Omega$ [eq. (X.30)]. Higher order terms may be generated by expanding all Boltzmann exponentials in powers of K/τ and H/τ.

$$\frac{Z(H)}{Z(0)} = \left\langle \prod_i \left(1 + \frac{H}{\tau} S_{i1} + \ldots\right) \prod_{i>j} \left(1 + \frac{K_{ij}}{\tau} \sum_\alpha S_{i\alpha} S_{j\alpha} + \ldots\right)\right\rangle_0 \tag{X.48}$$

The expansion must be such that each site (i) is associated with a second moment $S_{i\alpha}^2$. Thus, odd powers of H drop out, and we are left with graphs proportional to $(H/\tau)^{2p}$ ($p = 1, 2 \ldots$). In one such graph, we have p self-avoiding (and mutually avoiding) chains of lengths $N_1 \ldots N_p$. The total number of links are $N_t = N_1 + \ldots + N_p$. An example with $p = 2$ is shown in Fig. X.5.

Finally we may write

$$\frac{Z(H)}{Z(0)} = 1 + \sum_{p=1}^{\infty} \sum_{N_t=0}^{\infty} \left(\frac{H}{\tau}\right)^{2p} \left(\frac{K}{\tau}\right)^{N_t} \mathfrak{G}(p, N_t) \tag{X.49}$$

where $\mathfrak{G}(p, N_t)$ is the total number of ways to put p polymer chains of total

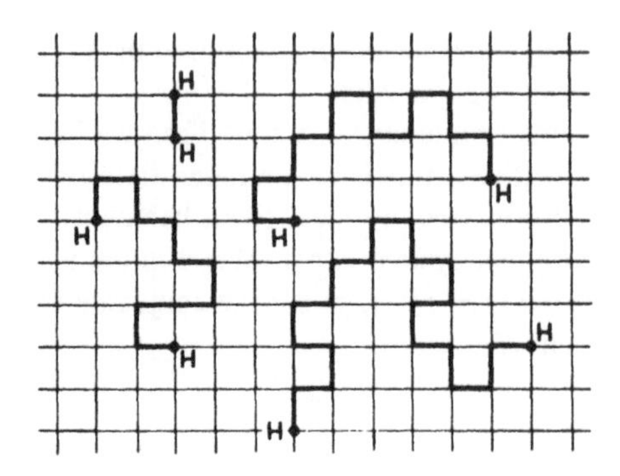

Figure X.5.

length N_t on the Flory-Huggins lattice, each site being occupied once at most. Thus, we may think of $\Xi = Z(H)/Z(0)$ as being the grand partition function for a system of mutually avoiding chains. For each chain we have a factor $(H/\tau)^2$ and for each monomer a factor K/τ.

Note that in this system the chains are polydisperse, but their average degree of polymerization $\bar{N}$ will be determined by a suitable choice of K and τ. Let us introduce a monomer concentration (per site) Φ and a polymer concentration (number of chains divided by the number of sites) Φ_p. Using the standard relations between concentrations and fugacities, we have

$$2\Phi_p + \Phi = \frac{1}{Q}\frac{\partial \ln \Xi}{\partial \ln(1/\tau)} \quad \text{(X.50)}$$

$$2\Phi_p = \frac{1}{Q}\frac{\partial \ln \Xi}{\partial \ln(H^2)} \quad \text{(X.51)}$$

(where Q is the total number of sites on the Flory-Huggins lattice).

We must impose a fixed value of $\bar{N}$:

$$\bar{N} = \frac{\Phi}{\Phi_p} \quad \text{(X.52)}$$

(Note that in practice $\bar{N}$ will be large, and we can replace $\Phi + 2\,\Phi_p$ by Φ in eq. (X.50)).

A typical thermodynamic quantity to be derived from the calculation is the osmotic pressure Π (divided by the polymer temperature T). A general thermodynamic theorem relates Π to the grand partition function

$$\frac{\Pi}{T} = \frac{1}{V}\ln \Xi \quad \text{(X.53)}$$

where $V = Qa^d$ is the total volume.

Let us now return to the magnetic aspect of the problem. We introduce a free energy (per site) as in eq. (X.8)

$$F(M) = F_o + \Gamma(M) - MH \qquad \text{(with } \Gamma(0) = 0 \text{ by convention)}$$

When this is minimized with respect to M we get an implicit equation for the equilibrium magnetization:

$$\frac{\partial \Gamma(M)}{\partial M} = H \tag{X.54}$$

If we insert the resulting value of $M = M(H)$ into $F(M)$, we reach the conjugate potential $G(H) = F[M(H)]$. It is this potential which is related to the partition function $Z(H)$

$$-\tau \ln \left(\frac{Z(H)}{Z(0)} \right) = Q[G(H) - G(0)] \tag{X.55}$$

Finally we may write

$$\frac{1}{Q} \ln \Xi = -\frac{1}{\tau}\left[\Gamma - M \frac{\partial \Gamma}{\partial M}\right]_{M = M(H)} \tag{X.56}$$

At this stage we have expressed all the polymer quantities (Φ, Φ_p, Π) in terms of the magnetic parameters (H, τ). We then write $\tau = \tau_c (1 + \epsilon)$ and note that near $\epsilon = 0$, $\partial\Gamma/\partial\epsilon$, for example, is more singular than Γ; this result allows us to reduce eqs. (X.50, X.51) to a simple form. We write first

$$-2\Phi_p = H \frac{\partial F}{\partial H} [M(H)] = H \left[\frac{\partial F}{\partial M} \frac{\partial M}{\partial H} + \frac{\partial F}{\partial H} \right]$$

Then we note that $\partial F/\partial M = 0$ at $M = M_{eq}(H)$ and also that $\partial F/\partial H$ is simply $-M$. We then arrive at

$$\Phi = \frac{\partial \Gamma}{\partial \epsilon} \tag{X.57}$$

$$\Phi_p = \frac{1}{2} M \frac{\partial \Gamma}{\partial M} \tag{X.58}$$

In the next section we insert the Widom scaling form for Γ into these equations.

X.3.2. Overlap concentration $\Phi^\star$ and related scaling laws

We use the Widom scaling form [eqs. (X.6), (X.8)] for Γ (M, τ) in the polymer eqs. (X.57, X.58, and X.52). This gives

$$\Phi = \epsilon^{\nu d-1} f_\Phi(x) \tag{X.59}$$

$$\Phi_p = \epsilon^{\nu d} f_p(x) \tag{X.60}$$

where $\nu d \equiv 2 - \alpha$ and where x is a dimensionless parameter proportional to $M/|\epsilon|\beta$. When ϵ is negative ($\tau < \tau_c$), a convenient definition of x is

$$x = \frac{M(\tau, H)}{M(\tau, 0)} \tag{X.61}$$

For the polymer problem, we can relate x to the length $\bar{N}$ of the chains through eq. (X.52); this gives:

$$\frac{f_\Phi(x)}{f_p(x)} = \bar{N}\epsilon \tag{X.62}$$

Thus x is a function only of $\bar{N}\epsilon = y$. We can give a simple meaning to y if we introduce the overlap concentration

$$\Phi^\star = \frac{\bar{N}a^d}{R_F{}^d} = \bar{N}^{(1-\nu d)} \tag{X.63}$$

Then:

$$\frac{\Phi}{\Phi^\star} = (\bar{N}\epsilon)^{\nu d-1} f_\Phi(x) = y^{\nu d-1} f_\Phi(x) \tag{X.64}$$

is only a function of y. Thus, any function of x (or y) only may be considered a function of $\Phi/\Phi^\star$. Let us apply this result to the osmotic pressure, Π. From eqs. (X.53, X.56) Π is given by:

$$\frac{\Pi a^d}{T} = \epsilon^{\nu d} f_\pi(x) \tag{X.65}$$

$$= \Phi_p \frac{f_\pi(x)}{f_p(x)}$$

$$= \Phi_p f_D \left(\frac{\Phi}{\Phi^\star} \right) \tag{X.66}$$

where f_D is another dimensionless scaling function. Eq. (X.66) is the fundamental des Cloiseaux result, stating that Π is equal to its perfect gas value, $(\Phi_p T/a^d)$, times a dimensionless function of $\Phi/\Phi^\star$.

X.3.3. Crossover between dilute and semi-dilute solutions

We know that solutions with $\Phi < \Phi^\star$ (dilute) are very different from solutions with $\Phi > \Phi^\star$ (semi-dilute). How does this appear in the magnetic analogy?

Consider the temperature versus magnetization diagram (τ, M) for the magnetic problem, shown in Fig. X.6. The accessible regions of this diagram are limited by the "coexistence curve"—i.e., the curve giving the spontaneous magnetization in zero field $M(\tau, 0)$. Eq. (X.52) or (X.63) that fixes the length of the chains gives one relationship between M and τ; for a given $\bar{N}$ the allowed points are on a certain line, which we call the *isometric line* (dashed line in the figure).

Consider first the dilute limit. Clearly, from the structure [eq. (X.49)] of the grand partition function, this corresponds to a case where $H \to 0$, so that the weights for two chains, three chains, etc. are much smaller than the weight for a single chain. Also the dilute limit must correspond to $\tau > \tau_c$

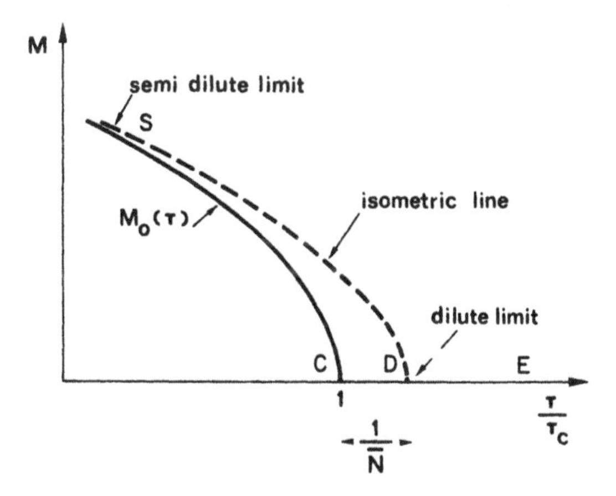

Figure X.6.

from what we have seen in the discussion of the single-chain problem. The conditions $\tau > \tau_c$ and $H = 0$ correspond to line CE on the figure. The isometric line intersects CE at some point D. From the results on a single chain we expect that the distance CD (which is proportional to ϵ) should be proportional to $\bar{N}^{-1}$.

We now proceed to the other end ($\Phi \gg \Phi^*$). All thermodynamic properties tend to become independent of $\bar{N}$ in this limit. Thus, we may set $\bar{N} = \infty$ or $\Phi_p = 0$ in calculating them. Looking back at eq. (X.58) where $\Phi_p = 1/2\ MH$, we see that this will again be obtained by taking $H \rightarrow 0$. Thus, the isometric line becomes very close to the coexistence curve. This limit corresponds to the region marked S in the figure.

There are many interesting features hidden in the algebra for this semi-dilute limit, which the reader can visualize in terms of blobs (defined in Chapter III).

(i) The size of a blob is $\xi \cong \epsilon^{-\nu} a$.

(ii) The number of monomers per blob is g, such that $g^\nu a = \xi$. This means that $g = \epsilon^{-1}$. Thus, the relationship between ϵ and N in the dilute limit becomes a relationship between ϵ and g for semi-dilute solutions.

(iii) The parameter $y = N\epsilon$ is the number of blobs per chain.

X.3.4. Correlations in the solution

There is no simple magnetic analog for the complete correlation $\langle \Phi(0)\Phi(\mathbf{r}) \rangle$. However, as shown by des Cloizeaux, the most accessible object is the correlation between the *extremities* of each chain.[8]

If all extremities are labeled, the corresponding correlation function is again related to the spin-spin correlations for the spin components S_1 *parallel* to the field. If only *one* chain is labeled at both ends, the associated correlation function can be related to a spin-spin correlation function for a component S_2 *transverse* to the field H. Of course the reader may shudder at the thought of discussing longitudinal and transverse correlations in a system of vectors with a number of components $n = 0$. However, a meaning can be given to these concepts; in fact, the first calculations of the end-to-end chain radius $R^2(\Phi)$ [eq. (III.33)] in the semi-dilute regime were based on these transverse correlations. However the direct argument based on blobs (given in Chapter III) is much simpler.

X.3.5. Current extensions

IMPROVEMENTS IN GOOD SOLVENTS

The main impact of the des Cloizeaux calculation has been to show that simple scaling laws do hold in polymer solutions. The main limitation, already mentioned, is related to the polydispersity of the chain ensemble which is considered. To be sure, we arrange that the average $\bar{N}$ is fixed and independent of concentration, but the distribution of N values is not fixed by the model and does not retain the same shape at all concentrations.

(i) For certain properties (and in particular for the region $\Phi \sim \Phi^*$) this is a serious defect. Direct calculations for monodisperse chains in the region $\Phi \lessdot \Phi^*$ have been constructed to circumvent this difficulty, but they are very complex.[9]

(ii) In the semi-dilute regime, all properties measured on the scale of one blob (and independent of $\bar{N}$) can be derived correctly from the des Cloiseaux approach. For example, the numerical coefficients of the osmotic pressure can be related to the known numerical approximations for the magnetic free energy at general n, after taking the limit $n = 0$.

POOR SOLVENTS

Near the theta point, there still exists a magnetic analog to a polymer solution—i.e., a so-called "tricritical point."[10] From this observation one can show that for a single chain, at $T = \Theta$, the mean field theory is approximately correct, except for some logarithmic corrections (explained in Chapter XI). If we go below $T = \Theta$ and study the consolute critical point, the des Cloiseaux procedure is no help because the inherent polydispersity upsets the picture seriously.

WALL EFFECTS

We have discussed some scaling properties of solutions near a wall, inside pores, etc. All these problems have their magnetic counterparts, but the state of the art for these magnetic problems is not very advanced, and the analogy does not help us very much.

X.3.6. What is the order parameter?

In the magnet problem, the onset of order is signaled by the appearance of a nonvanishing magnetization $\mathbf{M}(\mathbf{r})$. We call this the order parameter of the transition.

More generally, the order parameter is a quantity which shows diverging fluctuations (and a diverging susceptibility) near the critical point.* In

*There may be more than one physical observable with these features. For example, in a magnet both the magnetization and the energy have divergent fluctuations (related to the divergence of χ and C, respectively). The order parameter is then the observable with the *strongest* divergence (χ diverges more than C for the magnetic Curie point).

practice, to find the order parameter for a given physical transition is not easy. A classical example is superconductivity where the order parameter is a wavefunction for electron pairs. It took 50 years for this point to be understood.

A similar difficulty occurs when we compare polymer solutions with magnets. In a formal sense, the order parameter is the magnetization of a spin system with a number of spin components $n = 0$. This does not help very much. A more concrete statement, based on ideas of S. F. Edwards, is the following: the order parameter $\psi(\mathbf{r})$ for a polymer solution is similar to a quantum mechanical *creation* (or destruction) operator.[11] A factor $\psi(\mathbf{r})$ corresponds to the *initiation* of a chain (or to its termination).*

We can understand this statement through two approaches. In the grand partition function $\Xi = Z(H)/Z(O)$ for a polymer solution, we saw that each chain (with two ends) was associated with a factor H^2, where H is conjugate to the order parameter ψ. This means that each end has a factor H; thus, the concentration of end points must be proportional to ψ.

In Chapter IX we discussed one particular limit of the self-consistent field method, where the ground state was dominant. We then introduced a ground state wavefunction $\psi(r)$ with the following properties:

(i) The concentration of chain ends is proportional to $\psi(\mathbf{r})$ because only one weight factor $G_N(\mathbf{rr}')$ integrated over r' is involved.

(ii) The concentration of monomers is proportional to $|\psi|^2$ because two weight factors $G(\mathbf{r}_1\mathbf{r})$ and $G(\mathbf{rr}_2)$ are involved. These properties are represented in Fig. X.7.

The free energy functional $F\,[\psi(\mathbf{r})]$ obtained in eq. (IX.33) by a self-consistent field argument coincides with the mean field (Landau) approximation for the free energy in the magnetic problem (for which there are many excellent reviews[1]).

Thus at some points it is useful to think of the order parameter as an "initiator" (or "terminator") for a chain. (In our case with both ends equivalent we are not interested in distinguishing between the two.) However this viewpoint omits some important aspects related to the multicomponent feature of $\mathbf{M}(\mathbf{r})$. We saw, for example, (in connection with certain correlations) that it helps to distinguish between components of $\mathbf{M}$ parallel to the applied field $\mathbf{H}$ and perpendicular to $\mathbf{H}$. (Strangely enough, this distinction retains a meaning when we set $n = 0$.) These component aspects must be superimposed on the "initiator" aspect.

With these restrictions in mind, we can say that we have reached a certain level of qualitative understanding for the order parameter, but it is striking to see how slow our progress has been.

*The notation (ψ) is preferable to (M) because of the analogy with a wavefunction.

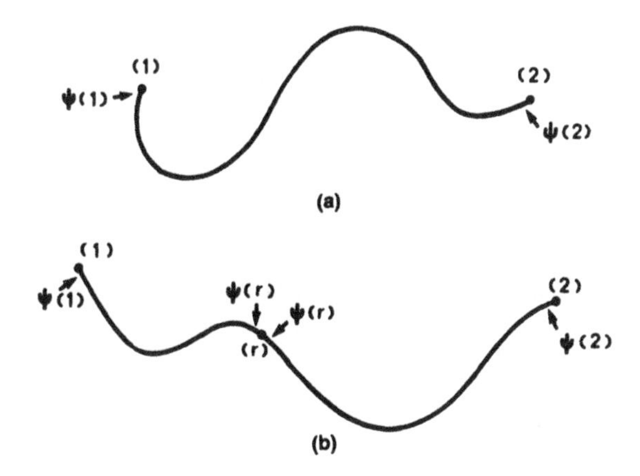

Figure X.7.

(a) The statistical weight for one chain, with ends at points (1) and (2), is proportional to $\psi(1)$, $\psi(2)$, where $\psi(\mathbf{r})$ is a "chain initiator" or "chain terminator." (b) The statistical weight for a chain crossing through point **r** contains the weights $\psi(1)$, $\psi(\mathbf{r})$, and $\psi(\mathbf{r})$, $\psi(2)$. Integration over the end points (1) and (2) gives a concentration at point **r** proportional to $|\psi(\mathbf{r})|^2$.

REFERENCES

1. H. E. Stanley, *Introduction to Phase Transitions and Critical Phenomena*, Oxford University Press, London, 1972.
2. P. G. de Gennes, *Magnetism*, G. Rado and H. Suhl, Eds., Vol. III, p. 115, Academic Press, New York, 1963.
3. G. Toulouse, P. Pfeuty, *Introduction au Groupe de Renormalisation*, Presses Universitaires de Grenoble, Grenoble, 1974. *Critical Phenomena*, J. Brey and R. Jones, Eds., Springer-Verlag, New York, 1976.
4. M. Fisher, *Rep. Prog. Phys.* **30,** 615 (1967).
5. M. Fisher, *Critical Phenomena*, M. Green and J. Sengers, Eds., p. 108, NBS Misc. Pub. No. 273, Washington, D.C., 1966. M. Fisher, A. Aharony, *Phys. Rev. Lett.* **31,** 1238 (1973).
6. M. Daoud *et al., Macromolecules* **8,** 804 (1975). For different presentations see: P. G. de Gennes, *Phys. Lett.* **A38,** 339 (1972). V. Emery, *Phys. Rev.* **B11,** 239 (1975). D. Jasnow, M. Fisher, *Phys. Rev.* **B13,** 1112 (1976).
7. See Refs. 14, 15, 16 of Chap. I.
8. J. des Cloizeaux, *J. Phys. (Paris)* **36,** 281 (1975).
9. See Ref. 5 of Chap. III.
10. P. G. de Gennes, *J. Phys. (Paris) Lett.* **36L,** 55 (1975).
11. A. Messiah, *Quantum Mechanics*, Vol. II, Dunod, Paris, 1957.

XI

An Introduction to Renormalization Group Ideas

XI.1. Decimation along the Chemical Sequence

Our aim here is to gain some insight in the calculation techniques initiated by K. Wilson,[1] without going into the intricacies of field theory. Fortunately, this is possible for polymers. The text in this chapter explains the principles of the method but does *not* prepare the reader for performing any elaborate calculations. For the latter purpose, there are some excellent courses and reviews.[2]

XI.1.1. A single chain in a good solvent

We start with a linear chain of N monomers, each of size a, in a space of arbitrary dimensionality d. (The size is defined in terms of the ideal chain radius R_0 through $a = R_0 N^{-1/2}$.) The interactions among monomers are described by an excluded volume parameter v:

$$v = a^d (1 - 2\chi) > 0 \qquad \text{(XI.1)}$$

In a continuous description, we say that for a local concentration profile $c(\mathbf{r})$ the repulsive energy is:

$$F_{rep} = \frac{1}{2} T v \int c^2(\mathbf{r}) d\mathbf{r} \qquad \text{(XI.2)}$$

If we prefer to specify the problem by a pairwise interaction U_{nm} between monomers, we can write this as a point repulsion

$$U_{nm} = Tv\delta(\mathbf{r}_{nm}) \tag{XI.3}$$

Parameters a, v, and N are the essential parameters of the one-chain problem. Instead of the excluded volume v, it is sometimes convenient to introduce a dimensionless coupling constant

$$u = \frac{v}{a^d} = 1 - 2\chi > 0 \tag{XI.4}$$

XI.1.2. Grouping the monomers into subunits

We do not directly attack the formidable problem of chain conformations with the repulsive interaction [eq. (XI.3)] included. More modestly, we begin by grouping g consecutive monomers into subunits, as shown in Fig. XI.1. We can choose the number g as we wish. For some problems it may be convenient to take a rather large g ($g = 10$); for others, $g = 2$ will be sufficient.

We have thus defined a number N/g of consecutive subunits. The idea is to compute the size a_1 and the excluded volume parameter v_1 for the subunits; this will require a direct calculation.

(i) All interactions inside one subunit must be taken into account for the determination of a_1.

(ii) All interactions between two subunits (which are widely separated on the chemical sequence) must be taken into account for v_1. This is summarized in Fig. XI.2, where the solid lines are chains, and the dashed lines are interactions.

Let us describe the structure of the new size parameter. If we were dealing with ideal coils, we would have simply, $a_1 = g^{1/2}\, a$. However, the interactions tend to swell the subunit. This swelling is described by a factor $1 + h$, where h depends on g and on the dimensionless coupling constant u

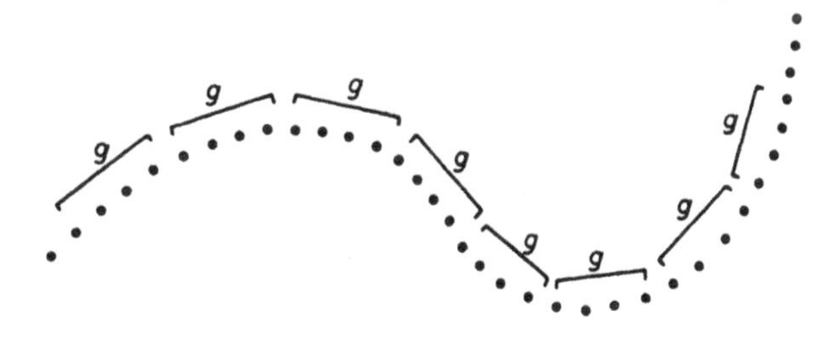

Figure XI.1.

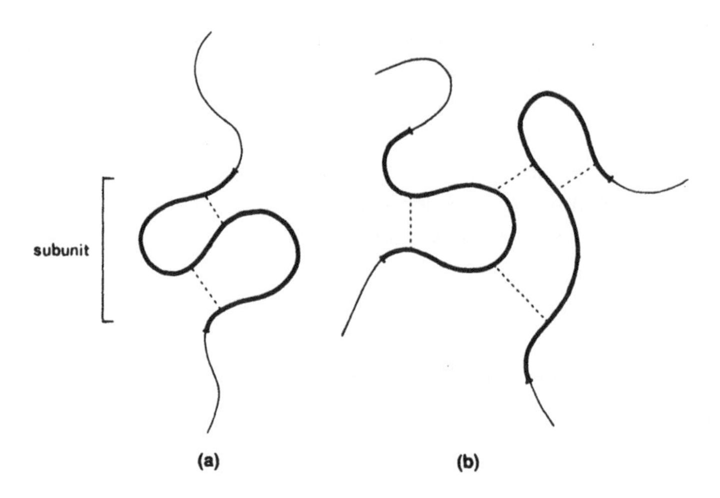

Figure XI.2.

(a) Principle for calculating the size of one subunit, incorporating all interactions inside this subunit. (b) Calculation of the effective interaction between two subunits.

$$a_1 = ag^{1/2}\,[1 + h\,(g,u)] \tag{XI.5}$$

Now consider the new excluded volume parameter. If the coils were nearly ideal, the subunit–subunit interaction would be g^2 times the monomer–monomer interaction (since there are g^2 possible interacting pairs). However, this is an overestimate. If the repulsion is strong, the two subunits do not interpenetrate each other well, and the number of interacting pairs is smaller than g^2. Thus, we must have

$$v_1 = vg^2\,[1 - l(g,u)] \tag{XI.6}$$

where l is another correcting factor. Introducing a reduced coupling constant $u_1 = v_1/a_1^d$, we have

$$\begin{aligned} u_1 &= ug^{2-d/2}\,\frac{1 - l}{(1 + h)^d} \\ &= ug^{2-d/2}\,[1 - k(g,u)] \end{aligned} \tag{XI.7}$$

where we have introduced the notation

$$k = 1 - \frac{1 - l}{(1 + h)^d} \tag{XI.8}$$

The numerical calculation of functions h and k is painful but feasible. The essential point is that the region of space concerned by the calculation (shown in Fig. XI.2) is comparable with the subunit size, and thus much smaller than the chain size. The worst part is the calculation of v_1, but this is a problem involving at most ($2\ g$) monomers, and it can be calculated exactly on a computer. Other methods of approximation for h and k can be devised, and we quote one of them later.

XI.1.3. Iterating the process

The idea of grouping the monomers into subunits is a classical idea of polymer physics. However, usually, with very few exceptions, polymer theorists performed the operation only once, to dispose of certain non-universal features of the monomer species.[3]

The essential idea of the renormalization group is to *repeat* the operation: at the next iteration, we have subunits of g^2 monomers, with size a^2 and reduced coupling constant u_2, and so on. In this way we generate a sequence

$$\begin{matrix} a \\ u \end{matrix} \rightarrow \begin{matrix} a_1 \\ u_1 \end{matrix} \rightarrow \begin{matrix} a_2 \\ u_2 \end{matrix} \rightarrow \ldots \rightarrow \begin{matrix} a_n \\ u_n \end{matrix}$$

Any pair (a_m, u_m) in this sequence represents one possible formulation of the single-chain problem.

The successive operations represented by the arrows can be associated with a mathematical group, called the *renormalization group* (actually a semi-group because we cannot define a unique inverse to the operations).

Successive steps are related by equations

$$a_m = a_{m-1}\ g^{1/2}\ [1 + h(u_{m-1})] \tag{XI.9}$$

$$u_m = u_{m-1}\ g^{2-d/2}\ [1 - k(u_{m-1})] \tag{XI.10}$$

(In these equations, to emphasize the main features, the dependence of h and k on g is not explicitly mentioned.)

At first sight, all that we have done is to replace the original insoluble problem (involving a, u, and N), by a sequence of equally insoluble problems (the m-th problem being based on a_m, u_m, and N/g_m). However, we have gained, as will become clear later.

XI.1.4. Existence of a fixed point

When the process has been iterated long enough (m large), the subunits are large, and we know from the properties of dilute solutions that two subunits *behave like impenetrable spheres.** This means that the excluded volume v_m must scale like the subunit volume a_m^d, or equivalently that the reduced coupling constant tends toward a finite limit:

$$u_m = \frac{v_m}{a_m^d} \rightarrow u^\star = \text{constant} \tag{XI.11}$$

The sequence of numbers $u_1, \ldots u_m$ approaches a fixed point $u^\star$ when $m \rightarrow \infty$.

The value of $u^\star$ is obtained from eq. (XI.10) and is a root of

$$g^{2-d/2}\,[1 - k(u^\star)] = 1 \tag{XI.12}$$

In most practical cases this fixed point is reached quickly. For example, with $g = 3$ and $m = 4$ we are already dealing with subunits of $3^4 \sim 100$ monomers, for which the hard sphere limit usually holds well.

XI.1.5. Scaling law for the chain size

The "fixed-point theorem" [eq. (XI.11)] is essential because when u_m reaches its limiting value $u^\star$, the relationship between a_m and a_{m-1} [eq. (XI.9)] becomes a simple geometric series

$$a_m = a_{m-1}\, g^{1/2}\,[1 + h(u^\star)] = \mu a_{m-1} \tag{XI.13}$$

(where μ is a known constant), and all properties on these large scales become "self-similar." The meaning of this word will become clearer if we consider a specific observable property, such as the root mean square end-to-end radius R of the chain. In the initial problem, we expect R to be of the form

$$R = af(N, u) \tag{XI.14a}$$

where the function f is unknown *a priori*. However, let us write R using the m-th formulation. We must have

$$R = a_m f\left(\frac{N}{g^m}, u_m\right) = a_{m-1} f\left(\frac{N}{g^{m-1}}, u_{m-1}\right) \tag{XI.14b}$$

with the *same* function f. Let us apply this identity in the region of high m where $u_m \cong u^\star$. Then we can drop u from the arguments of f, and we arrive at the condition

*This statement holds for dimensionalities $d \leqslant 4$. The case of $d \geqslant 4$ is discussed later.

$$\frac{f\left(\frac{N}{g^{m-1}}\right)}{f\left(\frac{N}{g^m}\right)} = \frac{a_m}{a_{m-1}} = \mu \tag{XI.15}$$

where we have used eq. (XI.13) defining μ. This imposes a power law structure on f:

$$f(N) = \text{constant } N^{\nu} \tag{XI.16}$$

Thus, in eq. (XI.16) we *prove* the existence of an exponent ν. Furthermore, we have an explicit value for ν, derived from eq. (XI.15)

$$\begin{aligned} g^{\nu} &= \mu \\ \nu &= \frac{\ln \mu}{\ln g} \end{aligned} \tag{XI.17}$$

Here we see the enormous advantage of renormalization groups, when compared with more classical methods, such as perturbation calculations (Chapter I.2). Even if we have only an approximate calculation of the h and k functions (and thus of $u^{\star}$ and μ), we obtain a nontrivial prediction for the exponent ν.

XI.1.6. Free energy of a single chain

We now show how the renormalization group transformation can be applied to a discussion of the statistical weight (or of the free energy) of the chain. Our discussion is purely qualitative, but it does give some insight into the origin of the "enhancement factor" presented in eq. (I.21).

Let us call $F(N, u)$ the free energy of a chain of N monomers with a dimensionless coupling constant u. Then we associate the monomers into subunits of g partners. We started with N independent vectors $\mathbf{r}_1 \ldots \mathbf{r}_N$. We end up with N/g independent vector $\mathbf{r}'_1 \ldots \mathbf{r}'_{N/g}$ defining the centers of gravity of the subunits. The original partition function was of the form

$$\exp\left[-\frac{F(N,u)}{T}\right] = \int d\mathbf{r}_1 \ldots d\mathbf{r}_N \exp(-\mathfrak{H}_u(\mathbf{r}_1 \ldots \mathbf{r}_N))$$

$\mathfrak{H}_u$ being the original interaction energy between all monomers. When we reduce the system to subunits, we integrate over $g - 1$ coordinates for each subunit. The result of the integration is to introduce a factor which we call $\exp(-\Delta F/T)$

$$\exp\left[-\frac{1}{T}F(Nu)\right] = \exp\left(-\frac{\Delta F}{T}\right)\int d\mathbf{r}'_1 \ldots d\mathbf{r}'_{N/g} \exp(-\mathfrak{H}_{u_1}(\mathbf{r}'_1 \ldots \mathbf{r}'_{N/g}))$$

$$= \exp\left(-\frac{\Delta F}{T}\right) \exp\left[-\frac{1}{T}F\left(\frac{N}{g}, u_1\right)\right]$$

or in terms of free energies

$$F(N,u) = \Delta F(N,u) + F\left(\frac{N}{g}, u_1\right)$$

What is the structure of ΔF? For an ideal chain we would get additive contributions from different subunits, each of them proportional to the number of integration variables $g - 1$

$$\frac{1}{T}\Delta F_{ideal} = A_0\frac{N}{g}(g - 1)$$

where $\exp(-A_0)$ would be the partition function of one monomer. When we switch on the interactions, two effects occur.

(i) A_0 is renormalized $A_0 \rightarrow A(u)$.

(ii) There is an end effect. The first and the last subunits in the chain feel smaller repulsions because they interact only on one side on the chemical sequence. This subtracts a contribution that is independent of N, which we shall call $B(u)$

$$\frac{1}{T}\Delta F = \frac{N}{g}(g - 1)A(u) - B(u) \tag{XI.18}$$

(Note that B vanishes for $u = 0$).

Let us now iterate eq. (XI.18); we get

$$F(N,u) = F\left(\frac{N}{g^m}, u_m\right) + (g - 1)\left\{\frac{N}{g}A(u_1) + \frac{N}{g^2}A(u_2) + \ldots \frac{N}{g^m}A(u_{m-1})\right\} - \{B(u) + B(u_1) + \ldots + B(u_{m-1})\}$$

The summation can be simplified if we assume that u_m is equal to $u^\star$ as soon as a few iterations have been performed. (The difference, $u_m - u^\star$, will introduce corrections in the form of a rapidly converging sum and is not important.) Then

$$\frac{1}{T}F(N,u^\star) = \frac{1}{T}F\left(\frac{N}{g^m}, u^\star\right)\frac{N}{g}(g-1)\left(1+\frac{1}{g}+\ldots\frac{1}{g^{m-1}}\right) + mB(u^\star)$$

(For large m the geometric series adds up to $g/g-1$). We now choose a value of m such that $g^m = N$. Then we have $m = \ln N / \ln g$ and

$$\frac{1}{T}F(N,u^\star) = \frac{1}{T}F(1,u^\star) + NA(u^\star,g) - [B(u^\star)/\ln g]\ln N$$

The first term, $F(1,u^\star)$, is a constant. Returning to $\Xi = \exp(-F/T)$, we see that Ξ has precisely the form described in eq. (I.21); it contains a factor $N^{\gamma-1}$ with

$$\gamma - 1 = \frac{B(u^\star)}{\ln g} \qquad \text{(XI.19)}$$

We see now that this factor reflects certain *end effects* in an open chain. This observation is interesting because it explains why the enhancement factor disappears when we consider closed polygons [eq. (I.28)]. A closed polygon has no ends!

A remark on the ζ parameter. In Chapter I we introduced a dimensionless parameter $\zeta \cong va^{-d} N^{2-d/2}$ and we stated the following property [eq. (I.42)]: a gaussian chain (unperturbed size $R_0 = N^{1/2}a$) subjected to a point repulsion between monomers $[Tv\delta(\mathbf{r}_{ij})]$ swells up by a factor R/R_0 which is a function of ζ only. One justification of this statement uses a decimation procedure which is somewhat different from the above, and is due to Kosmas and Freed.* We cite it here because of its great simplicity.

Using a continuous notation we write the weight factor $\mathfrak{H}_u/T$ in the form

$$\frac{\mathfrak{H}_u}{T} = \frac{3}{2a^2}\int_0^N dn\left(\frac{dr}{dn}\right)^2 + \frac{1}{2}\iint^N dn\, dm\, va^{-d}\,\delta(\mathbf{r}_n - \mathbf{r}_m)$$

where the first term describes ideal springs, and the second term describes interactions. We then perform a *single* decimation step by step switching to a new variable $n_1 = n/g$. We then find that we can maintain the same structure for $\mathfrak{H}_u$ if we change N into N/g and r_n into $g^{1/2}r_n$ The only new effect is then to change $u = va^{-d}$ into $ug^{2-d/2}$. Thus if we consider the dimensionless partition function Ξ/Ξ_0 (where Ξ_0 would correspond to the ideal chain) we can write

*M. Kosmas, K. Freed, *J. Chem. Phys.* **69**, 3647 (1978).

$$\Xi/\Xi_0 = f(N,u) = f\left(\frac{N}{g}, ug^{2-d/2}\right)$$

We may then choose $g = Nk$, where k is a constant, independent of N (and much smaller than unity, to ensure that the number of subunits $N/g = k^{-1}$ is still large enough to justify the continuous notation). We arrive at

$$\Xi/\Xi_0 = f(k, k^{2-d/2}\, uN^{2-d/2}) = f(k, k^{2-d/2}\zeta)$$

and conclude that Ξ/Ξ_0 is only a function of ζ. The same argument can be written for the swelling coefficient R/R_0. However, our readers should remain aware of the strong limitations of this theorem, which applies only to a special model: gaussian chains plus point-like pair-wise interactions.

If we start with a real chain, many other features will show up in the original hamiltonian, for instance chain stiffness, three-body interactions, etc. We must then decimate up to a point where all these nonpertinent features have dropped out. However, by this time, the coupling constant u will also have changed, and will be close to the fixed point value $u^\star$. Thus, for a real chain, the ζ-theorem boils down to the statement that R/R_0 is a function only of $u^\star N^{2-d/2}$ (that is to say of N) and is not very helpful.

XI.1.7. Calculations near four dimensions

In practice there are two main methods for calculating h and k in the renormalization group equations (XI.9, XI.10): 1) direct numerical calculations for small g, and 2) perturbation calculations for dimensionalities (d) "slightly below 4" ($d = 4 - \epsilon$).*

We discuss this second approach briefly. The crucial observation is that *for $d > 4$, the fixed point value $u^\star$ is equal to zero*. This is clear from eq. (XI.10) since

$$\frac{u_m}{u_{m-1}} \leqslant g^{2-d/2}$$

and when $d > 4$, the right side is smaller than unity. This suggests that for d slightly smaller than 4 ($d = 4 - \epsilon$), the fixed point value $u^\star$ is small. This guess can be confirmed by a self-consistent argument. If we assume u to be small, we can compute the corrections $h(u)$ and $k(u)$ by perturbation

*Dimensionalities d which are not integers can be manipulated by analytic continuation of the formulas for integral d. For example, the approximate Flory formula for $\nu = 3/(d+2)$ retains a meaning for nonintegral d.

methods. To first order they are linear in u. The fixed point equation then gives

$$1 - k(u^\star) = g^{d/2-2} = g^{-\epsilon/2} \cong 1 - \epsilon/2 \ln g$$

Thus $k(u^\star)$ (and $u^\star$ itself) are proportional to ϵ, and our assumption of small u was justified. Perturbation calculations along these lines have been carried out.[4] They give rather good results in three dimensions ($\epsilon = 1$).

All our presentation is based on a physical grouping into subunits. This is pedagogically convenient but suffers from certain defects. For example, interactions among *neighboring* subunits (Fig. XI.3) are not properly taken into account if the new coupling constant u is defined as in Fig. XI.2b through the interaction of subunits which are widely separated on the chemical sequence. This defect can be corrected by operating not on the

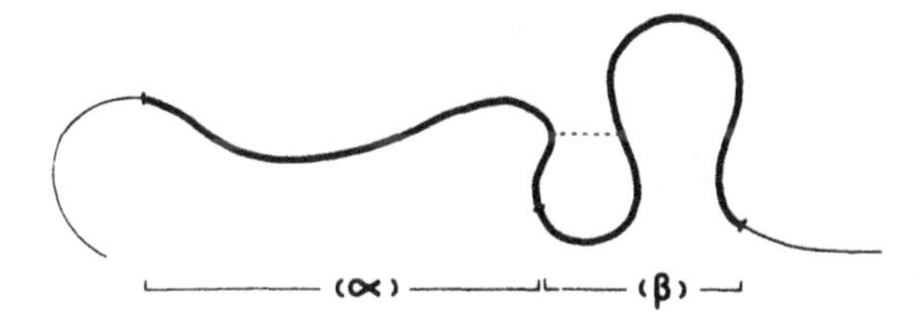

Figure XI.3.

Interactions between two subunits which are *consecutive* on the chemical sequence. Compare this drawing with Fig. XI.2b, which shows interactions between *nonconsecutive* subunits, and note the difference.

sequence of monomers 1, 2, . . . n . . . N but on Fourier transforms with respect to n. For the details of the procedure, see related discussions in Wilson and Kogut.[1]

XI.2. Applications

XI.2.1. Polyelectrolytes

Chains carrying ionizable groups (such as SO_3^- or CO_2^-) are frequent in nature and in many industrial applications requiring polymer–water systems. To each ionized group (say SO_3^-) is associated a counterion (say K^+) which moves rather freely in the surrounding water. General reviews on these "polyelectrolytes" are listed in Ref. 5. The effects of coulomb repulsions between the ionized groups are strong and complex. From a scaling point of view, the properties of polyelectrolyte solutions are not yet fully understood.[6] Here we restrict our attention to one soluble problem—namely, a *single* polyelectrolyte chain, with counterions dispersed very freely in the solution.

These assumptions are very restrictive because, as we shall see, polyelectrolytes are very stretched, and the resulting overlap concentration $c^\star$ is very low ($\Phi^\star$ is of order N^{-2} instead of $N^{-4/5}$ for neutral chains). Experiments at $c < c^\star$ are often unfeasible.

The assumption of dispersed counterions (negligible screening) is also very delicate. Manning[7] and Oosawa[5] showed that this assumption is correct only if the charge density along the chain remains below a certain threshold. If e is the electron charge and a is the length per monomer in a fully stretched chain, we must have (for monovalent ions)

$$\frac{e^2}{\epsilon\, aT} < 1 \tag{XI.20}$$

where ϵ is the dielectric constant of water.

If the inequality [eq. (XI.20)] is violated, the counterions "condense" on the chain and decrease the charge density, bringing it back to the threshold value. Another limit to the assumption of negligible screening is given by the residual ionic content of the water itself. If we use pure water (completely salt free) as the solvent, we still have a certain number of H^+ and OH^- ions that arise from spontaneous dissociation. These ions give a Debye screening length κ^{-1} which is large but finite (in the micron range). Our considerations hold only if the chain length is much smaller than κ^{-1}.

With all these difficulties theoretically eliminated, we ask: what is the conformation of one charged chain (of N monomers) in d dimensional space. The interaction between monomers (n) and (m) separated by a distance r_{nm} is of the general coulombic form

$$V(r_{nm}) = k_d \frac{e^2}{\epsilon r_{nm}{}^{d-2}} \tag{XI.21}$$

where k_d is a numerical constant.*

For $d = 3$, $V(r)$ is the usual $1/r$ interaction. For more general dimensionalities, eq. (XI.21) ensures that V satisfies the Poisson equation

*k_d is related to the area A_d of a d dimensional sphere of radius r: $k_d = 4\pi/(d-2)r^{d-1}/A_d$. This equation can be checked by applying the Gauss theorem to eq. (XI.22).

$$\nabla^2 V = \frac{4\pi e^2}{\epsilon} \delta(\mathbf{r}) \qquad \text{(XI.22)}$$

As usual we describe the strength of the interaction V by a dimensionless coupling constant u, obtained by comparing the coulomb repulsion between neighboring sites $V(a)$ with the thermal energy T

$$u = k_d \frac{e^2}{\epsilon\, a^{d-2}\, T} \qquad \text{(XI.23)}$$

The original problem of one chain with electric repulsions is then defined entirely in terms of a, u, and N.

To this problem we associate other problems obtained by grouping the monomers into subunits of lengths g, g^2, . . . g^m along the chemical sequence. The equation for the sizes $a_1 \ldots a_m$ of the units will still have the structure of eqs. (XI.5, XI.9) (although the precise structure of the correction function $h(g, u)$ will, of course, differ).

What is peculiar here is the equation for the coupling constants $u_1 \ldots u_m$. The charge of g units is *exactly* g times the charge of one unit. This implies that no other unknown function enters into the equation for u_m. Eq. (XI.10) is replaced by

$$u_m = u_{m-1}\, g^2 \left(\frac{a_{m-1}}{a_m}\right)^{d-2} \qquad \text{(XI.24)}$$

$$= u_{m-1}\, g^{3-d/2}\, [1 + h(u_{m-1})]^{2-d} \qquad \text{(XI.25)}$$

We can construct the relationship between u_{m-1} and u_m explicitly for the two essential limits, corresponding to weak coupling ($u_m \rightarrow 0$) and strong coupling ($u_m \gg 1$).

In weak coupling the chains are nearly ideal ($h \rightarrow 0$)

$$\frac{u_m}{u_{m-1}} = g^{3-d/2} \qquad (u_{m-1} \ll 1) \qquad \text{(XI.26)}$$

In strong coupling the chains are completely stretched, and

$$\frac{a_m}{a_{m-1}} = g \qquad \text{(XI.27)}$$

Inserting this into eq. (XI.24), we find

$$\frac{u_m}{u_{m-1}} = g^{4-d} \qquad (u_{m-1} \gg 1) \tag{XI.28}$$

From these two limiting properties [eqs. (XI.26, XI.28)] one can reconstruct the general structure of u_m versus u_{m-1} (Fig. XI.4). Depending on the dimensionality d, we find three cases: $d > 6$, $4 < d < 6$, and $d < 4$.

For $d > 6$ the plot of u_m versus (u_{m-1}) is always below the first bisector. Then, if we construct successive iterations, as shown in the figure, we find that the fixed point is $u^\star = 0$. Physically this means that a long chain is ideal.

For $4 < d < 6$ the plot of u_m versus (u_{m-1}) must cross the first bisector because the slope at the origin [given by eq. (XI.26)] is larger than unity, while the slope at infinity [eq. (XI.28)] is smaller than unity. Then successive iterations lead to a finite fixed-point value.

$$\lim_{m\to\infty} u_m = u^\star \tag{XI.29}$$

Returning to eq. (XI.24) we then find the behavior of the lengths a_m in the fixed point limit

$$\frac{a_m}{a_{m-1}} = g^{2/(d-2)} \tag{XI.30}$$

We could then derive the chain size using the detailed arguments of eqs. (XI.14–17). Here we use a simplified method, which is always used in these matters. We choose $u = u^\star$ for our original problem. This improves the convergence and simplifies the argument. Then the sequence of size $a_1, \ldots, a_m$ is a geometric series $a_m \cong a g^{2m/(d-2)}$. We now choose m so that at the m-th step, one subunit spans the entire chain ($g^m = N$). Then the chain size is

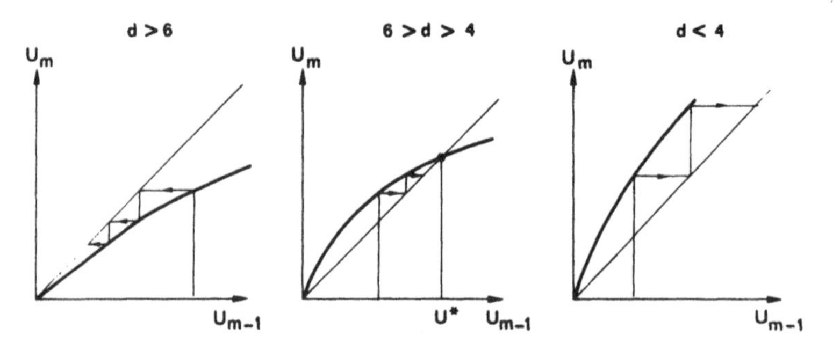

Figure XI.4.

$$R = a_m \cong aN^{2/(d-2)}$$

and we find an exponent[8]

$$\nu = \frac{2}{d-2} \tag{XI.31}$$

This result was obtained first by a more complex perturbation calculation near $d = 6$, but it is valid in the entire range $4 < d < 6$.

For $d < 4$ the iteration leads to $u_m \to \infty$, i.e., to rigid chain behavior. This applies particularly to the practical case for $d = 3$. A single polyelectrolyte chain (in the absence of all salts, etc.) should be fully stretched.

One case is of special interest. If we have a very weak charge density (one ionized group for N_c monomers, with $N_c \gg 1$), the first iterations will correspond to the weak coupling limit; it is only on large scales that the chain will behave as a rod. The corresponding picture is shown in Fig. XI.5.

The diameter D of the effective rod may be estimated as follows. At scales r smaller than D, the chain is ideal. Thus, a section of overall size D has a number of monomers

$$g_D \cong (D/a)^2 \tag{XI.32}$$

and a charge $g_D/N_c\, e$, corresponding to a coulombic energy

$$V_D \cong \left(\frac{g_D e}{N_c}\right)^2 \frac{1}{\epsilon D} \tag{XI.33}$$

$$\cong \frac{e^2}{\epsilon a}\left(\frac{D}{a}\right)^3 N_c^{-2} \tag{XI.34}$$

If D is the crossover size, the dimensionless coupling constant constructed with blobs of size D is of order unity, or V_D is comparable with T. This gives

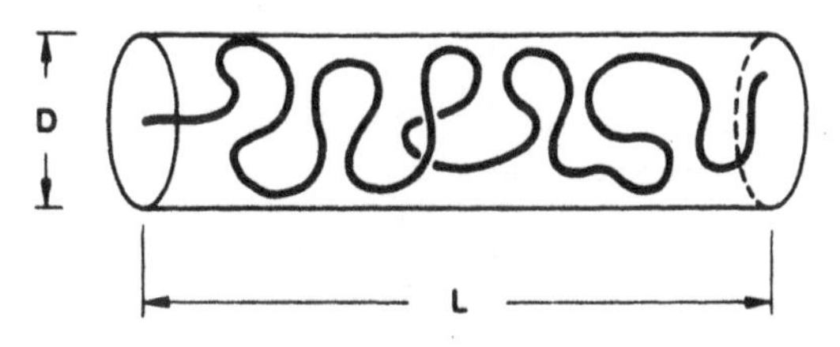

Figure XI.5.

$$\frac{D}{a} \cong N_c^{2/3} \left(\frac{e^2}{\epsilon a T}\right)^{-1/3} \qquad \text{(XI.35)}$$

The overall size of the chain is

$$R_{\parallel} \cong \frac{N}{g_D} D \qquad \text{(XI.36)}$$

This discussion could be refined by more detailed renormalization group calculations,* but it shows clearly the interest of diffraction studies on chains that carry only a few charges along their length. If the chain is not very long, we may reach a moment where $g_D = N$. Below this point, we return to an ideal chain, only weakly perturbed by coulombic effects.

XI.2.2. Collapse of a single chain

TWO COUPLING CONSTANTS

In Section IV.3.2 we gave a crude discussion of the properties of a single chain near the compensation temperature Θ. We saw that the chain was nearly ideal but that some delicate corrections were introduced by the three-body interactions. We shall now return to this problem using the decimation approach. Our starting point is a set of interactions described by eq. (III.9):

(i) A pair interaction $1/2\ vTc^2$.
(ii) A three body interaction $1/6\ w^2\ Tc^3$. (XI.37)

where c is the *local* concentration in the chain. We are interested in small values of the excluded volume v, and then the c^3 terms cannot be omitted as pointed out in Chapter IV. However, higher order terms ($c^4, \ldots$) remain negligible because $c \sim N^{-1/2}$ is small. Thus, we must introduce *two* dimensionless coupling constants

$$u = \frac{v}{a^3} = 1 - 2\chi$$
$$t = \frac{w^2}{a^6} \qquad \text{(XI.38)}$$

THREE TYPES OF CHAIN BEHAVIOR

We then construct a renormalization group, associating the monomers in subunits of g elements and iterating the process

*Also the effects of counterion screening should be incorporated.

$$\begin{pmatrix} u \\ t \\ a \end{pmatrix} \to \begin{pmatrix} u_1 \\ t_1 \\ a_1 \end{pmatrix} \ldots \to \begin{pmatrix} u_m \\ t_m \\ a_m \end{pmatrix} \qquad \text{(XI.39)}$$

It is convenient to show the successive values of u and t in a two-dimensional diagram (Fig. XI.6). Each step is represented by one point, and successive steps define a "trajectory." (We draw trajectories here in the limit where g is close to unity—i.e., where they are nearly continuous.)

Fig. XI.6. is drawn specifically for three dimensions. As explained in Section IV.3.2 the value $d = 3$ is very special. There are two types of trajectories: 1) trajectories converging toward the fixed point ($u = u^\star$, $t = 0$), and 2) trajectories going toward large negative u values. These two "basins" are separated by a dividing line, often called the "tricritical line" in analogy with certain problems in phase transitions.[2,9]

The physical implications are as follows. If we start with values of u and t corresponding to a point A in the first basin, and if we look to larger and larger subunits, we find that our subunits repel each other like hard spheres ($u_m \to u^\star$) and that the three-body interaction described by t becomes irrelevant ($t_m \to 0$). This means that on large scales our chain will be swollen and behave as a good solvent. If, on the other hand, we start at point B in the other basin, we go toward more and more negative (attractive) u values. This means that for large N (many iterations) we tend toward a collapsed situation.

THE RENORMALIZED Θ TEMPERATURE

If we choose an initial condition corresponding to point C (on the dividing line), the iteration leads to a universal behavior which is neither

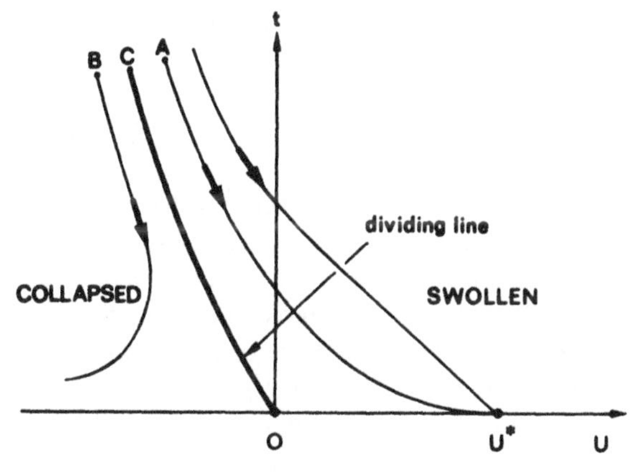

Figure XI.6.

swollen nor collapsed and which we call *quasi-ideal behavior*. Note that an arbitrary point C on the dividing line has a nonzero value of the u parameter $[u = u_d(t)]$; *quasi-ideal behavior is not associated with* $u = 0$.

In practice, for a given three-body interaction parameter w^2 (or t), we must distinguish between two temperatures: 1) the temperature at which $u = 1 - 2\chi$ vanishes, which we might call the bare Θ temperature, and 2) the temperature at which a large chain is quasi-ideal, which corresponds to the value of $u = u_d(t)$ on the dividing line. We call this temperature $\tilde{\Theta}(t)$ the renormalized Θ point. Most empirical definitions of a compensation point are based on $\tilde{\Theta}$. For example, one can define $\tilde{\Theta}$ from the requirement that the chain size R be close to the ideal value

$$\lim_{N\to\infty} [R^2/N] = \text{constant} \tag{XI.40}$$

Another practical definition of $\tilde{\Theta}$ can be proposed. We may call $\tilde{\Theta}$ the temperature at which the second virial coefficient between two very large coils vanishes. Fortunately, these two definitions *coincide*. When we are on the dividing line, the parameter u_m (at the m-th iteration) gives (in dimensionless units) the virial coefficient between two subunits. Since the dividing line ends at 0, where $u = 0$, this coefficient vanishes when the subunits are large enough. The distinction between Θ and $\tilde{\Theta}$ is essentially absent from the polymer literature (which has been written mainly on the mean field level).

CHAIN BEHAVIOR AT THE COMPENSATION POINT

We now sketch briefly the structure of the relationships between (u_m, t_m, a_m) and (u_{m-1}, t_{m-1}, a_{m-1}) near the fixed point O. In this region, u and t are small, and perturbation calculations can be used. The first essential result concerns the sizes, for which we have

$$a_m = g^{1/2}\, a_{m-1}\, [1 + k_u u_{m-1} + k_t t_{m-1} + O(u^2, t^2, ut)] \tag{XI.41}$$

Eq. (XI.41) can be understood easily when $g \gg 1$ (in which case the subunit is itself a small coil). The factor $g^{1/2}$ corresponds to ideal chain behavior. The u correction is derived from the perturbation calculation described in eq. (I.42). The second correction $k_t t$ results from a similar perturbation treatment applied to the three body interaction. The constants k_u and k_t depend on the value chosen for g.

Eq. (XI.41) applies in the entire neighborhood of O in the (u,t) plane. Let us now focus to the dividing line $u = u_d(t)$. On this line the chain must remain ideal. This means that the corrections linear in u and t cancel. Thus, the slope of the dividing line is defined by

$$u_d(t) = -\frac{k_t}{k_u}t \qquad \text{(XI.42)}$$

Then in eq. (XI.41) the only corrections which remain are of second order in u or t. As our parameter here we use the distance $\rho = \sqrt{u^2 + t^2}$ from the origin along the dividing line, and we write

$$\frac{a_m}{g^{1/2}\,a_{m-1}} - 1 = \text{constant}\ \rho_{m-1}^2 \qquad \text{(XI.43)}$$

Let us now turn to the equation for t_m (or for ρ_m, which is proportional to t_m) on the dividing line. On this line we have

$$t_m = t_{m-1} - kt_{m-1}^2 \qquad \text{(XI.44)}$$

In three dimensions the first term t_{m-1} on the right has a coefficient which is *exactly unity*. This can be seen from a regrouping in subunits in the three-body interaction [eq. (XI.37)]

$$\frac{1}{6}Tw^2c^3 \equiv \frac{1}{6}Tta^6c^3$$

$$= \frac{1}{6}Tt(g^{1/2}a)^6\left(\frac{c}{g}\right)^3 \qquad \text{(XI.45)}$$

All corrections to this result come from higher order interaction effects and thus are of order ρ^2 (or t^2) as indicated in eq. (XI.44) (k is a positive constant). The essential feature of eq. (XI.44) is that it does *not* give an exponential convergence of t_m toward the fixed-point values but only a slow convergence.*

The asymptotic law for t_m (or ρ_m) is found easily in the limit of g slightly larger than 1, where t_m and t_{m-1} are very close, and where m can then be treated as a continuous index. Instead of eq. (XI.44) we then have

$$\frac{\partial t_m}{\partial m} = -\,kt_m^2 \qquad \text{(XI.46)}$$

*For d smaller than 3 there are two fixed points near O, one describing gaussian chains and the other describing chains at Θ conditions. These two points merge when $d = 3$, and this confluence is the source of the slow convergence found in eq. (XI.44). This is called *marginal behavior* in renormalization language.

which integrates to

$$t_m = \frac{k}{m + m_0} \tag{XI.47}$$

where m_0 is an integration constant, related to the initial value of t

$$m_0 = \frac{k}{t_0} \tag{XI.48}$$

As explained, on the dividing line, the distance ρ to the origin is proportional to t. For large m, both scale like $1/m$. If we insert this result into the size equation [eq. (XI.43)], we arrive at

$$\frac{a_m}{g^{1/2}\, a_{m-1}} = 1 + k'm^{-2} \tag{XI.49}$$

or, introducing $b_m = g^{-m/2}a_m$, we have in the continuous limit

$$\frac{1}{b_m}\frac{\partial b_m}{\partial m} = k'm^{-2} \tag{XI.50}$$

This integrates to

$$\begin{aligned} b_m &= b \exp\left(\frac{k'}{m}\right) \cong b\left(1 + \frac{k'}{m}\right) \\ a_m &= g^{m/2}\, a\left(1 + \frac{k'}{m}\right) \end{aligned} \tag{XI.51}$$

We can now choose m so that $g^m = N$ (i.e., one subunit coincides with the whole chain), and we find a size:

$$R^2 = Na^2\left[1 + \frac{37}{363 \ln N}\right] \qquad (N \to \infty) \tag{XI.52}$$

In this final result, we have inserted the numerical coefficient derived in a complete calculation by M. Stephen.[10] (Note that in the final result g drops out; the properties of the chain are independent of our choice of subunits.)

The main conclusion is that at least for the size of the chain, quasi-ideal behavior is not very different from ideal behavior. The only corrections are proportional to $1/\ln N$ and are probably unobservable in practice. However, there are some other effects where they might show up, some of which are discussed below.

VICINITY OF THE COMPENSATION POINT: QUASI-IDEAL BLOBS

We consider now a temperature that is not exactly equal to the compensation temperature $\tilde{\Theta}$. In the (u,t) plane of Fig. (XI.6) this means that we start at a point (u_0,t_0) which is not exactly on the dividing line. For simplicity, we ignore the difference between Θ and $\tilde{\Theta}$; this corresponds to a dividing line that is *vertical*. Then we have

$$\left.\begin{aligned} u_0 &\sim \frac{T - \Theta}{\Theta} \\ t_0 &\sim \frac{w^2}{a^6} \sim 1 \end{aligned}\right\} \tag{XI.53}$$

If we choose a small, positive u_0, a long chain will show excluded volume effects at large scales. On the other hand, at small scales it will still be quasi-ideal. Our main purpose here is to find the boundary r_B between these two scales. We can also speak in terms of blobs, each containing g_B monomers and having size r_B. Each blob is quasi-ideal, but the necklace of blobs is a swollen necklace, with excluded volume effects. Note that the relationship between r_B and g_B must be of the type in eq. (XI.52); in this equation we have seen that the size corrections are negligible, and thus we can write

$$a^{-2}\,(r_B)^2 = g_B \tag{XI.54}$$

A similar problem occurs if we start with $u_0 < 0$. Then we can again define blobs, of size r_B and monomer number g_B. Each blob is still quasi-ideal, but the necklace of blob is a collapsed structure—the blobs fill the available space, with a certain filling density.

To determine the blob size, we construct a renormalization group equation for u. The quantity u measures the distance to the dividing line (since we ignore the difference $\Theta - \tilde{\Theta}$). This equation has the form:

$$u_m = g^{1/2}\,u_{m-1} - c u_{m-1}\,t_{m-1} + O(u^2) \tag{XI.55}$$

The first term $g^{1/2}$ corresponds to an ideal chain [as in eq. (XI.10)] and leads to an exponential increase in u. However, this increase is moderated slightly by the second term, and this will have some importance.* Note that c is *not* a concentration, but in a numerical constant, dependent on g. Iterating eq. (XI.55) we may write

$$u_m = g^{m/2} u_o [1 - \tilde{c}t_o][1 - \tilde{c}t_1] \ldots [1 - \tilde{c}t_{m-1}] \qquad \text{(XI.56)}$$

with $\tilde{c} = cg^{-1/2}$. We assume for the moment that we are very close to the dividing line, and that t_m is still given by eq. (XI.47). For large m we write

$$1 - \tilde{c}t_m = 1 - \tilde{c}k/m = 1 - p/m \cong \exp(-p/m) \qquad \text{(XI.57)}$$

where we have introduced an important numerical parameter $p = \tilde{c}k$. A complete calculation of k and $\tilde{c}$ shows that p is independent of g^{11} and is a purely geometric constant

$$p = \frac{4}{11} \qquad \text{(XI.58)}$$

Returning to eq. (XI.55) we have

$$u_m \cong g^{m/2} u_o \exp - \int_1^m dm' \frac{p}{m'} = g^{m/2} u_o \exp(-p \ln m)$$
$$\cong g^{m/2} \frac{1}{m^p} u_o \qquad \text{(XI.59)}$$

Thus, near the dividing line u_m increases a little more slowly than $g^{m/2}$. This expresses a correlation effect: the number of monomers in one subunit is $G = g^m$; in the ideal chain limit, counting all possible interacting pairs between two subunits, we would arrive at a coupling between them $u = G^2 / G^{3/2} u_o = G^{1/2} u_o$. However, the subunits do not interpenetrate each other freely because of the three-body interaction described by t, and this reduces the effective value of u.

We now return to the equation for the other coupling constant t. We wrote it on the dividing line to the form of eq. (XI.44). Now we must extend eq. (XI.44) to the vicinity of the dividing line (with our convention, to $u \neq 0$). It then has the form

*Note that there are no terms of order t^2 in eq. XI.55. If we start on the dividing line ($u = 0$), we never leave it.

$$\frac{dt}{dm} = -kt^2 + lu^2 + sut + \ldots \qquad \text{(XI.60)}$$

There are no terms linear in u. A two-point interaction cannot generate a three-point interaction in first order. The dominant new correction will be the u^2 term (since u increases exponentially, while t decreases). The constants l and s depend on g. We now retain only the k and l terms and estimate the contribution from the l term to $t(m)$, which we call $\delta t(m)$

$$\delta t(m) = l \int^m u_{m'} dm' \cong u_0^2 \int^m g^{m'} \frac{dm'}{(m')^{2p}}$$

$$\cong u_0^2 \, g^m \frac{1}{(\ln g) m^{2p}} \cong u_0^2 \frac{g^m}{m^{2p}} \qquad \text{(XI.61)}$$

For the integration we used the fact that $g^{m'}$ varies much faster than $1/m'$, and we replaced $(m')^{-2p}$ by m^{-2p}. We have to compare $\delta t(m)$ with the value of t on the dividing line [eq. (XI.47)]

$$t \cong \frac{t_0}{m} \qquad (m \gg 1) \qquad \text{(XI.62)}$$

From eqs. (XI.61) and (XI.62) we find that δt is negligible whenever the following inequality is satisfied

$$u_0^2 \, g^m / m^{2p} < t_0 \, m^{-1}$$

$$u_0^2 < t_0 \, g^{-m} / m^{1-2p} \qquad \text{(XI.63)}$$

Let us now use as our variable the number G of monomers per subunit $G = g^m$. The inequality in eq. (XI.63) may be transformed into

$$G < u_0^{-2} / |\ln u_0|^{1-2p} \qquad \text{(XI.64)}$$

In terms of a blob size g_B, this must be equivalent to $G < g_B$. Thus:

$$g_B = u_0^{-2} / |\ln u_0|^{1-2p} = u_0^{-2} / |\ln u_0|^{3/11} \qquad \text{(XI.65)}$$

Whenever G is smaller than g_B, the effect of u_0 (i.e., the departure from theta conditions) is negligible. Thus, eq. (XI.65) is the fundamental equation for the blob size. Related equations were derived originally (in connection with tricritical points and in a very different language) by Riedel and

Wegner,[12] Abrahams and Stephen,[13] and Wohrer.[11] Notice that if we omit the weak logarithmic factor $|\ln u_0|^{3/11}$, eq. (XI.65) agrees with the naive estimate of r_B (or g_B) discussed in Chapter IV—cf. eq. (IV.53) where $v/a^3 = u_0$. For many experiments the naive estimate is enough.

PHYSICAL PROPERTIES NEAR $T = \tilde{\Theta}$

The practical consequences of these delicate logarithmic effects have been analyzed in detail for the related problem of tricritical phase transitions.[11,12,13]

On the polymer side the understanding is less complete.[14,15,16] Recently the present author has proposed a qualitative picture of the transition based on the blob concept described above.[17] The idea is to estimate first the interaction constants u and t *between blobs*. They are obtained from eqs. (IX.59) and (XI.62), using the value of m which corresponds to the blob size ($g^m = g_B$). Then knowing the interactions between blobs u_B and t_B we can calculate the *size* R of a chain made of many blobs simply. For scales larger than the blob size, the blobs behave as would be predicted from the Flory theory; for instance, above the Θ point, where t_B becomes irrelevant, we obtain R from the Flory eq. (I.38) applied to N / g_B blobs of size r_B and excluded volume $u_B r_B^3$. Because of the logarithmic factor in eq. (XI.65) there also appear logarithmic factors in the size R. However, these factors seem extremely hard to see in any practical experiment. Another property of interest is the *specific heat*. A specific heat anomaly proportional to $\ln \Theta / \Delta T$ was suggested by M. Moore.[14] In Ref. 17 a slightly different structure is proposed: namely that the singular part of the free energy be simply *T per blob*. For the whole chain this then gives a free energy $F_{tot} = NT / g_B$, and from eq. (XI.65) one arrives at a specific heat proportional to $(\ln \Theta / \Delta T)^{3/11}$. The peak value would correspond to $g_B = N$ and is of order $(\ln N)^{3/11}$.

At the time of this writing, our main information on these properties is from numerical studies on a single chain on a lattice.[15] The chain is considered self-avoiding, but an attractive interaction (among monomers which are located on neighboring sites) is added. This gives a Θ point and a specific heat maximum (for $T \to \Theta$) which may well be logarithmic in N.[16] However, the possibility of fractional powers ($\ln^{3/11}$) has not been considered in the analysis.

From an experimental point of view, thermal measurements on very dilute chains are unpractical. What could be done is to study the *number N_c of contacts* between monomers by ultraviolet spectroscopy. N_c is essentially proportional to $\partial F_{tot}/\partial u_0$. The derivative $\partial N_c/\partial T$ should show an interesting anomaly near $T = \Theta$.

SUMMARY

A single chain at the compensation point $\tilde{\Theta}$ has a quasi-ideal behavior. The size R scales like $N^{1/2}$, and the pair correlation function $g(r)$ decreases like $1/r$ (for $r \ll R$). However, the three-body repulsive interactions remain effective even at $T = \tilde{\Theta}$. Their effect (in three dimensions) is to introduce some correlation between the monomers. The probability of contact between two (or three) monomers is reduced by certain logarithmic factors. These factors could show up in certain measurements which are sensitive to local properties (e.g., specific heat) and possibly in certain optical properties.

XI.2.3. Semi-dilute solutions and blobs

Renormalization groups can be applied to many chain problems. Here we take as an example a semi-dilute solution, with $d = 3$, in a good solvent. The dimensionless parameters characterizing the solution are a, u, and the volume fraction $\Phi = ca^3$. On each chain, we regroup the monomers into consecutive subunits, each made of g monomers. We reach new values a_1, u_1, and a number of subunits per volume a_1^3 which is

$$\Phi_1 = \frac{c}{g} a_1^3 \tag{XI.66}$$

We then iterate the process m times. The resulting Φ_m will be defined as the ratio between the average number of monomers in the volume $a_m^3 (ca_m^3)$ and the number of monomers in one subunit g^m. When Φ_m is less than 1 we are dealing essentially with a one-chain problem. When $\Phi_m > 1$, interchain effects become dominant.

The successive iterations may again be represented as trajectories in a (u, Φ) plane (Fig. XI.7). For $\Phi = 0$ the points of u axis converge exponentially toward the single-chain fixed point $u^\star$. For an initial Φ which is small but nonvanishing, the early iterations are much as they are for a single chain with regard to u, but Φ increases. We can then approach $u_m = u^\star$ if Φ_m is still smaller than 1 (single-chain behavior), but further iterations will modify the picture. When Φ_m becomes of order 1, we begin to deal with a dense system, for which the Flory theorem of Chapter II will hold; the effective iterations decrease, and u drops from a value close to $u^\star$ to values near zero. If we stop the iterations at the levels $\Phi_m = 1$ and $u_m \sim u^\star$, we have reached subunits that are identical to the "blobs" of Chapter III. Thus, the blob concept can be related to renormalization group trajectories. The example discussed here is somewhat trivial, but

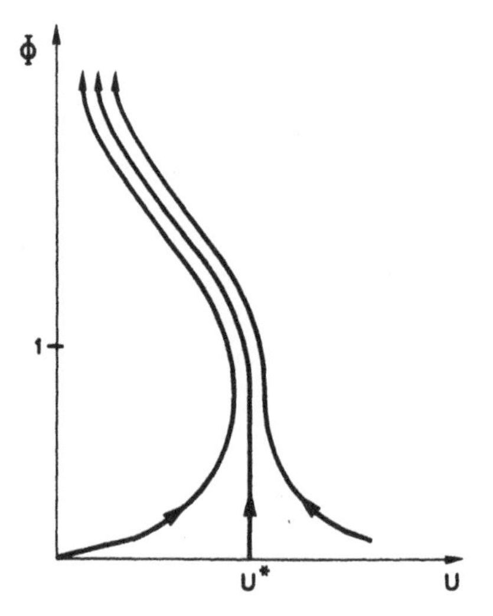

Figure XI.7.

similar calculations would be of interest for more complex systems such as confined chains.

REFERENCES

1. K. Wilson, J. Kogut, *Phys. Rep.* **12C,** 77 (1974).
2. S. K. Ma, *Rev. Mod. Phys.* **45,** 589 (1973). S. K. Ma, *Modern Theory of Critical Phenomena,* W. A. Benjamin, New York, 1976.
3. Ideas which are not far from the renormalization group procedure can be found in Z. Alexandrowicz, *J. Chem. Phys.* **49,** 1599 (1968).
4. M. Gabay, A. Garel, *J. Phys. (Paris) Lett.* **392,** 123 (1978). Y. Ooro (unpublished).
5. F. Oosawa, *Polyelectrolytes,* Marcel Dekker, New York, 1971. E. Selegny, M. Mandel, U. Strauss, *Polyelectrolytes,* Reidel, Dordrecht, 1974.
6. P. G. de Gennes, *et al., J. Phys. (Paris)* **37,** 1461 (1976).
7. G. Manning, *J. Chem. Phys.* **51,** 924, 934, 3249 (1969).
8. P. Pfeuty, R. Velasco, P. G. de Gennes, *J. Phys. (Paris) Lett.* **38L,** 5 (1977).
9. See Ref. 10 of Chap. X.
10. M. Stephen, *Phys. Lett.* **53A,** 363 (1975).
11. The most detailed reference on these calculations is M. Wohrer, Ph.D. Thesis, Paris, 1976. Available from CEA Saclay, Orme du Merisier, BP no. 2, 91 Gif s/ Yvette, France.
12. E. Riedel, F. Wegner, *Phys. Rev.* **B7,** 248 (1973).
13. E. Abrahams, M. Stephen, J. P. Straley, *Phys. Rev.* **B12,** 256 (1975).
14. The following reference proposes a different power for the logarithmic singularity in the specific heat: M. A. Moore, *J. Phys. (Paris)* **A10,** 305 (1977).

For a different viewpoint, see Y. Oono, T. Oyama, *J. Phys. Soc. Japan* **44,** 301 (1978).
15. J. Mazur, F. McCrackin, C. M. Guttman, *Macromolecules* **6,** 859 (1973). D. C. Rapaport, *Macromolecules* **7,** 64 (1974). C. Domb, *Polymer* **15,** 259 (1974).
16. D. C. Rapaport, *Phys. Lett.* **48A,** 339 (1974). D. C. Rapaport, *J. Phys.* **A10,** 637 (1977).
17. P. G. de Gennes, *J. Phys. (Paris) Lett.* **39L,** 299 (1978).

Author Index

Subject Index